UNDERSTANDING U

Understanding Urban Warfare

LIAM COLLINS AND JOHN SPENCER

Howgate Publishing Limited

First published in 2022 by
Howgate Publishing Limited
Station House
50 North Street
Havant
Hampshire
PO9 1QU
Email: info@howgatepublishing.com
Web: www.howgatepublishing.com

British Library Cataloguing-in-Publication Data
A catalogue record for this book is available from the British Library

ISBN 978-1-912440-35 1 (pbk)
ISBN 978-1-912440-42 9 (hbk)
ISBN 978-1-912440-36 8 (ebk – PDF)
ISBN 978-1-912440-37 5 (ebk – ePUB)

The views expressed in this publication are those of the authors and do not necessarily reflect the official policy or position of the Department of Defense or the U.S. government.

The public release clearance of this publication by the Department of Defense does not imply Department of Defense endorsement or factual accuracy of the material.

Contents

PART II: OPERATIONAL CASE STUDIES

Figures and Tables

Acknowledgements

The authors would like to thank Mr. Vincent Viola. This book would not have been possible without his support.

The authors would also like to thank all the contributors that helped make this book possible. They graciously donated their time to conduct interviews for the Urban Warfare Project Podcast series and later reviewed and updated those interviews to provide the chapters in this book. They include David Kilcullen, Richard Norton, Saskia Sassen, Sokwoo Rhee, Patrick J. Mahaney Jr., Daphné Richemond-Barak, Jayson Geroux, Lee Van Arsdale, Larry Perino, Kyle Lamb, James Rainey, Leonard DeFrancisci, Louis DiMarco, Rob MacMillan, Roger Noble, and Charles Knight.

Finally, the authors would like to thank John Amble for his help editing the initial podcasts from which these chapters were produced, Madeline Koch for transcribing the podcasts so that they could be edited and updated for this volume, and Jayson Geroux and Edward Coss for their help in editing this book.

Author Biographies

Liam Collins is the executive director of the Madison Policy Forum, a senior fellow at New America, and a permanent member with the Council on Foreign Relations. He retired as a colonel from the U.S. Army after 27 years of service. As a career special forces officer, he conducted multiple operational and combat deployments to Afghanistan, Iraq, Bosnia, South America, and the Horn of Africa. He was the founding director of the Modern War Institute and a former director of the Combating Terrorism Center at West Point. He is co-editor of the *Routledge Handbook of U.S. Counterterrorism and Irregular Warfare Operations*. He earned a Bachelor of Science degree in Mechanical Engineering (Aerospace) from the United States Military Academy, and a master in public affairs and a PhD from Princeton University's School of Public and International Affairs.

John Spencer is an award-winning scholar, professor, author, combat veteran, and internationally recognized expert and advisor on urban warfare and other military related topics. Considered the world's leading expert on urban warfare, he served as an advisor to the top four-star general and other senior leaders in the U.S. Army as part of strategic research groups from the Pentagon and the United States Military Academy. Spencer is the Chair of Urban Warfare Studies at the Modern War Institute at West Point, co-director of the Urban Warfare Project, and host of the Urban Warfare Project podcast. He is also the Chair of Urban Warfare Studies with the Madison Policy Forum and a Colonel in the California State Guard with assignment to the 40th Infantry Division, California Army National Guard as the Director of Urban Warfare Training. Serving over twenty-five years in the active Army as an infantry soldier, Spencer has held ranks from private to sergeant first class and second lieutenant to major. His assignments as an officer included combat deployments to Iraq as a platoon leader and company commander; an instructor in the U.S. Army's Ranger School; a Joint Chief of Staff and Army staff intern; fellow with the Chief of Staff of the Army's Strategic Studies Group; and co-founder, strategic planner, and deputy director of the Modern War Institute at West Point. He is the author of the book *Connected Soldiers: Life, Leadership, and Social Connection in Modern War*.

Guest Biographies

John Amble is the editorial director of the Modern War Institute at West Point and co-director of the Urban Warfare Project. He is also a military intelligence officer in the U.S. Army Reserve and a veteran of the wars in Iraq and Afghanistan. Previously he served as managing editor of *War on the Rocks*. He has been featured in print and broadcast media in the United States, Canada, and the Middle East, and his work has appeared in academic journals and books published by the academic press. He is the co-editor of *Strategy Strikes Back: How Star Wars Explains Modern Military Conflict* and *Winning Westeros: How Game of Thrones Explains Modern Military Conflict.*

Leonard J. DeFrancisci is a retired colonel who served for 32 years in the United States Marine Corps. He spent 14 years in civil affairs, conducting operational deployments to Iraq, Afghanistan, Nicaragua, Korea, Liberia, Albania, Jordan, and the Horn of Africa. He served in many different positions within civil affairs to include serving as commander of the 2nd Civil Affairs Group and commander of the 4th Civil Affairs Group. In 2004 he served as a civil affairs detachment commander for Regimental Combat Team - 1 during the second battle of Fallujah.

Louis DiMarco is a professor of military history at the United States Army Command and Staff College and specializes in urban warfare history. He is a retired U.S. Army lieutenant colonel. DiMarco is the author of the book, *Concrete Hell: Urban Warfare from Stalingrad to Iraq.*

Jayson Geroux is an infantry officer with The Royal Canadian Regiment of the Canadian Armed Forces. He is an urban operations instructor, urban warfare historian, and scholar who has done extensive research on the 1943 battle of Ortona. Geroux acquired his master's degree from the Brigadier Milton Gregg V.C. Centre for the Study of War and Society at the University of New Brunswick, Canada, in which his thesis focused on the battle of Ortona.

David Kilcullen is a best-selling author and leading scholar in the field of urban warfare, unconventional warfare, and guerilla warfare. He served 24 years as an infantry officer in the Australian Army. He worked in the U.S. State Department, where he was chief strategist in the Counterterrorism Bureau, Senior Counterinsurgency Advisor to Multi-National Force Iraq, and Senior Advisor for Counterinsurgency in the office of Secretary of State Condoleezza Rice. He is the author of numerous books to include *The Accidental Guerilla: Fighting Small Wars in the Midst of a Big One*, *Counterinsurgency*, and *Out of the Mountains: The Coming Age of The Urban Guerrilla*.

Charles Knight is a senior lecturer in terrorism, asymmetric conflict, and urban operations at Charles Stuart University, and a senior researcher at the University of New South Wales. Knight researches urban operations, an interest sparked when he was tasked to develop training for confronting the Soviets in a German city as a young parachute regiment officer. His research is informed by fieldwork in Lebanon and Cambodia as well as by his uniformed service. Knight spent over 30 years in the Australian Army, serving in the 1 Commando Regiment; commanding 2/17 Battalion, Royal New South Wales Regiment; and a decade in special operations development driving reform of close combat training and writing the army's urban doctrine.

Rob MacMillan is a collective trainer at the Fort Carson mission training complex. He is a retired lieutenant colonel with 30 years of service in the U.S. Army with a career spanning from a scout platoon team leader in Alaska to the manager of foreign military sales programs in Iraq. MacMillan served in infantry, airborne infantry, cavalry, and armor units. He was the operations officer for the 1st Combined Arms Battalion, 68th Armored Regiment, during the battle of Sadr City.

Patrick J. Mahaney Jr. retired from the U.S. Army as a colonel after 32 years of service. As a career special forces officer, his service included nine combat tours and extensive operational assignments worldwide. In 2014-2015, Mahaney served as the director of the Chief of Staff of the Army's Strategic Studies Group while the group researched the implications of megacities and urbanization for the U.S. Army. A native of New York City and member of the Council on Foreign Relations, his work focuses on the innovation, resilience, and adaptation needed to meet emerging challenges and complex operating environments for public safety and national security. He is the co-founder and director of the National Center for Urban Operations.

Kyle Lamb is the founder and president of Viking Tactics, a tactical training and gear business. He retired as a sergeant major after 21 years of service in the U.S. Army, most of it in special operations. Lamb conducted operational and combat deployments to Bosnia, Iraq (Operation Desert Storm and Operation Iraqi Freedom), and Somalia. In Somalia he was part of the assault force during Operation Gothic Serpent made famous by the book and movie, *Black Hawk Down*. He has received numerous valorous awards and decorations including the Silver Star.

Roger Noble serves as the Ambassador for Counter-Terrorism within Australia's Department of Foreign Affairs and Trade. He retired as a major general from the Australian Army. During the battle of Mosul in 2016, Noble served as the deputy commander for the Combined Joint Forces Land Component Command.

Richard Norton is a professor of national security affairs at the U.S. Naval War College and a retired U.S. Navy commander. While in the Navy, he served extensively at sea on cruisers and destroyers, as well as shore tours in the Navy's Office of Legislative Affairs and the Navy Staff. Norton has authored numerous articles including "Feral Cities" in the *Naval War College Review* and "Feral Cities: Problems Today, Battlefields Tomorrow?" in the *Marine Corps University Journal*.

Larry Perino retired as a colonel from the U.S. Army after nearly 25 years of service. During his career, he served in various positions within multinational, joint, conventional, training, and special operations commands. He was a platoon leader with the 75th Ranger Regiment during Operation Gothic Serpent in Somalia in 1993, where he was awarded a Silver Star for his actions. He currently resides in the Pensacola, Florida area with his wife and works as an assistant vice president at Navy Federal Credit Union.

James Rainey is a lieutenant general in the U.S. Army and currently serves as the Headquarters, Department of the Army, deputy chief of staff for operations, plans, and training (G-3/5/7). During the second battle of Fallujah in November 2004, Rainey served as the commander of the 2nd Battalion, 7th Cavalry Regiment within the 1st Cavalry Division.

Sokwoo Rhee was the associate director of the Cyber-Physical Systems Program at the National Institute of Standards and Technology at the time he was interviewed. In that position, Rhee was responsible for the smart city program.

Daphné Richemond-Barak is an assistant professor at the Lauder School of Government, Diplomacy, and Strategy at Reichman University where she heads the International Program in Government. She also serves as senior researcher and head of the international law desk at the International Institute for Counter-Terrorism, adjunct scholar at the Modern War Institute at West Point, and senior fellow at the Institute of Law and Land Warfare. Richemond-Barak is one of leading scholars of subterranean or underground warfare, author of *Underground Warfare*, and founder of the International Working Group on Subterranean Warfare.

Saskia Sassen is the Robert S. Lynd Professor of Sociology at Columbia University, and a member of its Committee on Global Thought, which she co-chaired until 2015. She has been called "the leading urban theorist of the global world." She is the author of eight books including *Cities in a World Economy* and *The Global City*. She is the editor or co-editor of three books, with her most recent being *Cities at War: Global Insecurity and Urban Resistance*.

Lee Van Arsdale retired as a special forces colonel from the U.S. Army. In the course of his 25-year army career, he served in three combat zones and was awarded the Silver Star for valor in combat and the Purple Heart for wounds received in combat. He also participated in numerous classified operations across the globe. Following his military career, Van Arsdale was a business executive for several organizations and now serves on the boards of select companies and does volunteer work for veterans' organizations. He was a lieutenant colonel within Task Force Ranger's headquarters during Operation Gothic Serpent in Somalia in 1993.

List of Abbreviations

AFP	Armed Forces of the Philippines
AQI	al Qaeda in Iraq
ASCOPE	areas, structures, capabilities, organization, people, and events
COP	combat outpost
DARPA	Defense Advanced Research Projects Agency
EFP	explosively formed penetrator
HMMWV	high mobility multipurpose wheeled vehicle (Humvee)
IED	improvised explosive device
IS-P	Islamic State in the Philippines
ISIS	Islamic State of Iraq and al Sham
JAM	Jaish al-Mahdi
KIA	killed in action
MAGTF	Marine Air-Ground Task Force
MILF	Moro Islamic Liberation Front
MICLIC	mine clearing line charge
MNLF	Moro National Liberation Front
MRAP	mine resistant ambush protected
MSR	main supply route
NIST	National Institute of Standards and Technology
NVG	night vision goggles
PMESII-PT	political, military, economic, social, information, infrastructure, physical environment, and time
RCT	regimental combat team
RPG	rocket-propelled grenade
SIGACT	significant activity
SLICE	strategize, locate, isolate, constrict, and eliminate
SWEAT MSO	sewage, water, electricity, academics, trash, medical, safety, and other
TRADOC	Training and Doctrine Command
UNISOM	United Nations Operations in Somalia

Introduction

Militaries seek to accomplish the strategic goals set by political leaders. Accomplishing these goals may require defeating the enemy's military, conquering key territory, eroding the enemy's will to resist or capacity to fight, or completing a number of other objectives. They may also involve degrading the enemy's military power, political power, economic power, and informational power. Much of this power is now concentrated in cities.

Cities are home to national and regional governance and the centers of political power. They are the economic engines of nations. They are the gateways for transit, logistics, and trade out of, into, and through countries, especially for militaries campaigning into a foreign land. Cities are population centers that support or contest a nation's activities in war.

Thus, cities often become the tactical, operational, or strategic focus of militaries during war and, as such, they are unavoidable. Yet, too often, militaries find themselves unprepared for the complex challenges of urban warfare.

To help understand the urban environment, the U.S. Army breaks it into three parts that the army calls the urban triad: complex man-made physical terrain, a population of significant size and density, and the supporting infrastructure that services that population.[1] While the military uses the urban triad to help describe an urban environment, it does not provide a distinct requirement for what constitutes an urban area. In other words, it does not provide a minimum population, population density, minimum number of buildings, or building density to determine when a settlement is to be considered an urban environment or a city.

---

[1] Department of the Army, *Urban Operations*, ATP 3-06 (Washington, DC: Headquarters, Department of the Army, 2017), 1-3, https://armypubs.army.mil/epubs/DR_pubs/DR_a/pdf/web/ARN6452_ATP%203-06%20FINAL%20WEB.pdf.

Likewise, there is no consensus definition for what constitutes a city across nations or academic disciplines.[2] Nations use different criteria to classify an urban area as a city. One criterion is population size, which is not surprising given that, according to urban geography researcher Tim Hall, a "fundamental precondition for the emergence of cities is the existence of a civilization."[3] The words "city" and "civilization" even share the same Latin root. Hall defines "civilization" as a "complex sociocultural organization that contains formal institutions and that organizes strangers into a cohesive community under the control of a centralized authority."[4]

Beyond population size, other criteria used to classify a city include administrative function, population density, urban characteristics, or a combination of these criteria with other standards such as economic function. Even nations that use population thresholds to determine when a settlement is classified as a city use numbers that vary greatly. For example, the minimum threshold for Denmark and Iceland is 200 inhabitants, for the Netherlands and Nigeria it is 20,000, for Mali it is 30,000, and for Japan it is 50,000.[5] U.S. Army doctrine defines a city as an urban area with over 50,000 residents.[6]

At its most basic level, a city can be thought of as a type of terrain: urban terrain. There are many types of terrain that armies must prepare to operate in: jungle, mountainous, desert, and arctic among others. Each type of landscape poses its own unique challenges: the thick jungle canopy negates the effect of many airborne intelligence, surveillance, and reconnaissance platforms; rugged mountains make mounted movement impossible; soft desert sands and swamps may be difficult to traverse in vehicles; and bitter arctic cold negatively impacts the performance of humans, vehicles, and weapons systems.

Urban terrain presents many of these same problems but also has additional, inherent challenges. The roofs of buildings act like a jungle canopy that negate intelligence, surveillance, and reconnaissance assets from knowing what is inside. Rubbled structures, until cleared, can make

[2] John Spencer, "The Destructive Age of Urban Warfare; Or, How to Kill a City and How to Protect It," Modern War Institute, March 28, 2019, https://mwi.usma.edu/destructive-age-urban-warfare-kill-city-protect/.

[3] Tim Hall, *Urban Geography*, 2nd ed. (London: Routledge, 2002), 28.

[4] Ibid., 28-9.

[5] United Nations Human Settlements Programme, *What Is A City?* (Nairobi: United Nations Human Settlements Programme, 2020), 3, https://unhabitat.org/sites/default/files/2020/06/city_definition_what_is_a_city.pdf.

[6] Department of the Army, *Urban Operations*.

vehicular movement impossible, and even when streets are not blocked with rubble, every street has the potential to be a three-dimensional ambush zone. Enemy can hide in and emerge from subterranean tunnels, they can hide inside rooms at ground level and engage targets from windows or loopholes,[7] and they can engage targets from several stories up while avoiding armored vehicles' main guns that have limited angles of elevation.

The density of buildings in urban areas also negates the range and effects of many weapon systems, ensuring that most combat is fought at extremely close quarters. An enemy can be located one room away, yet "invisible" to the soldier located on the other side of the wall. Buildings must be deliberately cleared one room at a time, requiring thousands of rooms to be cleared before a city can be secured. To make matters worse, if soldiers leave a building after clearing it, they may have to re-clear it because the enemy might have reoccupied the structure through a tunnel or a mousehole.[8]

Smaller caliber weapons cannot penetrate thick concrete building material. As a result, high explosive munitions (bombs, artillery, mortars) may be necessary to attack an enemy inside a strong concrete building, but this can produce significant destruction. No other terrain presents these challenges.

In its totality, urban man-made physical terrain consists of many complex dimensions which may include maritime spaces; subterranean networks of transportation tunnels, utility tunnels, basements, parking garages, and even underground malls; super-surface spaces like rooftops and building facades; and the labyrinths of interior spaces that may stretch dozens of stories into the sky.

Yet as complex as the physical terrain is, the population of urban environments is perhaps even more intricate. Civilians live in all types of terrains, but their density in urban terrain is an order of magnitude higher. These populations have, according to U.S. Army doctrine, "underlying relational, social, and cultural patterns" that form networks that may be

---

[7] A loophole is a hole in the wall that is used as a firing port. See, U.S. Marine Corps, *Military Operations on Urbanized Terrain (MOUT)*, MCWP 3-35.3 (Washington, DC: Headquarters: U.S. Marine Corps, 1998), https://www.marines.mil/Portals/1/MCWP%203-35.3.pdf.

[8] Mouseholes are "openings that are made from the interior or exterior of a structure (walls, floors, ceilings, roofs) to facilitate inter- and intra-building communications and movement." Department of the Army, *Combined Arms Operations in Urban Terrain*, ATTP 3-06.11 (Washington, DC: Headquarters, Department of the Army, 2011), B-1, https://www.globalsecurity.org/military/library/policy/army/attp/attp3-06-11.pdf.

friendly, neutral, or adversarial.[9] Trying to distinguish who is friendly, neutral, or adversarial is often difficult.

Trying to avoid civilian casualties is a significant constraint for armies that have a moral compass. The law of armed conflict, also called the law of war or international humanitarian law, was established to balance between military requirements and humanitarian concerns.[10] After World War II, international humanitarian law was strengthened in hopes that civilian populations would never have to pay such a heavy cost in war again.[11]

Thus, when a military engages in combat in cities, they often limit their use of force, much more than for any other type of terrain. The rules of engagements that militaries develop—outlining what they can and cannot do—restrict the types of weapons they can use, the targets they can engage, and the methods they can employ to achieve their military missions.

Historically, commanders emplace stricter rules of engagements than required by the law of war. For example, during the battle of Manila, General Douglas MacArthur repeatedly denied requests by U.S. forces for air bombardment while fighting to liberate the city because he wanted to spare the Philippine civilians and the city itself from the excessive damage of urban warfare.[12] MacArthur understood that when the war was over, the population, the city, and the city's infrastructure would remain—in one form or another—and it would be best in the post-conflict phase for the death and destruction to be minimized during the conflict phase.

Cities and civilizations are comprised of many complex adaptive systems that evolve over time. Civil, social, and economic conditions fuel a city's development. Cities grow, constantly evolve, and continuously achieve a natural stasis that could be optimal or suboptimal for its inhabitants based on the conditions. Like other complex systems, when there is a major change in the environment, such as warfare, the entire system is impacted. Thus, when a military enters a city, it affects the natural stasis of that urban area and changes it. Given the complexity of a city's systems, it is nearly impossible to accurately predict the second- or

---

[9] Department of the Army, *Urban Operations*, ATP 3-06, 1-3.

[10] International Committee of the Red Cross, *What is International Humanitarian Law?* (Geneva: International Committee of the Red Cross, 2004), 1, https://www.icrc.org/en/doc/assets/files/other/what_is_ihl.pdf.

[11] "Geneva Conventions," History.com, August 21, 2018, https://www.history.com/topics/world-war-ii/geneva-convention.

[12] William Robertson and Lawrence Yates, *Block by Block: The Challenges of Urban Operations* (Fort Leavenworth, KS: U.S. Army Command and General Staff College Press, 2003), 112-113, https://cgsc.contentdm.oclc.org/digital/collection/p16040coll3/id/88/.

third-order effects of any military operation. Moreover, given the increased global nature and connectedness of many cities, military operations in one city often have global impacts.

Combat operations often disrupt and destroy many parts of a city, damaging or destroying as much as 95 percent of a city's buildings in some cases.[13] Beyond the physical damage to buildings, urban combat often destroys political power systems, social processes, and the natural flows within a city.[14] This destruction can negatively impact millions of people and result in billions of dollars in damage and economic loss.[15]

The financial cost to repair Mosul and its surrounding areas following its 2017 battle was estimated to have been $100 billion dollars and this does not include other costs such as the loss of employment, wages, and productivity.[16] The effect of warfare on the urban environment—its people, terrain, and infrastructure—has no parallel, in scope or magnitude, to any other environment. When combat occurs in wooded, mountainous, or jungle environments for example, the damage is largely absorbed by the natural terrain because the density of man-made structures is so much lower or non-existent.

While urban warfare has always been challenging, it is arguably even more so today given technological advancements that have impacted warfare. The most obvious is that weapons are more lethal and prolific. The accessibility and speed of information technology is another important change. Any combatant or civilian on the battlefield possesses the ability to upload footage from their cellular phone and transmit it to the global community on social media. Thus, in urban terrain a military can expect to have their actions watched and potentially recorded by hundreds if not thousands of cameras and shared globally.

During Russia's 2022 invasion of Ukraine, any person in the world could watch a live video feed from the capital.[17] This aspect of the

[13] Ervin Malicdem, "Aftermath of the Battle of Marawi," *Shadow Expeditions*, November 30, 2017, https://www.s1expeditions.com/2017/11/223-marawi-battle-structures.html.

[14] John Spencer, "The City is Not Neutral: Why Urban Warfare is So Hard," Modern War Institute, March 4, 2020, https://mwi.usma.edu/city-not-neutral-urban-warfare-hard/.

[15] See for example, John Spencer and Margarita Konaev, "Sticker Shock: The World Just Got the Bill for Rebuilding War-Ravaged Cities in Iraq," Modern War Institute, February 21, 2018, https://mwi.usma.edu/sticker-shock-world-just-got-bill-rebuilding-war-ravaged-cities-iraq/.

[16] "Iraq's Mosul struggles to rebuild without funds," *France 24*, August 29, 2021, https://www.france24.com/en/live-news/20210829-iraq-s-mosul-struggles-to-rebuild-without-funds.

[17] See, for example, e.hocamm web, "Ukraine live camera - Kyiv live camera 22 MARCH multi camera - 24/7," YouTube video, March 23, 2022, https://www.youtube.com/watch?v=ZaE2INoAgJ0.

information domain in urban combat can influence local, regional, and global communities. In 2004's first battle of Fallujah, American forces were forced to stop only nine days into their operation because of Iraqi and global perceptions of what was considered excessive methods after media images of injured civilians were broadcasted from within the city.[18]

To be sure, there are advantages and disadvantages to fighting in urban environments. For a weaker military, or a military attempting to hold terrain, the urban environment can offer immense defensive advantages to the force that can get there first. Some have called the urban terrain "a great equalizer" because it reduces an attacking military's intelligence, surveillance, and reconnaissance and strike capabilities.[19] The dense physical terrain can make an attacker's weapons and maneuver tactics less effective. An attacker can be canalized, separated, and defeated in detail by the unique multidimensional features of the urban terrain.

Yet, despite urban combat being a common feature in modern wars and while it is fairly common for militaries to develop schools or specialties for other challenging terrains, there are no such schools for urban warfare in the United States.[20] As a result, the United States military, like many militaries, often finds itself unprepared for urban fighting and having to learn techniques in the heat of combat that could have been learned, practiced, and perfected in training.

This is particularly surprising given that militaries frequently seek to understand terrain so that they can leverage ground to their advantage. Any military practitioner knows, for example, that it is preferable to place artillery on the reverse slope of a hill because it affords better protection than the forward slope. Likewise, troops understand the inherent dangers involved with fording rivers so they will check the depth and speed of the water prior to crossing a stream or river.

Militaries teach their officers the importance of understanding "dirt" so they understand whether a vehicle can traverse the terrain based on

---

[18] James Farwell, *Information Warfare: Forging Communication Strategies from Twenty-first Century Operational Environments* (Quantico, VA: Marine Corps University Press, 2020), https://www.usmcu.edu/Portals/218/InfoWarfare_Web_1.pdf.

[19] Robert H. Scales Jr., "The Indirect Approach: How U.S. Military Forces Can Avoid the Pitfalls of Future Urban Warfare," *Armed Forces Journal International* 136, no. 3 (1988), https://apps.dtic.mil/sti/pdfs/ADA399275.pdf.

[20] John Spencer, "The Army Needs An Urban Warfare School and It Needs it Soon," Modern War Institute, April 5, 2017, https://mwi.usma.edu/army-needs-urban-warfare-school-needs-soon/.

the composition and moisture content of the soil.[21] If not, they risk getting stuck, which is an unenviable position to be in when the enemy is trying to kill you.

For decades the U.S. Army has had a jungle school, originally located in Panama and now located in Hawaii, to teach its troops how to operate in jungle terrain.[22] Likewise, the U.S. Army has a mountain warfare school to teach troops how to traverse, survive, lead, and operate in mountainous terrain.[23]

The army, however, has yet to establish an urban warfare school, which is puzzling given the challenges that the urban environment presents and the fact that the world is only becoming more urban. In 1950, there were only 83 cities with populations over one million; in 2018, there were 548.[24] In 1950, 30 percent of the world's population lived in cities, by 2020, it had nearly doubled to 56 percent. Thus, the likelihood of future battles being waged in urban areas seems more, not less, likely.

From a military perspective, controlling a city can often offer tactical, operational, or strategic value. Cities rarely grow randomly; they grow because they are located on a coast that is suitable for a port; they were built along a river for trade; or they were built on a hill to be defendable. They may also grow along a major trade or supply route, or, like Nagorno-Karabakh's Shusha, control a mountain corridor. In other words, cities are often built at locations that have value at multiple levels.

A city may have strategic value. They may house the nation's government or serve as an economic center that drives a nation's economy and warfighting effort. As such, a particular city might be a country's center of gravity—"the source of power that provides moral or physical strength, freedom of action, or will to act."[25]

Unlike other terrain that is fairly static, the urban environment is constantly changing and growing larger and more complex each day. How can one even conceptualize a city? Is it useful to think of a city as an

---

[21] All cadets at the United States Military Academy are required to take EV203 Physical Geography in which they learn about different soils. The nickname for the course is 'Dirt.'

[22] See, "Lightning Academy," U.S. Army Garrison Hawaii, https://home.army.mil/hawaii/index.php/25thID/units/lightning-academy.

[23] See, "Army Mountain Warfare School (AWMS)," Fort Benning, https://www.benning.army.mil/infantry/amws/.

[24] Population Division, *The World's Cities in 2018* (New York: United Nations, 2018), 2, https://digitallibrary.un.org/record/3799524?ln=en.

[25] Joint Chiefs of Staff, *Joint Planning*, JP 5-0 (Washington, DC: Joint Chiefs of Staff, 2017), GL-6, https://www.jcs.mil/Portals/36/Documents/Doctrine/pubs/jp5_0.pdf.

organism? Is it more appropriate to think of a city as a closed system? Or is it better to think of a city as an open and incomplete system?

A city is essentially a system of flows—commodities, people, power, energy, information, fuel, water—that move into, through, and out of it. If military operations disrupt a city, the urban area cannot effectively process these natural flows and similar to an organism, toxic buildup may result in the form of sewage, waste, pollution, crime, or violence. Entering and occupying a city without having at least a basic understanding of how a city functions risks disrupting these natural flows and increasing these toxic byproducts and making the military operation significantly harder.

How a military can best operate in a city depends on its mission. For example, in an urban counterinsurgency, military forces must understand the human geography to achieve their mission of finding, fixing, and finishing the enemy. They must understand the city as a system to identify the requirements of the people they wish to safeguard and those on whom they are dependent for intelligence. [26]

This book's aim is to mitigate the gap in the general lack of understanding of the urban environment when it comes to military operations. As such, this book is divided into two parts. In part I, "Understating the Complex Operational Environment," experts provide different ways to analyze and understand a city. In part II, "Understanding Past Urban Battles," experts who have conducted extensive field research and practitioners who fought, describe a number of urban battles that derive lessons that are presented in the book's conclusion.

This is an interview style book, meaning that each chapter was conducted as an interview with the relevant expert instead of being authored by the expert. Chapters are updated versions of interviews that were conducted as part of the Modern War Institute's *Urban Warfare Project Podcast* series.

Part I: Understanding the Complex Operational Environment

Part I offers a number of different ways to conceptualize or understand a city. As cities grow larger and become increasingly complex, the

[26] John Spencer, "The City is Not Neutral."

lexicon—feral, global, smart, mega—used to describe and understand them is also expanding.

In "Understanding the City," Dr. David Kilcullen discusses the four emerging trends of population growth, urbanization, littoralization, and networked connectivity. These trends have expanded the concept of the city from an "urban triad" to an "urban quad," one that must take information, which may originate from hundreds or thousands of miles away, into account.

Dr. Richard Norton introduced the term feral city—a metropolis that remains within the territory of a sovereign state, but one in which the government has lost the ability to maintain the rule of law within the city's boundaries—to describe the phenomena that he was witnessing in some cities.[27] In "Feral Cities," Norton discusses these types of cities and provides a taxonomy to classify cities as healthy, feral, or somewhere on the spectrum in-between. Norton identifies five variables to examine the health of cities: governance, economy, services, security, and civil society. This taxonomy can be useful to identify failing cities, the ones that militaries are most likely to find themselves engaged in.

In "Global Cities," Dr. Saskia Sassen discusses a new type of city that emerged in the 1980s. Global cities shifted the power, ownership, function, and the relationship or stature of cities and their host nations. She describes how some global cities have become more important economically and politically than the nation in which they reside. She focuses on the important role of financial power and the flow of information and capital that emanates from global cities. Sassen also introduces the analogy of thinking of a city as an open and incomplete system as opposed to a closed organism in a final state of being.

The concept of "smart cities" dates back to the 1970s, but it took off in the early 2000s with the explosive growth of information technology and smart devices. This type of city uses information and communications technology to improve efficiency and the quality of life of its residents. In "Smart Cities," Dr. Sokwoo Rhee discusses the four layers of the Internet of Things that allow cities to become smart: the physical systems layer, the communications layer, the data analytics layer, and the human and services layer. He also discusses how smart cities might impact urban operations for military forces.

[27] See chapter 2, "Feral Cities," in this book.

Colonel (retired) Patrick Mahaney discusses the unique challenges that cities with populations exceeding ten million residents present to militaries in "Megacities and the Military." Mahaney argues that cities are one of the most complex operational environments that a military can operate in. Yet, the U.S. Army, and many other armies, remain woefully unprepared to function in large urban areas even though dozens of major city battles have been fought in the past two decades. He also offers some recommendations as to how the U.S. Army could better train and educate its force to be prepared for the challenging future urban environment that is megacities.

In part I's final chapter, "Beneath the City," Dr. Daphné Richemond-Barak discusses the unique challenges that the subterranean environment presents to militaries. As the author of *Underground Warfare* and founder of the International Working Group on Subterranean Warfare, she has conducted groundbreaking work on this type of warfare to include the legal challenges that operating in the subterranean environment presents to military forces. She also describes the different types of tunnels that were used by Hezbollah and Hamas in wars against Israel.

Part II: Understanding Past Urban Battles

History is ripe with examples of militaries learning techniques and tactics that are not new, but instead have been learned and practiced many times before. This comes to light in part II which explores some of the most relevant urban battles in modern history. These battles are examined by military officers who fought in, or experts who have conducted extensive research on, these battles.

The first one examined is World War II's battle of Ortona. While an older battle, we included Ortona to demonstrate that some facets of urban combat, such as the importance of combined arms operations, are just as relevant today as they were nearly eight decades ago. In "The Battle of Ortona," John Spencer interviews Major Jayson Geroux who is an urban operations instructor, urban warfare historian, and scholar who has conducted extensive research on the battle. Historically referred to as "Little Stalingrad," this particularly violent battle took place in the town of Ortona, Italy from December 20-27, 1943, and became the seminal urban battle for the Canadian Army during World War II. The Canadians learned many urban lessons while dislodging the defending and heavily barricaded Germans from the city to support the Allies' continued advance up the eastern side of the Italian peninsula.

In "The Battle of Mogadishu," Liam Collins interviews Colonel (retired) Lee Van Arsdale, Colonel (retired) Larry Perino, and Sergeant Major (retired) Kyle Lamb about their experiences as part of Task Force Ranger in Mogadishu, Somalia in 1993. All three took part in the battle that was made famous by Mark Bowden's book *Black Hawk Down*, which was later turned into a feature film with the same title. This chapter is an important addition because it demonstrates the challenges that the urban environment presents for even limited military operations. While other chapters discuss operations to seize or control a city, this one describes the challenges that urban areas present for a force that is simply trying to conduct operations within the city without attempting to control it.

The next four chapters discuss American urban battles during the Iraq War. In the first, John Spencer questions Lieutenant General James Rainey about "The Second Battle of Fallujah." Rainey served as a battalion commander during the November to December 2004 battle. At the time, Fallujah was an enemy stronghold, having been controlled by al Qaeda in Iraq since the United States was forced to prematurely end its first attempt to rid the city of insurgents in April 2004. Unlike Mogadishu, the city was largely devoid of civilians, as 70-90 percent of the city's 300,000 residents fled prior to the start of the battle.[28] Yet it would still require several weeks for coalition forces to clear the city of its estimated 3,000 insurgents.[29]

In "Rebuilding Fallujah," John Spencer discusses the second battle of Fallujah, from a civil affairs perspective, with Colonel (retired) Leonard DeFrancisci. During the battle, DeFrancisci served as a U.S. Marine Corps civil affairs detachment commander. Prior to the assault, his civil affairs team built lists of key individuals, mapped out critical infrastructure, supported displaced residents, and facilitated the division's deception operations. During the assault, his team conducted humanitarian assistance infrastructure assessments, and on one occasion, figured out how to stop the

[28] Timothy S. McWilliams and Nicholas J. Schlosser, *U.S. Marines in Battle: Fallujah, November-December 2004* (Quantico, VA: U.S. Marine Corps History Division, 2014), 7, https://www.usmcu.edu/Portals/218/FALLUJAH.pdf.

[29] Dexter Filkins and James Glanz, "With Airpower and Armor, Troops Enter Rebel-Held City," *New York Times*, November 8, 2004, https://www.nytimes.com/2004/11/08/international/with-airpower-and-armor-troops-enter-rebelheld-city.html.

city's flooding so the assault could continue. After the battle, his civil affairs team helped get the city functioning again and facilitated the recovery of more than 2,000 dead enemy personnel.

In "The Battle of Ramadi," Dr. Louis DiMarco, a retired lieutenant colonel and professor of history at the U.S. Army Command and General Staff College, considers the 2006 battle. At the time, Ramadi was regarded as "the deadliest city in Iraq," so the United States-led coalition set out to clear the insurgent stronghold.[30] DiMarco explains how Colonel Sean MacFarland applied counterinsurgency principles to clear, hold, and rebuild the city. Some of the techniques that MacFarland applied, such as getting his force out of large forward operating bases on the edge of town and into small combat outposts throughout the city, would be implemented across Iraq in 2007 when President George W. Bush implemented the surge.

"The Battle of Sadr City" is the fourth and final chapter from the Iraq War. John Spencer, who also fought in the battle, questions Lieutenant Colonel (retired) Rob MacMillan, a battalion operations officer during the 2008 battle. The previous three chapters involve American forces seizing a city. This chapter, instead, examines how the army was able to control Baghdad's Sadr City neighborhood, and its population of nearly two million, without physically entering it. MacMillan describes how his battalion had to quickly transition from stability operations to high-intensity offensive operations and how they were able to build a giant concrete barrier, 20 feet high and nearly five kilometers long, to seal off the southern side of Sadr City during Operation Gold Wall.

In "The Battle of Mosul," Major General (retired) Roger Noble explains his role in liberating Mosul, Iraq from the Islamic State of Iraq and al Sham (ISIS). Noble, an officer in the Australian military, served as a deputy division commander during his 2016 deployment. Unlike the battles described in the previous four chapters in which coalition forces led operations, the Iraqis led combat operations in Mosul with the coalition providing only advice and limited assistance. As such, Noble depicts a different approach that the Iraqis used to seize the city, executing the operation in a very Iraqi way. The Iraqis often moved slower than many thought they should, but their technique ultimately proved successful as they were able to push ISIS out of the city. Noble also explains how the

[30] Mark Kukis, "The Most Dangerous Place in Iraq," *Time*, December 11, 2006, http://content.time.com/time/world/article/0,8599,1568724,00.html.

Iraqis effectively leveraged information operations to support their lethal operations.

In chapter 14, John Spencer interviews Dr. Charles Knight for "The Battle of Marawi." Knight spent nearly three decades in the British Army and is presently a senior lecturer in terrorism, asymmetric conflict, and urban operations. He conducted extensive field research of the battle and describes how the Philippine Army was completely unprepared for urban operations in 2017. Most of the city's residents fled during the opening week of the battle. That clearly saved civilian lives and made the fighting easier for the Armed Forces of the Philippines (AFP), although they still faced a daunting task. The AFP faced a brutal fight as the Islamic State in the Philippines established strong point defenses throughout the city. Knight illustrates how the AFP had to systematically clear every building to rid the city of the insurgents. Similar to Mosul, the AFP effectively leveraged information operations to support their clearing operation.

The final battle discussed, the Second Nagorno-Karabakh War's 2020 "Battle of Shusha," is also the most recent urban battle other than urban conflict during Russia's invasion of Ukraine that had just started as we were completing this project. John Spencer delineates the strategic value that a city, even a small one, can have in war: the six-week war ended once Shusha fell, resulting in Azerbaijan reclaiming seven regions that it had lost to Armenia in its first war. In a reversal of roles, the Azerbaijani forces played the attacker for a city that they had failed to defend nearly three decades earlier.

In the conclusion, we discuss military applications that emerge from various chapters in part I and urban operations lessons that emerge from an examination of the various urban battles contained in part II. In an increasingly urban world, the future character of warfare will become steadily more and more focused on cities. This book sets out to help understand that future.

PART I

UNDERSTANDING THE COMPLEX OPERATIONAL ENVIRONMENT

1

Understanding the City

In its most basic form, the city can be thought of as a type of terrain. Thus, just as militaries have invested in understanding other types of terrains—desert, swamp, mountainous, jungle, or arctic—they must understand urban terrain. This knowledge allows military practitioners to distinguish between severely restrictive, restrictive, and unrestrictive terrain.[1] It allows planners to draw meaningful arrows on maps. It allows militaries to maneuver and accomplish their tactical, operational, and strategic objectives.

Urban terrain is more challenging than other types of terrain. It changes and grows at a much faster rate and is usually becoming more complex. This leads to it being much more difficult to understand or master. Urban terrain has existed for centuries, and militaries have been fighting in cities for centuries, but cities have grown exponentially bigger and exponentially more complex in recent decades. Mosul in 2017 looks nothing like Stalingrad in 1942. By contrast, the deserts that Lawrence of Arabia crossed in 1917 are not much different from how they look today.

Thus, it is critical for military practitioners to gain a sufficient understanding of this type of terrain, especially considering that urban areas often represent key terrain—terrain "whose seizure or retention affords a marked advantage to either combatant."[2]

---

[1] Department of the Army, *Intelligence Preparation of the Battlefield*, ATP 2-01.3 (Washington, DC: Headquarters, Department of the Army, 2019), 4-12 to 4-13, https://home.army.mil/wood/application/files/8915/5751/8365/ATP_2-01.3_Intelligence_Preparation_of_the_Battlefield.pdf.

[2] The U.S. Army defines key terrain as "an identifiable characteristic whose seizure or retention affords a marked advantage to either combatant." Department of the Army, *Offense and Defense*, ADP 3-90 (Washington, DC: Headquarters, Department of the Army, 2019), 3-10, https://armypubs.army.mil/epubs/DR_pubs/DR_a/ARN18010-ADP_3-0-000-WEB-2.pdf.

Controlling a city might lead to the control of a critical mountain pass or a crucial line of communication (supply route). Cities often have tactical, operational, and strategic value. Cities are more than just a terrain feature, however. They are a complex system of communities, social groups, networks, people, and physical spaces that are constantly changing.

Cities have been compared to living organisms or complex adaptive systems that ingest and process critical flows of resources such as commerce, energy, information, fuel, water, and waste.[3] A city needs fresh air, water, food, and power and a way to expel its wastes such as sewage and trash. Cities are also critical nodes in a complex global economy.

Rapid population growth and urbanization result in a world that is becoming increasingly urban and the likelihood that militaries will find themselves operating in urban areas—for either combat or noncombat operations—will remain and likely increase. The challenge for military forces is gaining a sufficient understanding of this complex and evolving environment so that it can successfully accomplish its mission.

In this chapter, John Spencer interviews Dr. David Kilcullen, a best-selling author and leading scholar in the field of urban warfare, unconventional warfare, and guerrilla warfare. Kilcullen served for 24 years as an infantry officer in the Australian Army and worked in the U.S. State Department where he was chief strategist in the Counterterrorism Bureau, senior counterinsurgency advisor to Multi-National Force Iraq, and senior advisor for counterinsurgency in the office of Secretary of State Condoleezza Rice.

He is the author of numerous books including *The Accidental Guerilla: Fighting Small Wars in the Midst of a Big One*, *Counterinsurgency*, and *Out of the Mountains: The Coming Age of The Urban Guerrilla*, that have influenced the United States and other militaries around the world. In this chapter, Kilcullen discusses the four megatrends of population growth, urbanization, littoralization, and networked connectivity and why they are relevant to the military practitioner.

John Spencer: I would like to start by discussing your book, *Out of the Mountains*. When it came out in 2013, it served as a wakeup call about the future of urban warfare. It has been fairly influential, and I routinely see

[3] John Spencer and John Amble, "A Better Approach to Urban Operations: Treat Cities Like Human Bodies," Modern War Institute, September 13, 2017, https://mwi.usma.edu/better-approach-urban-operations-treat-cities-like-human-bodies/.

the book cited in articles that I read discussing urban warfare. You discuss four emerging megatrends in the book: population growth, urbanization, littoralization, and networked connectivity.[4] Could you briefly discuss these trends?

David Kilcullen: The first megatrend is population growth. The world's population has been growing exponentially. It took from the start of human history until 1960 for the planet to reach three billion people. Yet it only took 40 more years for the world's population to double from three billion to six billion.[5] In the past 20 years, the world's population has grown by another 1.65 billion people and is on pace to grow from six to nine billion in fewer years than it took to grow from three to six million.[6] Thus, by 2035 we could have triple the number of people on this planet than in 1960 (see figure 1.1). That is a pretty big deal, not only as it relates to conflict and security, but for every aspect of life.

The second megatrend is urbanization. At the same time that the world's population is growing, the proportion of that population that is living in cities, not just the total population, is growing. In 1900, about ten percent of the world's population lived in urban areas. In 2007, the world's population was split evenly between urban and rural areas (see figure 1.2).[7] In 1950, there were only 83 cities with populations over one million. In 2018, there were 548, and by 2030 the United Nations estimates there will be 706.[8] By 2050, the United Nations also predicts that two-thirds of the world's population will be living in urban areas, with many of those located near a coast.[9]

The third megatrend is littoralization. People, economic activity, and a variety of other phenomena tend to cluster on coastlines. Presently, about 40 percent of the world's population lives within 100 kilometers of the

---

[4] David Kilcullen, *Out of the Mountains: The Coming Age of the Urban Guerrilla* (London: Hurst & Company, 2013), 17, https://doi.org/10.1080/23296151.2017.1384277.

[5] The world actually hit six billion in 1999. Department of Economic and Social Affairs, Population Division, *Total Population – Both Sexes* (New York, United Nations, 2021), https://population.un.org/wpp/Download/Standard/Population/.

[6] At current rates, the world's population will reach nine billion by 2035. Ibid.

[7] "The Urban Planet," *World Atlas of Desertification*, European Commission Joint Research Center, https://wad.jrc.ec.europa.eu/urbanplanet.

[8] Kilcullen, *Out of the Mountains*, 29; and Population Division, *The World's Cities in 2018* (New York: United Nations, 2018), 2, https://digitallibrary.un.org/record/3799524?ln=en.

[9] "Urbanization and the Megacity," World Population History, https://worldpopulationhistory.org/urbanization-and-the-megacity/.

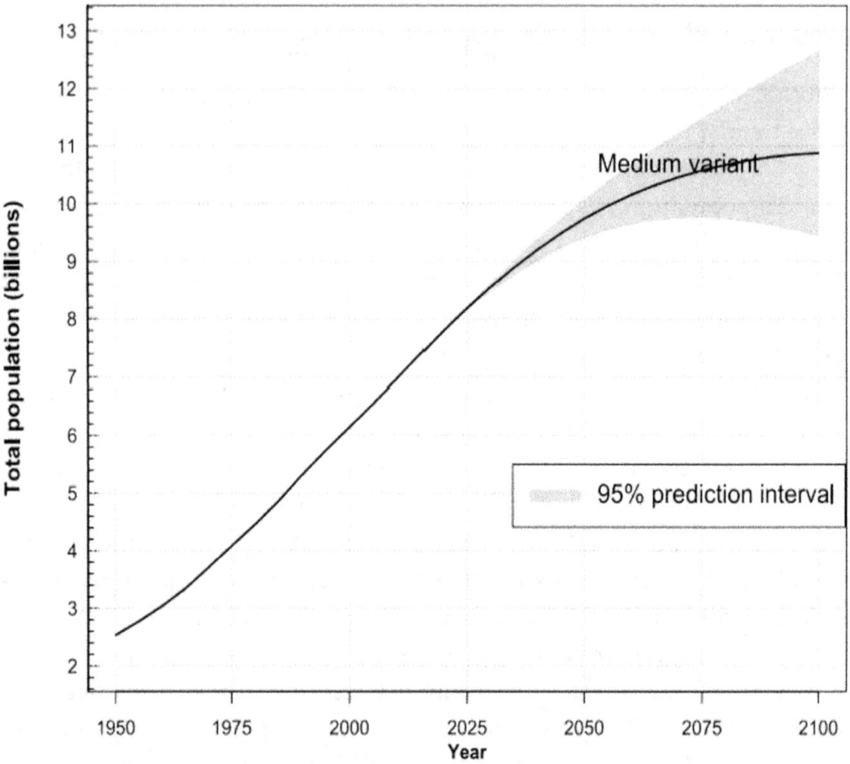

Figure 1.1. World population, 1950-2100.
Source: World Population Prospects 2019, Volume II: Demographic Profiles by the Department of Economic and Social Affairs, Population Division, © 2019 United Nations. Reprinted with the permission of the United Nations, https://population.un.org/wpp/Graphs/1_Demographic%20Profiles/World.pdf.

coast.[10] Since many cities are located near the coast, the dual megatrends of population growth and urbanization means that an increasing share of the world's population is clustering on coasts.

The fourth megatrend is networked connectivity. The first three megatrends have been around for centuries, but networked connectivity

[10] Data comes from the Center for International Earth Science Info. See, "Populations in Coastal Zones," Resource Watch, https://resourcewatch.org/data/explore/Populations-in-Coastal-Zones?section=Discover&selectedCollection=&zoom=2.11381382844091&lat=14.51107327694869&lng=-37.427202005158065&pitch=0&bearing=0&basemap=dark&labels=light&layers=%255B%257B%2522dataset%2522%253A%2522995ec4fe-b3cc-4cf4-bd48-b89d4e3ea072%2522%252C%2522opacity%2522%253A1%252C%2522layer%2522%253A%25222e533124-3258-4ec2-a777-beed1712d1a0%2522%257D%255D&aoi=&page=1&sort=most-viewed&sortDirection=-1.

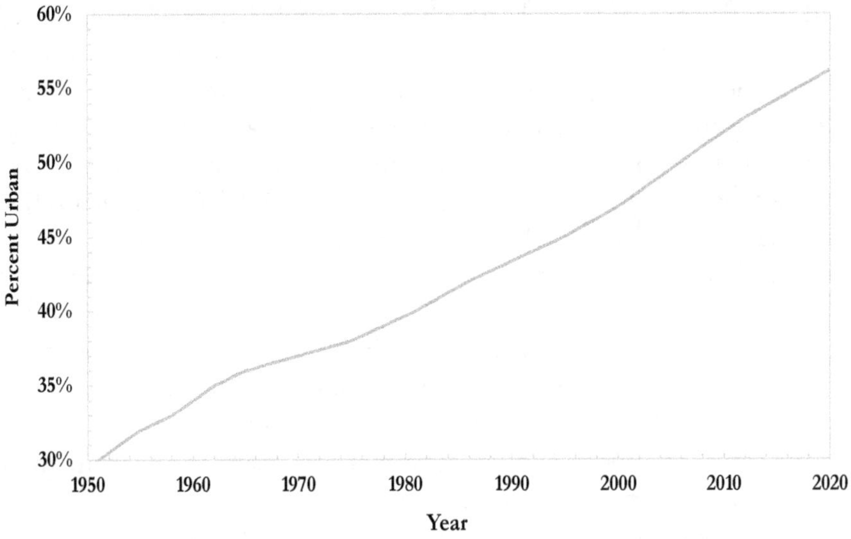

Figure 1.2. Percent of the world's population living in urban areas, 1950-2020.
Source: Graph was created by Liam Collins using data from Worldometer, "World Population by Year," https://www.worldometers.info/world-population/world-population-by-year/.

is a more recent phenomenon. There has been a massive explosion in electronic connectivity since around 2000, with the dramatic expansion of mobile devices, smart phones, the internet, access to television, and satellite communications. It is changing the degree of connectivity and, I would argue, is altering what we mean by a city. The traditional way of thinking about a city is in terms of its population, its complex manmade physical terrain, and its supporting infrastructure, and those are the three components that United States military doctrine considers to be part of the "urban triad."[11]

But if you also consider connectivity, a city's network footprint is far more than its terrain footprint or what is simply inside its city limits. For example, Mogadishu, the capital of Somalia, has nearly 2.5 million residents.[12] But there is a large Somali diaspora spread all over the world with clusters in Minneapolis and in cities in Norway, Switzerland, Sweden,

[11] Department of the Army, *Urban Operations*, ATP 3-06 (Washington, DC: Headquarters, Department of the Army, 2017), 1-3, https://armypubs.army.mil/epubs/DR_pubs/DR_a/pdf/web/ARN6452_ATP%203-06%20FINAL%20WEB.pdf.
[12] "Population of Cities in Somalia (2021)," World Population Review, https://worldpopulationreview.com/countries/cities/somalia.

and other European cities.[13] These Somali populations are linked by cell phones, the internet, financial transfers, and a bunch of other things back to their home city of Mogadishu. This produces a kind of overlapping of cities where the urban footprint today is dramatically different from the concept of a city in terms of its physical terrain and resident population of the past.

Spencer: Are these four megatrends bad? For many, urbanization has increased their economic opportunity and livelihood. Some countries seem to have thrived as a result of these megatrends. For example, the four "Asian Tigers"—Hong Kong, Singapore, South Korea, and Taiwan—experienced remarkable growth starting in the 1960s. At the time, they were considered developing countries. Now they are considered fully developed countries. Thus, these megatrends do not seem to be inherently bad or good and to some extent, it is what a nation makes of them.[14] So why are these megatrends concerning for the military?

Kilcullen: In the book I actually discuss how these megatrends are, on the whole, a good thing. The rise of urbanization has coincided with dramatic improvements in public health, in women's freedom, in economic prosperity, and many other things that are undeniably very good.[15] I think urbanization is very much a good thing. Urbanization is not necessarily a problem; it is only a problem when rapid or unplanned urban growth exceeds a city's ability to handle that growth.

This rapid change can create stresses and strains on an urban system that can lead to corruption, violence, a lack of economic opportunity, and a variety of other problems. The military needs to be concerned, not because urbanization is bad, but because urban areas are where more and more people are living. That means there is an increasing chance that the military could find itself in an urban environment for combat or noncombat operations.

Spencer: *Out of the Mountains* came out several years ago, and since that time we have seen urban battles all over the world: Islamic State of Iraq and al Sham (ISIS) in cities across Syria and Iraq, Russia in Ukraine, the Philippine Army in Marawi, or the Azerbaijanis in Shusha, Nagorno-Karabakh. You detail some of those events in your 2016 book *Blood Year: The*

[13] Phillip Connor and Jens Manuel Krogstad, "5 Facts about the Global Somali Diaspora," Pew Research Center, June 1, 2016, https://www.pewresearch.org/fact-tank/2016/06/01/5-facts-about-the-global-somali-diaspora/.

[14] Charlotte Severns, "The Development of the Four Asian Tigers," *Borgen Magazine*, March 6, 2021, https://www.borgenmagazine.com/development-of-the-four-asian-tigers/.

[15] Kilcullen, *Out of the Mountains*, 246.

Unraveling of Western Counterterrorism, but since writing *Out of the Mountains* and *Blood Year,* what has changed in regard to these four megatrends and the implications from them, or have things played out as you predicted?

Kilcullen: The genesis of *Out of the Mountains* grew out of a series of lectures that I used to give at the NATO Defense College in Rome. I started giving those lectures around 2010 when we were in the process of withdrawing from Afghanistan. I got very tired of lecturing students who would say two incompatible things. On the one hand, they would say that the war in Afghanistan was ending—which obviously turned out to be premature. On the other hand, they would say that the future of war was going to look a lot like Afghanistan.

I would explain that Afghanistan is landlocked, it is remote, it is one of the least urbanized countries on the planet, and until about 2015, most of the conflict in Afghanistan had been rural, not urban. So, I would ask: "What makes you think that Afghanistan is typical? What makes you think Afghanistan is going to be representative of what we see in the future?" No one could provide a good answer to that question, and that partly drove me to study the four megatrends in greater detail.

Looking at *Out of the Mountains* six years after having finished it, I definitely called some things correctly, but I also got some things wrong. I was right about the growth of urban conflict. A vast majority of conflict since that time has taken place in or around urban areas, to include Raqqa, Mosul, Fallujah, Tikrit, Ramadi, and Marawi—some of those I know you examine in this book.

There have been at least a dozen battles involving ISIS that have been fought in urban areas. Even in Libya, a predominantly desert country, a vast majority of the fighting since 2011 has occurred in coastal, urban areas.[16] This should not come as a surprise given that 80 percent of Libya's population lives in urban areas, most of which are located on the Mediterranean Sea.[17]

I was wrong in my prediction that megacities would be least able to cope with rapid urbanization. Instead, it was the smaller cities in the developing world—particularly in coastal Africa, Latin America, and Asia—that were least able to cope with rapid growth.[18] I think the book's

[16] Approximately 90 percent of Libya is considered desert. See, Abby Martin and Eva Lambson, "The Amazing Libya," Story Maps, December 11, 2020, https://storymaps.arcgis.com/stories/64cd1121b4e64c3090cf49f0b6cf927f.

[17] "Libya," *The World Factbook,* Central Intelligence Agency, https://www.cia.gov/the-world-factbook/countries/libya/#environment.

[18] Kilcullen, *Out of the Mountains,* 32.

analysis of megacities is still relevant, they just are not the most concerning right now.

Megacities—cities with populations exceeding ten million—pose an extremely difficult environment for military operations. What you might call mega-urban warfare requires entirely different skill sets from those of urban warfare in a smaller city for a variety of reasons including the "edgelessness" of megacities and the resulting inability to invest (or encircle) objective areas. It is important to understand and to be prepared to operate in a megacity, but as data has shown since 2013, it is not the most likely environment where militaries will find themselves engaged. Instead, recent conflicts have been fought in medium and smaller cities, so that seems the most likely trend for the foreseeable future.

One megatrend that I absolutely did not emphasize enough in the book was the speed of the connectivity explosion. I thought that I grasped it and discussed how it contributed to the Arab Spring[19] and how military techniques were diffusing because of the connectivity explosion. Yet, I still underestimated how fast and transformative it would be. I also predicted the application of military-grade precision and the emergence of what you might call info-kinetic maneuver—the ability to maneuver simultaneously in electronic and physical space—as long-term possibilities, yet they have already happened. So, I underestimated how quickly this would advance.

Spencer: I study some of this myself. Some large cities, such as Cairo have been around for millennia. They have experienced problems in the past, but I do not think to the level that exist today. For example, Rio de Janeiro has had ungoverned spaces for a very long time, since at least the creation of the first *favela* (slum) in the 19th century.[20] But what was the tipping point that caused security in so many of these cities—like Rio in 2018 or like many cities during the Arab Spring—to spiral to the point where military action was necessary? Do you think connectivity and the access to information has contributed?

Kilcullen: The question of tipping points is really important, particularly when we are discussing complex adaptive systems which, by

[19] The Arab Spring "was a series of pro-democracy uprisings that enveloped several largely Muslim countries, including Tunisia, Morocco, Syria, Libya, Egypt and Bahrain." See "Arab Spring," History.com, January 17, 2020, https://www.history.com/topics/middle-east/arab-spring.

[20] Bruce Douglas, "The story of cities #5: the rise and ruin of Rio de Janeiro's first favela," *The Guardian,* April 5, 2016, https://www.theguardian.com/cities/2016/apr/05/story-cities-15-rio-de-janeiro-first-favela-providencia-2016-olympic-games.

definition, involve nonlinear dynamics. Connectivity is a critical factor for tipping points, but I think one of the key issues is how cities adapt to stress. There are different ways a city can respond to extreme stresses. A city can deal with stresses through its existing resources, maybe by redistributing resources or becoming more efficient and effective in the way it uses current resources.

But a city may reach a tipping point where that is no longer feasible, prompting the city to undergo a rapid shift and a restructuring in the way it operates. That can be a positive thing if it leads to the city adapting and growing more robust through stress, what some people would now call antifragile. But a city may also fail to effectively respond to that stress, which leads to collapse and the emergence of what Naval War College professor Richard Norton describes as a feral city, a city that is still connected but one that is largely ungoverned and only partially functioning.[21]

Spencer: I love Professor Norton's feral city analogy, and I used that term at a conference that I recently attended. For some reason, the term offended many academics who were attending.

Kilcullen: Really!

Spencer: Yes, because a feral cat somehow offended this group. I was really surprised. It was like 200 people!

Kilcullen: That is interesting. For a biologist or a zoologist, the concept of ferality is not derogatory. It is simply a description of what happens to a domesticated organism when it goes back into a condition of wildness. It often gets bigger, shaggier, and more aggressive—there are observable changes in the creature's physical appearance and behavior. I think it is worth examining why people find that word so offensive when applied to a city. Many people describe the city as being an organism, so it seems that feral would be an appropriate word in that application.[22]

Returning to *Out of the Mountains,* I would have liked to have included the trend and impact of climate change in the book, but, at the time, there was a lack of good data or consensus as to how it would impact the earth in the 2030-2050 timeframe. I discuss it in the book, and I separately wrote

[21] See chapter 2, "Feral Cities," in this book.

[22] See, for example, Elizabeth Rapoport, "Interdisciplinary Perspectives on Urban Metabolism: A Review of the Literature," University College London Environmental Institute Working Paper, October 27, 2011, https://www.alnap.org/system/files/content/resource/files/main/um-lit-review-final.pdf; and Paola Pluchino, "The city as a living organism is circular by nature," *Green Biz,* August 6, 2021, https://www.greenbiz.com/article/city-living-organism-circular-nature.

a paper on climate impacts for NATO's urbanization project, but I do not describe it as a megatrend and do not make any projections about its likely impact. People were studying climate change, of course, and there was a lot of discussion around sea-level rise, but there was a lack of consensus as to what its specific urban impacts would be in the 2030–2050 timeframe, which is where the book is focused. People were discussing the possibility of a one-meter rise in sea-level rise by 2100, but we lacked good predictions based on firm data as to what it would be in 2040 or 2050.[23]

Since then, the data has become much better. The effects of climate change at high latitudes, such as the melting of the polar ice caps, is pretty obvious.[24] Climate change has impacted weather patterns, agriculture, and desertification in many developing countries. Whether or not those changes are caused by humans is somewhat irrelevant to this discussion, because the data is very clear that something is happening.[25] Since writing the book, I have conducted research into how climate change might affect urban development, and I think water shortage is the principal thing that military planners should consider. It is pretty clear from the conflict literature that water shortages in urban areas are a major trigger of urban violence.[26]

In *Out of the Mountains*—and I was one of the first academic researchers to write about this—I discussed the linkage between drought, water riots in Syria, and the outbreak of the Syrian civil war.[27] Shortly after the book's release, some data came out which tended to confirm the anecdotal evidence that I had heard from Syrians as I wrote the book.[28] For military professionals concerned with the urban environment and conflict, it is

[23] See, for example, Nobuo Mimura, "Sea-level rise caused by climate change and its implications for society," *Proceedings of the Japan Academy, Series B Physical and Biological Sciences* 89, no. 7 (2013), 281-310.

[24] See, for example, Sophie Coulson et al., "The Global Fingerprint of Modern Ice-Mass Loss on 3-D Crustal Motion," *Geophysical Research Letters* 48, no. 16 (2021), https://doi.org/10.1029/2021GL095477.

[25] "Climate Change," Food and Agriculture Organization of the United Nations, https://www.fao.org/climate-change/en/.

[26] See, for example, Jurgen Scheffran, "The Geopolitical Impact of Climate Change in the Mediterranean Region: Climate Change as a Trigger of Conflict and Migration," in *IEMed. Mediterranean Yearbook 2020* (Barcelona: European Institute of the Mediterranean, 2000), 55-61, https://www.iemed.org/wp-content/uploads/2021/01/The-Geopolitical-Impact-of-Climate-Change-in-the-Mediterranean-Region-Climate-Change-as-a-Trigger-of-Conflict-and-Migration.pdf.

[27] Kilcullen, *Out of the Mountains*, 220.

[28] See, for example, Brad Plumer, "Drought helped cause Syria's civil war. Will climate change bring more like it?" *Washington Post*, September 10, 2013, https://www.washingtonpost.com/news/wonk/wp/2013/09/10/drought-helped-caused-syrias-war-will-climate-change-bring-more-like-it/.

important to consider how climate change might impact water availability, security, and conflict.

Spencer: It seems to me that natural disasters, such as hurricanes, have increased in recent years. Is that simply my perception or is that supported by evidence? It is relevant because the military periodically deploys for humanitarian missions following natural disasters. For example, the United States military provided disaster relief to Puerto Rico after Hurricane Maria in 2018 and Haiti after its 2021 earthquake.[29]

Kilcullen: There seems to be fairly conclusive evidence that extreme weather events are "becoming more frequent and more severe."[30] For the sake of argument, let us take climate change out of the equation. The megatrends of population growth, urbanization, and littoralization have resulted in 267 million people living on land that is less than two meters above sea level and three billion living within 200 kilometers of a coastline.[31] Thus, millions of people are at risk from natural disasters, and the problem only gets worse with climate change. For example, one study found that a one-meter rise in sea level, even with no population growth, would increase the number of people living on land that is less than two meters above sea level from 267 million to 410 million.[32]

A recent *Wall Street Journal* article discussed the rising cost of hurricanes hitting the United States. It demonstrated that the data is very clear: the cost of coastal weather events is increasing and much of that can be attributed to the fact that there are simply more people living near the coast.[33] Littoralization is happening everywhere. The total population of the coastal regions of Mediterranean countries, for example, grew from

[29] Megan Friedl, "Service Members Continue Helping Puerto Rico after Hurricane Maria," Department of Defense, September 20, 2018, https://www.defense.gov/News/News-Stories/Article/Article/1640884/service-members-continue-helping-puerto-rico-after-hurricane-maria/; and "U.S. Military Support to Haiti Earthquake Relief," SOUTHCOM.mil, September 3, 2021, https://www.southcom.mil/HaitiEarthquakeSupport/.

[30] Rocio C. Labrador and Amelia Cheatham, "U.S. Disaster Relief at Home and Abroad," Council on Foreign Relations, August 24, 2020, https://www.cfr.org/backgrounder/us-disaster-relief-home-and-abroad.

[31] A. Hooijer and R. Vernimmen, "Global LiDAR land elevation data reveal greatest sea-level rise vulnerability in the tropics," *Nature Communications* 12, no. 3592 (2021), https://doi.org/10.1038/s41467-021-23810-9; and Liz Creel, *Ripple Effects: Population and Coastal Regions* (Washington, DC: Population Reference Bureau, 2003), 1, https://www.prb.org/wp-content/uploads/2020/12/RippleEffects_Eng.pdf.

[32] Hooijer and Vernimmen, "Global LiDAR land elevation data."

[33] Kara Dapena, "The Rising Costs of Hurricanes," *Wall Street Journal*, September 29, 2018, https://www.wsj.com/articles/the-rising-costs-of-hurricanes-1538222400.

100 million in 1980 to 150 million in 2005.[34] In just 25 years, an additional 50 million people and their property are now at risk, without even considering the effects of climate change.

Spencer: I want to return to understanding the city. *Out of the Mountains* turned me onto the analogy of looking at a city as an organism, but I often get pushback when I use that analogy.[35] Some argue that a city cannot be considered an organism because it cannot be killed. Throughout history, there have been attempts to destroy cities, but they almost always survive. Whether using an analogy or not, how can people better understand cities?

Kilcullen: I am a big fan of the idea, pioneered by Abel Wolman in 1965, of analyzing a city as a biological system. A city is essentially a system of flows—commodities, people, energy, information, fuel, or water, for example—that move into, through, and out of an urban environment. Thinking about these flows and viewing the city as an organism, leads to the idea of urban metabolism.[36] A city, like any system or organism, seeks homeostasis. A city (or any organism) tries to preserve itself. Thus, it seeks balance by effectively absorbing all its inputs, transforming them into biomass and energy, and then getting rid of waste products. If the city is unable to effectively process its inputs, toxic buildups result: pollution, crime, or violence, for example.

There is a subtle yet significant distinction between thinking about a city as an organism and thinking about it as an ecosystem. An ecosystem has multiple competing or collaborating organisms within it. It has a food web with predators and prey, for example. It includes a number of different systems. It is not necessarily a closed system where the bigger the input the bigger the pushback. Instead, it is a complex adaptive system where small inputs can have giant impacts, one where you can sometimes see a phase-change across the entire system.

I would tend to disagree with those who say you cannot kill a city. There are actually plenty of examples of cities being killed, such as in ancient times with Carthage, Jerusalem, or the ancient Mayan city of Naranjo, and

---

[34] Environment Programme, *2017 Mediterranean Quality Status Report* (New York: United Nations, 2017), https://www.medqsr.org/sites/default/files/inline-files/2017MedQSR_Online_0.pdf.

[35] Kilcullen, *Out of the Mountains*, 41-42.

[36] Kilcullen discusses the concept of urban metabolism in *Out of the Mountains*, but the idea originally comes from Abel Wolman who wrote about it in 1965. Abel Wolman, "The Metabolism of Cities," *Scientific American* 213, no. 3 (1965), 179-190, http://irows.ucr.edu/cd/courses/10/wolman.pdf.

more modern examples like Lidice in Nazi-occupied Czechoslovakia and Oradour-sur-Glane in Nazi-occupied France during World War II, or the city of Pripyat following the Chernobyl nuclear disaster.[37]

But I think it is a valid observation that it is a lot harder to kill an ecosystem than it is to kill any one organism within the ecosystem. Thus, I think it is useful to think of a city as an ecosystem. The most important thing about using any biological analogy, though, is to understand the limits of that analogy. If a city really is in some meaningful sense an actual ecosystem, then it should be possible to do what you might call a dynamic urban "intelligence preparation of the environment" to map, study, and observe that ecosystem in real time, and figure out how to affect it.[38]

The military must think about how to protect, damage, or control a city because the scale of these urban environments now is at the point where trying to control many cities by occupation is simply impossible. Thus, militaries must think in terms of how to control, protect, or influence a city, not through occupation, but by interdiction—through affecting the flows into and out of the city. That in turn means dealing with it as an ecosystem rather than a single undifferentiated block of urban terrain, which is how older doctrine tends to think about it.

Spencer: This brings up a really good point that I wanted to discuss. You said that "cities are sponges that soak up troops," not to steal the headline of one of your articles.[39] I think most experts and practitioners recognize this, and many conclude that the military should avoid cities altogether and never go there.[40] Since standing armies are not getting bigger and definitely not keeping pace with population and urban growth, how can militaries prepare for operating in large cities?

Kilcullen: I will provide a couple examples to illustrate how cities are sponges. When the Germans attacked the Red October factory in Stalingrad

---

[37] John Spencer, "The Destructive Age of Urban Warfare; Or, How to Kill a City and How to Protect It," Modern War Institute, March 28, 2019, https://mwi.usma.edu/destructive-age-urban-warfare-kill-city-protect/.

[38] U.S. Army doctrine defines intelligence preparation of the battlefield as "the systematic process of analyzing the mission variables of enemy, terrain, weather, and civil considerations in an area of interest to determine their effect on operations." See, Department of the Army, *Intelligence Preparation of the Battlefield*, Glossary-5.

[39] David Kilcullen, "Urban Combat: 'Cities Are Sponges That Soak Up Troops'," *Cipher Brief*, October 22, 2017, https://www.thecipherbrief.com/urban-combat-cities-are-sponges-that-soak-up-troops.

[40] See, for example, Robert H. Scales Jr., "The Indirect Approach: How U.S. Military Forces Can Avoid the Pitfalls of Future Urban Warfare," *Armed Forces Journal International* 136, no. 3 (1988), https://apps.dtic.mil/sti/pdfs/ADA399275.pdf.

in 1942, they sent an entire brigade against a single building complex.[41] When the Russians attacked Berlin in 1945, they assembled a force of five armies—nearly 500,000 men—and pushed an entire battalion down each side of some streets on the main axis of advance.[42] The traditional approach to deal with an urban environment has been to "invest" (encircle or cordon) the city, break into it, advance toward key objectives, and then clear it, block by block.

This technique requires throwing very large troop formations into relatively small pieces of physical terrain because there are so many spaces in that three-dimensional terrain that have to be cleared and held. Every single room or interior space needs to be garrisoned in case the enemy counterattacks or reinfiltrates into it, and that soaks up troops. This was challenging enough during World War II when cities were a fraction of their current size and militaries exceeded ten million uniformed members.[43] But with smaller armies and larger cities, that technique is no longer possible, so people have been trying to figure out how to control cities with fewer troops.

One of the cases that I examined in *Out of the Mountains* was the 2008 Mumbai attack. The population of Mumbai exceeds 20 million people.[44] You could put the entire U.S. Army (active component, reserve component, and National Guard), totaling over one million soldiers into Mumbai, and many residents would probably not even know that they were there.[45] There simply are not enough troops in the world to secure these cities using the traditional approach of putting small groups of soldiers on every street corner.

The question becomes how to control or influence a city without requiring massive troop formations numbering in the millions. Al Qaeda in Iraq provided one answer to that question when it was able to exert significant control over Baghdad with only a few thousand fighters. When Abu Musab al-Zarqawi, the head of al Qaeda in Iraq, was killed in 2006, one of the pieces of pocket litter allegedly recovered from his body was a

---

41 Antony Beevor, *Stalingrad: The Fateful Siege: 1942-1943* (New York: Penguin Books, 2007), 187.

42 David Porter, "Seelow Heights and Berlin," *Military History*, June 11, 2020, https://thepast.com/feature/seelow-heights-and-berlin/.

43 At peak strength, the Russians had 13.2 million, the United States 12 million, Germany had 9.5 million, Japan 7.5 million, and China 5.7 million. See, "Top Ten Largest Armed Forces," Erenow, https://erenow.net/ww/the-history-buffs-guide-to-world-war-ii/9.php.

44 "Mumbai Population 2022," World Population Review, https://worldpopulationreview.com/world-cities/mumbai-population.

45 For fiscal year 2020, the regular force included 480,000 soldiers, the National Guard 336,000, and the Army Reserve 189,500. See Mark Cancian, "U.S. Military Forces in FY2020: Army," CSIS, October 15, 2019, https://www.csis.org/analysis/us-military-forces-fy-2020-army.

hand-drawn sketch map of Baghdad. The map depicted Zarqawi's battle for the "belts" around Baghdad.

The belts were "residential, agricultural, and industrial areas that encircle[d] the city, and networks of roadways, rivers, and other lines of communication that [lay] within a 20 or 30 mile radius of Baghdad and connect[ed] the capital to the rest of Iraq."[46] The map depicted Zarqawi's campaign plan and how he prioritized control of the belts—and, in particular, of entry and exit points through those belts along main supply routes—to control the city.[47] Thus, al Qaeda in Iraq sought to influence and control access to the city, not by a large presence within the city, but instead by effectively occupying and penetrating the peri-urban space around the city.

When I served as a counterinsurgency advisor during the surge in Iraq in 2007, I worked with Colonel J.B. Burton who commanded a brigade in northwest Baghdad. One of the things that Burton and other commanders were trying to do was to figure out the relationships between rural villages outside Baghdad and the population within the city itself. If you examined different parts of Baghdad, you could draw links between sections within the city and specific villages or towns outside Baghdad where people lived before moving into the city. This sort of flow modeling was difficult back in 2007, but it has become easier now with the connectivity explosion, which allows us to track those relationships much easier. What Burton and others were doing was looking for ways to influence the city without having to put troops on the ground.

Thus, we may be approaching a time where we have several military options for influencing or controlling a city, some of which do not involve putting troops on the ground at all. Imagine a situation where the city's major airport or port is serviced by a labor force and that labor is predominantly drawn from the poorer residents of the city. Many of the workers are more recent immigrants who reside in the poorer neighborhoods and these neighborhoods are often controlled by gang leaders. That is a common circumstance in many developing world cities.

Now imagine that you can figure out that the gang leader who controls a particular workforce that runs the port emigrated from a remote village 20 or 30 years ago. Maybe instead of trying to use force at the port, you can instead exert influence or control by inserting a special forces detachment

[46] "Baghdad Belts," Institute for the Study of War, https://www.understandingwar.org/region/baghdad-belts.
[47] Ibid.

in the village that he emigrated from which is 100 miles away. Maybe the detachment can speak with the gang leader's mother or grandmother to convince him to work with the military. Or maybe you can exert influence or control even more remotely. Rather than a special forces detachment 100 miles away, you could reach out remotely over the internet or through diaspora populations to engage with people who can affect the way that a city operates.

Yet none of these examples can prevent the military from having to be prepared to physically operate in an urban environment. To do so would be equivalent to a fire department saying, "the solution lies in fire prevention, so instead of worrying about firefighting, all we need to worry about is fire prevention." The military cannot say, "It is too hard to operate in an urban environment, so we are not going to do it." That approach never works.

The United States military took that approach with counterinsurgency following the Vietnam War, but when the United States became involved in a counterinsurgency fight in Iraq in 2003, it was poorly prepared. The military may want to avoid urban areas, but it still must be prepared to operate within them if needed. Thus, while different ways may be emerging to control a city without entering it, that does not absolve the military from the responsibility of being able to enter and operate effectively within a city.

Spencer: Do you think the United States military or other militaries have grasped these changes and are effectively preparing themselves to operate in contemporary urban environments?

Kilcullen: I think the United States is doing a decent job, as are the other Five Eyes nations (United Kingdom, Australia, Canada, and New Zealand), as are France and Poland.[48] Nobody wants to fight in an urban environment, but I think these militaries are doing a decent job of embracing it. Russia is another country that has really done a lot of thinking about urban warfare and, frankly, is pretty good at it. The Russians had a very searing experience in the 1990s in Chechnya with two major urban battles in Grozny in 1994 and 1999. They learned a ton in the five years between those two battles. The traditional Russian approach, of course, is totally

[48] The Five Eyes is an intelligence-sharing alliance that traces its origins back to World War II. See, Raghav Bikhchandani, "What is 'Five Eyes,' the intelligence alliance US wants South Korea, India Japan to be part of," *The Print*, September 9, 2021, https://theprint.in/world/what-is-five-eyes-the-intelligence-alliance-us-wants-south-korea-india-japan-to-be-part-of/730475/.

unlike ours and not one we would or should seek to copy—in effect it has been to just remove the city: shell it and then bulldoze it.[49]

Spencer: Yes, I agree. That was one lesson that the Russians learned after the first battle of Grozny: avoid getting sucked in. Instead, prepare the city with plenty of artillery fire before entering.[50]

Kilcullen: Correct. And the Russians also learned some lessons about operating in urban terrain from their limited urban battles in Ukraine in 2014-2015. And since 2015, they have engaged in urban conflict in Syria. Much can be learned from their experiences, and they are certainly one of the more advanced nations in terms of thinking about urban conflict.

The Israelis have also done a huge amount of work on understanding urban conflicts and understanding how to operate in the urban environment. Israel is unique for several reasons (particularly the country's small size) so not all their practices may transfer well, but many of their techniques do. The Israelis pioneered reconnaissance techniques and the use of combat engineers in the urban environment. They move fully inside the urban structure through a technique they call infestation. If you want to understand operating in an urban environment, you must study the Israelis.[51]

Spencer: So that was actually my next question. Do you see enough study going into this problem, or should there be more, both from within and outside the military?

Kilcullen: I think that there is a decent effort to study this, and it often involves academic rigor. You can see a lot of stuff in the urban studies community, law enforcement, and research agencies of various kinds.[52] Many think tanks are also studying it.[53] I first became heavily engaged in urban study about 25 years ago, when the "Black Hawk Down" battle in Mogadishu took place.[54] I was a captain at the time. I was training for military advisory work in Southeast Asia and many of us were transfixed by that battle. That battle also prompted a major look at urban operations

[49] Kilcullen, *Out of the Mountains*, 109.

[50] Olga Oliker, *Russia's Chechen Wars 1994-2000* (Santa Monica, CA: RAND, 2001), https://www.rand.org/pubs/monograph_reports/MR1289.html.

[51] Eyal Weizman, "Lethal Theory," *Log* 7 (2006), 53–77, http://www.jstor.org/stable/41765087.

[52] See, for example, Jocelyn Fontaine et al., *Updated: Views of the Police and Neighborhood Conditions* (Washington, DC: Urban Institute, 2019), https://www.urban.org/sites/default/files/publication/100706/2019.11.11_ni_community_survey_brief_final_0.pdf.

[53] See, for example, Gian Gentile, *Reimagining the Character of Urban Operations for the U.S. Army: How the Past Can Inform the Present and Future* (Santa Monica, CA: RAND, 2017), https://www.rand.org/pubs/research_reports/RR1602.html.

[54] See, chapter 8, "The Battle of Mogadishu," in this book.

in the United States.[55] So you can go back to the mid-1990s to see that many of the things that we are discussing today are the same things people were discussing 25 years ago.

By the late 1990s, the U.S. Army was studying Chechnya, Bosnia, Mogadishu, and other places, and thinking about how to conduct urban operations. There was even an urban operations journal.[56] The United States Marine Corps published a report in 2000, *Chaos in the Littorals*, that examined urban littoral environments.[57] So, this was something that militaries were studying in the late 1990s. But then 9/11 happened and the military largely directed its focus to unconventional war, counterterrorism, and counterinsurgency.

The two battles of Fallujah in 2004 reminded the military that the urban environment was a critical operating space, but the Defense Department as an institution, finding itself thoroughly engulfed in a counterinsurgency campaign, could not dedicate sufficient effort to studying urban warfare. It was not until 2010 or 2011 that the institution re-engaged with the need to study the urban environment.

In the intervening ten years, while military experts were focused on other things, urbanists, urban planners, and urban ecologist academics continued to study the urban environment (which of course developed rapidly during that time as well), so urban knowledge continued to advance. Bringing these researchers and their research together with the military has been important to advance where we are now. But collaboration or understanding is not always easy because these are two very different groups that have different political outlooks, viewing cities and urbanization through very different lenses, and this often results in misunderstanding. Thus, it is important for the military planner and urban ecologist to overcome their suspicions and instead recognize

---

[55] See, for example, *United States Forces, Somalia: After Action Report and Historical Overview: The United States Army in Somalia, 1992-1994* (Washington, DC: Center of Military History, 2003), https://history.army.mil/html/documents/somalia/SomaliaAAR.pdf.

[56] The website Small Wars Journal (https://smallwarsjournal.com/content/about) was originally started as the *Urban Operations Journal* by the founder Dave Dilegge in the mid-1990s. See also, Howard Altman, "Dave Dilegge, the 'grandfather of urban warfare studies,' has died," *Military Times*, May 6, 2020, https://www.militarytimes.com/news/your-military/2020/05/06/dave-dilegge-the-grandfather-of-urban-warfare-studies-has-died/.

[57] Francis Donovan, *Chaos in the Littorals: Anarchic Wars and the United States Marine Corps* (Quantico, VA: Marine Corps University, 2000), https://apps.dtic.mil/sti/pdfs/ADA526292.pdf.

their differences to learn from one another. That is how we can take our understanding to the next level.

Spencer: As you say, much work has been done, but there is still much to do. What are some research areas or specific research questions that need to be examined more?

Kilcullen: I will provide my top three. The first relates to something we discussed earlier: Are cities better understood as organisms or as ecosystems? This is a modeling question. How do you understand a city under stress? One approach could be building a dynamic urban "intelligence preparation of the battlefield" staff ride that allows you to examine a city. A traditional intelligence preparation of the battlefield involves building various overlays: operational graphics, threat overlays, modified combined obstacle overlays, for example. A dynamic urban intelligence preparation of the battlefield would involve building a different set or different types of overlays. It might involve video or "ultrasound" that allows us to see, in real time, dynamic changes in an urban environment. Or maybe it might be better to think about it as a graphic equalizer—if I dial this up over here, what happens over here?

The second area that needs more research relates to the connectivity explosion and info-kinetic maneuver that I mentioned earlier. Until maybe five to seven years ago, most western theorists considered cyber war as a stand-alone form of conflict, something that was separate from physical conflict. But in recent years, we have seen how conflict in these two domains have merged.

In Ukraine in 2014, the Russians effectively combined lethal operations, cyber operations, and information operations when they sent text messages to Ukrainian soldiers prior to an artillery strike with messages such as, "retreat and live."[58] Commanders currently have fire control officers to control artillery fires. In the future they are likely to have cyber control officers as well so they can simultaneously maneuver in the informational and physical domains.

Earlier, I mentioned that the military thinks of a city in terms of an urban triad: its population, its physical infrastructure, and the terrain. The informational layer should be added to this triad, and remember, much of that information layer may be emanating from outside the city limits.

[58] Oleksandar Golovko, "Ukrainian Frontline: Cyber + EW + Psyops," PowerPoint (Kyiv: General Staff of the Armed Forces, 2018).

NATO's urban operations capstone concept now talks about an "urban quad" that includes information.[59]

Thus, this is a critical area of examination. Some good work has already been started in this area, such as *On Cyber: Towards an Operational Art for Cyber Conflict*, by Greg Conti and David Raymond, but more is still needed.[60] *On Cyber* is a brilliant book that everyone should read. I think this is an urban issue because connectivity tends to be urban. That is where cell phone networks and other systems are densest.

The third area of research that needs to be studied more is how to operate inside a city with a small-sized force without getting sucked in and defeated in detail. That involves intelligence preparation, airborne maneuver, autonomous systems, and technologies that allow you to secure a space without having to put a human in it. There are technologies, techniques, and concepts that must be developed.

We are still thinking about urban operations as an updated version of 2004's Operation Phantom Fury (The second battle of Fallujah).[61] We cannot afford to do that. Those kinds of operations do work, but they take a very long time, they soak up a lot of troops, and they are very casualty intensive. Rather than an updated version of old techniques, modern urban warfare requires entirely new approaches that consider emerging technologies.

Spencer: I often point to the first battle of Fallujah to remind people of the political pressure that often accompanies urban battles. The attack was terminated prematurely due to political pressure from the Iraqi government. The perceived damage to the city contributed to this decision.[62] Thus, military planners must consider political constraints during planning or risk developing a plan that is not politically feasible.

Kilcullen: Right. It is extraordinarily interesting when you contrast operations in Raqqa and Mosul in 2017 with operations in Fallujah thirteen years earlier. Politicians think much differently about urban operations now than they did at the beginning of the Iraq War. Many now realize that fighting in urban areas will likely involve significant casualties

---

[59] NATO, *Capstone Concept: Joint Military Operations in an Urban Environment*, NATO SH/SDP/J5/STP1/VP/18-000452 (Brussels: NATO, 2018).

[60] Greg Conti and David Raymond, *On Cyber: Towards an Operational Art for Cyber Conflict* (New York: Kopidion Press, 2017).

[61] For more on the operation, see chapter 9, "The Second Battle of Fallujah," in this book.

[62] Joel D. Rayburn and Frank K. Sobchak, eds., *The U.S. Army in the Iraq War, vol. 1* (Carlisle Barracks, PA: U.S. Army War College Press, 2019), 291-292, https://digitalcommons.usmalibrary.org/cgi/viewcontent.cgi?article=1018&context=books.

and significant collateral damage. But it is the military's job to provide politicians with options and ensure they clearly communicate the potential costs of those options.

Given the megatrends of population growth, urbanization, littoralization, networked connectivity, and climate change, not to mention an examination of recent history, it seems likely that the military will find itself conducting operations—either combat or noncombat—in urban areas in the near future. Thus, it is imperative for the military to continue to invest in urban operations.

2

Feral Cities

In 2003, Dr. Richard Norton published an article titled "Feral Cities" in the *Naval War College Review*.[1] At the time, there was a great deal of scholarly attention paid to the phenomenon of failed or failing states, but little attention paid to failed or failing cities.[2] Dating back to at least the 1960s, the idea of a city being "ungovernable" had been bantered around, but there was a relative lack of deep analysis into what this actually meant.[3] Norton, thus, coined the term "feral cities" and offered a taxonomy to examine and classify cities as healthy, feral, or somewhere in between.

Feral cities are "a metropolis in a nation-state where the government has lost the ability to maintain the rule of law within the city's boundaries."[4] Yet these cities "remain connected to the greater international system through such avenues as trade and communication" so they cannot be ignored.[5] The taxonomy that Norton uses to classify cities includes five variables: governance, economy, services, security, and civil society. As global urbanization and population growth increases, less developed counties will continue to struggle to extend governance over their rapidly

---

[1] Richard J. Norton, "Feral Cities," *Naval War College Review* 56, no. 4 (2003), 97, https://apps.dtic.mil/sti/pdfs/ADA523703.pdf.

[2] See, for example, James F. Miskel and Richard J. Norton, "Spotting Trouble: Identifying Faltering and Failing States," *Naval War College Review* 50, no. 2 (1997), 79–91, https://digital-commons.usnwc.edu/cgi/viewcontent.cgi?article=2875&context=nwc-review.

[3] See, for example, Nathan Glazer, "Is New York City Ungovernable?" *Commentary*, September 1961, https://www.commentary.org/articles/nathan-glazer-2/is-new-york-city-ungovernable/.

[4] Richard J. Norton, "Feral Cities: Problems Today, Battlefields Tomorrow?" *Marine Corps Journal* 1, no. 1 (2010), 51, https://www.marines.mil/Portals/1/Publications/Marine%20Corps%20University%20Journal%20Vol%201,%20No%201%20PCN%2010600001400.pdf.

[5] Ibid.

growing cities and megacities.[6] Thus, a taxonomy to identify at-risk cities can be useful, from a policy perspective, to help prevent cities from becoming feral.

In this chapter John Spencer interviews Dr. Norton to discuss feral cities. Norton is a professor of national security affairs at the U.S. Naval War College and a retired U.S. Navy commander. While in the Navy, he served extensively at sea on cruisers and destroyers, as well as shore tours in the Navy's Office of Legislative Affairs and the Navy Staff. In addition to his 2003 article, Norton published the article "Feral Cities: Problems Today, Battlefields Tomorrow?" in the *Marine Corps University Journal* in 2010.[7]

John Spencer: Can you tell us how you started studying cities and how you coined the term "feral cities"?

Richard Norton: My interest in cities dates back to 2000. I was attending a conference with my wife in South Africa when I observed something that I had not seen before. The South Africans had recently moved their stock exchange, the Johannesburg Stock Exchange, from its downtown location to Sandton, 17 kilometers north of its previous location.[8] That move would be the equivalent to the New York Stock Exchange moving its headquarters from its Wall Street location in the heart of New York City to Mahwah, New Jersey where its data center is located. The South Africans made the move for security reasons.

As I looked around, it became more apparent that businesses were abandoning Johannesburg's central business district. I started to think that we could see cities slip out of the control of their owning states, yet still play a significant role in the global community. I decided that feral seemed an appropriate word, because it was descriptive of something that had been domesticated but was now wild. Also, to be honest, it was a great hook, and it was one of those titles that kind of sells itself.

Spencer: For sure. I am familiar with international relations research into failed states, but a feral city is different. What is a feral city? How do

---

[6] Department of Economic and Social Affairs, Population Division, *The World's Cities in 2018: Data Booklet* (New York: United Nations, 2018), https://digitallibrary.un.org/record/3799524?ln=en.

[7] Norton, "Feral Cities," 50-77.

[8] "The JSE celebrates 20 years in Sandton Central," *Construction World*, November 10, 2020, https://www.crown.co.za/construction-world/property/14738-the-jse-celebrates-20-years-in-sandton-central.

you define it? How do you categorize it? And how is it different from a failed state?

Norton: My interest in feral cities came at the high-water mark of other scholars' interest in failed states. A feral city is a metropolis that remains within the territory of a sovereign state, but the government has lost the ability to maintain the rule of law within the city's boundaries, something else governs the city. For example, some parts of Baltimore might meet this definition. In some cases, a feral city results because the forces that were indigenous to the city were capable of overcoming those of the state. In other cases, a feral city results because, over time, the state has lost the political will or the capacity to govern. And the more I looked, the more I became convinced that this was actually happening. Although, much like failed states, we have not seen many truly feral cities, we do see cities that are approaching feral status just as we see fragile and faltering states.

Spencer: Just to be clear, when a state loses control of a city, you mean the state's government has lost control?

Norton: Correct. If for some reason Paris were to become a feral city, it would still fall within the territory of France, but the French government would be incapable of executing governance over some or all of Paris.

Spencer: I find it interesting that even a feral city remains connected to the international community or what you have described in your work as "globally connected."[9] Can you describe what you mean by globally connected?

Norton: If you think back to the past, some of the great pirate cities—Port Royal, Jamaica, for example—the city was connected to the larger global commerce because it was a port. Ships could come in and out, but Jamaica had little to no part in the governance and security of most of the city. Today, cities that are feral or trending feral remain connected to the rest of the world by air, some by road and rail, some by port, and always by the internet. When I first wrote about this in 2003, the idea was that anyone with a panel truck and a satellite phone could run a world-wide enterprise from a feral city. Today all you need is a briefcase and a laptop. Thus, a city, although it may lack formal governance and security, remains connected.

An example is Mogadishu. For many years, Mogadishu was the poster child for feral cities. Somalia was a failed state and Mogadishu lacked formal governance and security. Clans and warlords ran the city.

[9] Norton, "Feral Cities," 97.

Yet daily flights continued into the Mogadishu airport, professors still taught classes at the university—although students and professors had to periodically take cover due to automatic weapons fire—and the city's population remained connected by road, rail, and the internet. That is an example of how once you are on the grid, it is really hard to get off the grid. The days of Tiananmen Square, where the Chinese could pretty well isolate Beijing, are over.

Spencer: In the midst of the current COVID-19 pandemic, I find it fascinating that when you wrote your 2003 article, you identified the risk of pandemics emanating from feral cities, with a feral city "serving as a pathogen incubator and point of origin of an intercontinental epidemic."[10] Can you discuss why feral cities pose a concern for pandemics or epidemics?

Norton: The toxic environment of a feral city poses uniquely severe threats. They could function as Petri dishes where a new illness or strain of an existing disease could mutate and spread within the city without detection, before spreading throughout the world that the feral city remains connected to. By contrast, a functioning city can often detect and trace the point origin of a new disease and isolate it before it has the chance to spread globally. This was the case for the 2003 SARS epidemic.

COVID-19 did not originate from a feral city, yet functioning cities still offer better hope to mitigate the effects of the pandemic through better record keeping, better medical facilities, and better schools and governance that allow learning to happen and many to maintain their livelihood even if they cannot physically go to school or work. Given the lack of government resources, the lack of health services, and the lack of control, feral city cities remain particularly susceptible to pandemics.

Spencer: Yes, I have seen some interesting research about how increased global urbanization is increasing the number of COVID-19 types of diseases that are occurring. The urban sprawl and urban expansions bleed more into the natural habitats of wild animals and the resulting co-relation serves as an incubator for diseases to mutate and flourish.[11] We have to beat COVID-19, but the research concludes that we better be prepared for more COVID-19 type pandemics.

[10] Ibid., 100.

[11] Karin Brulliard, "The next pandemic is already coming, unless humans change how we interact with wildlife, scientist say," *Washington Post*, April 3, 2020, https://www.washingtonpost.com/science/2020/04/03/coronavirus-wildlife-environment/.

Norton: That is correct. The United Nations, for example, reports that 68 percent of the humans on earth will live in an urban environment by the year 2050, and there are already parts of the world where over 80 percent of the population are urban dwellers.[12] When people are so densely packed, unless they have an extraordinarily well-administered and very robust health system, it just stands to reason that a new disease that gets loose is going to wreak havoc, and I think that is what we are seeing with COVID-19. And you are absolutely right that as bad as COVID-19 is, we do not know what the next pandemic will be like.

Spencer: Last year, I had the opportunity to travel to Mumbai, India to study the 2008 attack. Mumbai is by no means a feral city, but the sanitation and essential services in that city are huge issues and the city has over one million people living in slums.[13] Reflecting back on what I saw there, I could imagine that disease could spread like a wildfire in an environment like that.

Norton: I agree. Mumbai is a great example in many ways. As you point out, the Dharavi slums contain more than one million residents. The state of India has effectively given up some sovereignty and allowed a kind of a people's court to be established in some of these areas. Nongovernment organizations often play an active role in these areas, and you could argue that it is a good deal for the state, but if the state is letting others handle social and health issues, it cannot help but to weaken the social contract between state and population. I would agree that Mumbai is not a feral city, but if you grouped cities into green, yellow, and red categories, Mumbai would probably trend to the yellow category. It will be really interesting, not to sound macabre, to see which cities handle the pandemic better than others. Mumbai would definitely be on the list of the cities that I would like to see how it handles the pandemic at the granular level, but I am not an epidemiologist.

Spencer: No, neither am I. I do love the study of cities, even though I am not an urbanist. I just listened to an interview with a good friend of mine, Max Brooks. He is getting a lot of attention for some of the similarities between the zombie apocalypse in his book *World War Z* and

[12] Department of Economic and Social Affairs, Population Division, "68% of the world population projected to live in urban areas by 2050, says UN," United Nations, May 16, 2018, https://www.un.org/development/desa/en/news/population/2018-revision-of-world-urbanization-prospects.html.

[13] Yue Zhang, "Building a Slum-Free Mumbai," Wilson Center, March 3, 2016, https://www.wilsoncenter.org/article/building-slum-free-mumbai.

the current pandemic.[14] He explained that if you look at some of the most vicious diseases in history that still remain, some of them can be controlled because they are found in remote areas, in African villages, for example, that are not globally connected. Thus, it is much easier for the World Health Organization to go in and contain the disease.[15] But because of urbanization and globalization, containment is nearly impossible in most of the world today.

Norton: I think you are right. I think your mention of Max Brooks is on point but also humorous. Several years ago, as president of the Naval War College, Phil Wiscup brought Max in to talk about *World War Z*. He provided a great presentation and the immediate linkages between a pandemic as a threat for which we were unprepared, and his book was excellent.

Spencer: After watching the video of that presentation, I asked him to come and speak to the Chief of Staff of the Army's Strategic Studies Group in 2014, where I was working at the time.[16] Later, when I was assigned to the Modern War Institute at West Point, I asked him to come and talk to cadets and faculty at the United States Military Academy.[17] He has been a non-resident fellow with the Modern War Institute since its inception, so he is a close friend and like you said, he has a very brilliant mind.

Norton: You are lucky to know him that well, that is super.

Spencer: Returning to the discussion of controlling a disease or controlling a city in general, much of your definition of feral cities is about control, what might also be described as security or power. You discussed how Mumbai has allowed some of its slums to have its own judicial system. Is the lack of governmental control or the delegation of some functions a bad thing? For instance, Rio de Janeiro was in the news recently because some

---

[14] Terry Gross, "'All of This Panic Could Have Been Prevented' Author Max Brooks on COVID-19," *NPR Fresh Air*, March 24, 2020, https://www.wbur.org/npr/820601571/all-of-this-panic-could-have-been-prevented-author-max-brooks-on-covid-19.

[15] "Factors that contributed to undetected spread of the Ebola virus and impeded rapid containment," World Health Organization, January 2015, https://www.who.int/news-room/spotlight/one-year-into-the-ebola-epidemic/factors-that-contributed-to-undetected-spread-of-the-ebola-virus-and-impeded-rapid-containment.

[16] U.S. Naval War College, "Max Brooks: World War Z," YouTube video, April 5, 2012, https://www.youtube.com/watch?v=-nGG5E04cog.

[17] Modern War Institute, "Max Brooks on the Courage to Champion Creativity in the Military," YouTube video, January 31, 2017, https://www.youtube.com/watch?v=Rgq04T9YOCc.

of its gangs have been enforcing COVID-19 social distancing directives and they have been very vicious about their "follow the rule or else."[18]

Norton: I think that is one of the more interesting things that we are going to see as a result of this pandemic. The *favelas* (shanty towns or slums) in Rio and elsewhere in Brazil have their own informal governance structures that produce what I call patchwork governance. Brazil controls most of Rio, but there are neighborhoods where the state's ability to provide security and policing services is nearly nonexistent. You have everything from the theft of electricity to gang members sanctioning the limited medical services.[19]

If you live in an area like that, you have to make some pretty stark choices. When a local gang member comes to you and says, "We have the clinic on the corner, and you are safe as long as you follow our rules," it is pretty easy to follow those rules. It reminds me of a conversation that I had with a colleague whose daughter goes to one of the worst *favelas*—I think it is *Complexo do Alemão*—in Rio every year. He told me that she has never been molested. When I asked, "Why is that?" He responded, "It is because the local warlord, basically the drug lord, put the word out that nobody touches her, or they will have to answer to his wrath. Because he likes the idea that the population sees him, as opposed to the government, providing these health services."

I think the idea of the gangs policing raises a really interesting question about how a city works, and it may not be a bad thing if you consider the alternative of no policing. A city that might be viewed as very dangerous by outsiders could be potentially very safe and very human-security oriented to its residents, although I have yet to see that. There are other ramifications. You are an urban warfare specialist, and you know as well as anybody how difficult combat is in an urban environment. If 68 percent of the world's population lives in urban environments and war remains a human activity—humans fighting humans—even if the army, Marines, or any other service wants to avoid cities, it seems like they are

[18] Caio Barretto and Tom Phillips, "Brazil gangs impose strict curfews to slow coronavirus spread," *The Guardian*, March 25, 2020, https://www.theguardian.com/world/2020/mar/25/brazil-rio-gangs-coronavirus.

[19] Clarissa Lins, "Providing electricity to Rio de Janeiro's favelas," *The Guardian*, March 18, 2014, https://www.theguardian.com/sustainable-business/providing-electricity-rio-de-janeiro-favelas; and John Sullivan and Robert Bunker, "Third Generation Gangs Strategic Note No. 38: Cape Town Extortion Gangs Target Paramedics," *Small Wars Journal*, June 1, 2021, https://smallwarsjournal.com/jrnl/art/third-generation-gangs-strategic-note-no-38-cape-town-extortion-gangs-target-paramedics.

going to have to fight in cities at some point in the future. I know that the army and the Marine Corps have looked at combat in megacities and consider it a daunting challenge.[20]

Spencer: The complexities of understanding the powers within a city is a topic that keeps coming up on the Modern War Institute's *Urban Warfare Project Podcast* series and I recently published an article about my own personal experience with the gangs in Baghdad. I discussed how we co-opted them to achieve some short-term security gains just to get security functions jumpstarted.[21] Knowing that we only achieved success in Baghdad by using the gangs made me think of how badly it could get in a feral city if a military tried to take power or attempt to exert its control onto the city without understanding the existing power structures. It could get very ugly and ultimately turn into a situation where the military would have to "destroy the city to save it," as we sometimes say.

Norton: I think that is exactly right, which speaks directly to the need for cultural expertise and the diplomatic and politics part of understanding feral cities. For example, the Marines successfully re-took Fallujah—which at the time clearly met the definition of a feral city (and cities in war often become feral)—during the second battle of Fallujah. But in order to retake the city, the Marines, by some estimates, seriously damaged over 60 percent of the buildings in the city.[22] That was a pretty steep price to pay to "liberate" the city. Unfortunately, this level of destruction is not unusual. It can be even worse. By some estimates, over 80 percent of the buildings in Warsaw, Poland, and Hue, Vietnam were destroyed during World War II and the Vietnam War respectively.[23]

Spencer: Something that I think about often is that we have a lot of people focused on the country level, but little at the city level. Most of our diplomatic relations and understanding of other environments usually start at the nation level, and most of our military's foreign area officers are

[20] See, for example, Marc Harris et al., *Megacities and the United States Army: Preparing for a Complex and Uncertain Future* (Arlington, VA: Office of the Chief of Staff of the Army, 2014), 3, https://api.army.mil/e2/c/downloads/351235.pdf.

[21] John Spencer, "Gangs of Baghdad," Modern War Institute, January 21, 2020, https://mwi.usma.edu/gangs-of-baghdad/.

[22] Kendall Gott, ed., *Eyewitness to War Volume I: The US Army in Operation AL FAJR: An Oral History* (Fort Leavenworth, KS: Combat Studies Institute Press, 2006), 14, https://apps.dtic.mil/sti/pdfs/ADA460458.pdf.

[23] "The Destruction of Warsaw: The Nazi Plan to Obliterate a City," *History Channel*, September 1, 2019, https://www.history.co.uk/article/the-destruction-of-warsaw-the-nazi-plan-to-obliterate-a-city; and Olga Dror, "Learning From the Hue Massacre," *New York Times*, February 20, 2018, https://www.nytimes.com/2018/02/20/opinion/hue-massacre-vietnam-war.html.

aligned to regions and countries, but never cities. When I was on the Chief of Staff of the Army's Strategic Studies Group, we sent a team to Dhaka, Bangladesh. It was clear that the United States embassy country team did not really understand how Dhaka worked, because they were prioritized on the country level. My experience in Mumbai was similar. They did not have a Mumbai expert within the country team. How do you think we are doing on developing expertise on major urban areas, feral or not?

Norton: My quick answer is that I do not know, but I am not encouraged by what I have encountered. Unfortunately, I do not think your experiences with embassies are unique. In some ways it reflects international relations and its three levels of analysis: the system, the state, and the individual. The systemic level of analysis explains a state's behavior by examining the international system, and there are theories, such as realism, that help to make sense of a complex world. The state level of analysis examines state behavior by focusing on the characteristics of the state: the nature of the government and how bureaucracies within the government work. The individual level of analysis examines leaders and important personages to explain state behavior.

But maybe now there is another level of analysis, between the state and the individual, and that could be the megacity. I do not think this has been examined enough. First off, my apologies to urbanists and futurists who specialize in cities, but I think the larger international relations discipline is still coming to grips with the impact of a megacity. The army, obviously, has specialized interest in them, but understanding the unique challenges of megacities applies to everything from sanitation to health, to education, yet despite urban hypertrophy and explosive demographic growth, the number of experts remains comparatively few.

Spencer: Some of the United States military's geographic combatant commands have committed people to study large cities that pose risks within their area of responsibility. For example, when General Bob Brown was at U.S. Army Pacific Command, he would often remark that 21 of the world's 41 megacities were located within his command's area of responsibility. He recognized their importance, but I am not sure the command knew what to study or in which cities to focus their effort.[24]

[24] TRADOC G-2 OE Enterprise, "TRADOC Mad Scientist 2018 Megacities: 1.07 The Warfighter & Megacities-USARPAC / GEN Brown," YouTube video, April 12, 2018, https://www.youtube.com/watch?v=bJD2ORr0VVA.

In the course of your research, did you create a metric or a taxonomy that could give a warning as to whether a city was becoming feral? This would have been a tool that General Brown could have used to focus his staff.

Norton: I did, and I have continued to refine it over the years. But I would push back at you a little on your comment about combatant commanders trying to get a grasp on this. U.S. Army Pacific Command is a great example of a commander who understood the issue, but I would bet that despite the command's interest in megacities, they still had very little true understanding. If you asked them, "Tell me, who is important in Manila? How does Manila really work? If I had to occupy Manila, where are the gangs, the unions, or the leaders who would be able to put Manila back together?" I do not think you would get much of an answer. Maybe we know more than we did when we rolled into Basra or Baghdad in 2003, but I am not confident we have learned that much.

To return to your question of the diagnostic tool, as I was thinking about how to recognize a feral city, I initially came up with four important categories—governance, security, services, and industry—and later added a fifth: civil society (see table 2.1). A healthy city has a stable government; hopefully it is democratic, but it does not have to be. The key is that the people trust the government and the government feels a responsibility to its people. In terms of the economy, a healthy city has domestic and foreign investment, a stable tax base, good employment opportunity, and low unemployment.

A healthy city has robust and resilient services, which include critical services like trash pickup, sewage, education, and ambulance services, but also some of the more exotic things that do not immediately come to mind when thinking about city services like parks and museums. In a healthy city, the police provide adequate security and largely behave ethically. In a healthy city, civil society and civic organizations work with the government, they support and reinforce each other, they perform watchdog functions, and there is a healthy dialog.

In feral cities, formal governance is weak or non-existent, and the limited governance is instead provided by gangs, clans, terrorist organizations, or village councils. Services must be contracted privately, or residents go without. The security forces are often gangs vying for power among many other gangs. There is very little civil society; instead, society is based on clan, tribe, ethnicity, or some other identity.

	Government	Economy	Services	Security	Civil Society
Healthy ("Green")	Enacts effective legislation, directs resources, controls events in all portions of the city all the time. Not corrupt.	Robust. Significant foreign investment. Provides goods and services. Possesses a stable and adequate tax base.	Complete range of services, including educational and cultural, available to all city residents.	Well regulated by professional, ethical police forces. Quick response to a wide spectrum of requirements.	Rich and robust. Constructive relationships with the government.
Marginal ("Yellow")	Exercises only "patchwork" or "diurnal" control. Highly corrupt.	Limited to no foreign investment. Subsidized or decaying industries and growing deficits.	Can manage minimal level of public health, hospital access, potable water, trash disposal.	Little regard for legality or human rights. Police are often matched/ stymied by criminal "peers."	Relationships with the government are confrontational. Civil societal organizations fill governmental voids.
Going Feral ("Red")	At best it has negotiated zones of control; at worst it does not exist.	Either local subsistence industries or industries based on illegal commerce.	Intermittent to nonexistent power and water. Those who can afford to will privately contract.	Nonexistent. Security is attained through private means or paying protection.	Civil Society fractured along clan, ethnic, or other lines. Local elites in control. Security-oriented civil society organizations may become criminal.

Table 2.1. Feral city taxonomy.
The table comes from Norton, "Feral Cities: Problems Today, Battlefields Tomorrow?" 57.

I thought the economy would be similar, but I was wrong. I thought the only businesses in a feral city would be illicit ones such as drugs, arms, or smuggling. What I learned is that some legitimate businesses find the atmosphere within a feral city conducive enough to remain. A form of economy remains, and it is not just a subsistence or an illegal one, but the economy does lack major manufacturing because the risk to the owners is simply too high.

Turning to Mogadishu again, there are cell phone towers all over the city, and these are not fly-by-night companies, they are reputable companies. In some ways, operating in Mogadishu may even be liberating—there is no Occupational Safety and Health Administration and there are very few controls—so there is still money to be made. The ways in which money is moved may be a combination of modernity and very old traditional patterns. Another thing that surprised me about Mogadishu was the university. Despite meeting most of the requirements of a feral city, it had a functioning university. Feral cities are not empty, people are social creatures and parents want their children to have better lives. So, Mogadishu University continued to educate young adults even through some of the city's darkest times. Thus, even the most feral of cities have some functioning parts.

Spencer: I love your taxonomy, and the red, yellow, green categories. I am an old infantry soldier, and that is very easy to understand. Where do you get the data for your taxonomy? One source that comes to mind is Transparency International's corruption perception index, but I believe that only provides data at the country level.[25]

Norton: That is correct. There are some databases, but none have the level of granularity that I need. The data on crime rates is one of the exceptions. Crime rate alone is not enough to determine that a city is feral, but it is always kind of interesting to look at the 50 most dangerous cities. A high crime rate suggests weak security, a breakdown of civil authority, a loss of trust in the government by the civil society, and potentially, the loss of some investment, all of which contribute to moving a city from the healthy end of the spectrum to the feral end.

After writing my first article on feral cities, I came away thinking you could never have a feral city in the United States. But as people read the article, they would tell me, "Rick, East Los Angeles [or they might name some other city] is a feral city." I would say it was not because the mayor has plenty of forces to deal with East Los Angeles; and if the city did not have the forces, then the state does; and if the state does not have them, then the federal government does. So, I thought never in the United States.

Then Hurricane Katrina hit New Orleans in 2005. And for at least five days the United States could not extend the rule of law into parts of the city, and it was not because of gangs. Instead, it was a product of the devastation wrought by the hurricane. It impacted services in ways that people never considered. The New Orleans Police Department used radios with rechargeable batteries because that was cheaper than buying

---

25 "Corruption Perceptions Index," Transparency International, https://www.transparency.org/en/cpi/2021.

thousands of batteries each month. But once the charging stations went under water, they lost their communications.[26] Hurricane Katrina was really kind of an eye opener for me. So, certain circumstances—perhaps a giant earthquake in California or the Yellowstone Caldera blowing—could result in a healthy city becoming feral, at least for a short period of time.

Spencer: So, it is a live assessment?

Norton: Absolutely. If you look at it as a snapshot, then you are bound to make erroneous conclusions. Think of it as a spectrum where dark red is really bad and bright green is really good and a city can move along that spectrum. If you look at New York, parts of the city may have been "yellow" during "The War Years" in the 1960s and 1970s. This was a period of dwindling prosperity where many property owners turned to arson to burn down their property for the insurance payouts and residents routinely attacked firefighters that arrived to put out the fires.[27]

In 1980, President Ronald Reagan gave a very inspirational speech from one of the more devastated and blighted portions of New York City.[28] Big portions of the city were trending in and out of "yellow" or possibly even "red" or feral. The city government even negotiated with local gangs.[29] I am told that part of the city is now wonderful. New York is much "greener" than it was during the War Years or when President Reagan spoke there. I mentioned Johannesburg at the beginning of our discussion. That city has shown significant improvement since 2000 when they pulled their stock exchange out of the city, but improvements have recently stagnated. So, cities are constantly moving along the spectrum.

Spencer: I am sure you get this question all the time, but beyond Mogadishu, are there any other cities that you would describe as having fallen into the red category and seem likely to remain there?

---

[26] *Hurricane Katrina: Managing Law Enforcement and Communications in Catastrophe: Hearing before the Committee on Homeland Security and Governmental Affairs*, Senate, 109th Congress, 2nd sess., February 6, 2006. https://www.govinfo.gov/content/pkg/CHRG-109shrg27025/html/CHRG-109shrg27025.htm.

[27] Celeste Neill, "The New York City Fire Department: A Timeline of the City's Firefighting History," HistoryHit.com, September 9, 2021, https://www.historyhit.com/new-york-fire-department-history/.

[28] Hezakya Newz & Film, "Ronald Reagan Goes to the South Bronx and Gets Heckled, Loses His Temper," YouTube video, February 5, 2019, https://www.youtube.com/watch?v=hHRSGDpMGGQ.

[29] "50th anniversary of Bronx gang peace treaty seen as opportunity to quell violence," ABC7NY.com, December 10, 2021, https://abc7ny.com/50th-anniversary-bronx-gang-peace-treaty-south-gangs/11320339/.

Norton: I think Port-au-Prince, Haiti falls into that category. That is not necessarily an indictment against the Haitian government, but it is really difficult to recover from the double whammy of corrupt governance and earthquake devastation, particularly when informal systems of governance are in place. I think Lagos, Nigeria is a city to watch, just because it grew so fast. The northern tier of Mexico is also a place to watch due to the cartels and drug violence. The drug cartels have had an extraordinarily destabilizing influence in the region, and I think we do not give the Mexican government and the Mexican people enough credit for combating that threat. They have lost the equivalent of several divisions over the last decade or so trying to combat that threat.[30] I think it is easier for us to just say that it is sad, and it is their problem, but I think it is a transborder problem that is getting worse. I think cities such as Nuevo Laredo and Tijuana continue to trend in the feral direction despite the continuing efforts of the Mexican government to change that.

Spencer: In your first article you listed Mexico City in the yellow zone.[31]

Norton: Mexico City is one of the cases that led me to say something else is going on here, because when I wrote the article in 2003, there were nearly 90,000 police in Mexico City, and that is nearly as many police in one city as Canada has soldiers in its entire army.[32] While large, the public and Mexico's own government lacked confidence in it. Mexican politicians often denounced the police as simply another gang.[33] But Mexico City has not yet slid into the red, and I think that is a testament to the strength of Mexicans, of Mexico City's civil society, and Mexico's efforts to improve the lives of the average resident. Mexico City is much better today than what I would have predicted in 2003.

Part of the joy of research is saying, "That did not work out the way that I thought it would. Why?" That led me to include "civil society" as a fifth component of my taxonomy. Again, people often ask about the

---

[30] Carlos Galina, "Mexico's Security Dilemma," Council on Foreign Relations, January 24, 2020, https://www.cfr.org/in-brief/mexicos-security-dilemma.

[31] Norton, "Feral Cities," 103-4.

[32] Benjamin Reames, "Police Forces in Mexico: A Profile" (working paper, Mexico Policing Project, 2003),https://escholarship.org/content/qt1sq4g254/qt1sq4g254_noSplash_e7293b3668ff369291a6400097f9f7b5.pdf?t=krnbgkC. Canada has 68,000 Regular Forces in their Army and 27,000 Reserve Forces for a total of 95,000. See, "Canadian Armed Forces 101," Government of Canada, March 2020, https://www.canada.ca/en/department-national-defence/corporate/reports-publications/transition-materials/defence-101/2020/03/defence-101/caf-101.html.

[33] See, for example, Robert Siegel, "Mexico's Police Corruption Investigation," *All Things Considered*, NPR News, June 26, 2007, https://www.npr.org/templates/story/story.php?storyId=11433234.

term feral city and some take offense, arguing that it is a loaded term. My response is that it may be provocative, but it is also accurate.

Spencer: Feral is both provocative and controversial. Why so?

Norton: It is provocative because it means that nation-states are losing control. The government is surrendering part of its sovereign territory to someone or something else. It is controversial because people say that I undervalue the law and order that does exist; or that I undervalue the ability of people who may live in slums to have productive, positive human interactions, and lead a fairly fulfilling life; or that cities themselves might not be the be-all and end-all that some people claim them to be.

I remember a debate that I had with a very pro-cities guy, and we agreed that one of the things that keeps a modern city of chrome, steel, and rebar functioning is the bond that the government or city has with its citizens in terms of protecting its citizens, providing services to its citizens, and thinking about the human security of its citizens. But if that bond is broken or the city becomes incapable of being in control, bad things happen. The glue that holds the city together begins to crumble.

Spencer: In your 2003 article, you discussed pollution. How is pollution related to the taxonomy of a city?

Norton: Pollution often results from rapid urban hypertrophy. Some cities that were doing well experienced a huge influx of people—either because jobs have suddenly opened or because other pressures are pushing populations to the city—and existing services get overwhelmed before they can absorb the influx. Sewers may become overwhelmed and the next thing you know, you have a situation where a river, which is normally fairly clean, produces fecal blooms in the offshore waters after heavy rains. This is a breeding ground for disease. There are parts of some cities that are almost no-man's-land, not because of gangs vying for control, but because the toxicity of the area is so high that people cannot live there.

In the United States we take so much for granted: clean tap water, consistent power, and sewage treatment, for example. Sure, there are exceptions, such as the Flint, Michigan water crisis in 2014, but by and large, well developed nations with robust services can enact measures to reduce pollution.[34] Yet for other nations, pollution remains a significant problem. For example, if someone is struggling to get by, they may not have

[34] Merrit Kennedy, "Lead-Laced Water in Flint: A Step-by-Step Look at the Making of a Crisis," NPR, April 20, 2016, https://www.npr.org/sections/thetwo-way/2016/04/20/465545378/lead-laced-water-in-flint-a-step-by-step-look-at-the-makings-of-a-crisis.

a reliable source of fuel, so they turn to what they can—such as charcoal and wood products—to produce heat, and those products produce much more pollution. Colonel (retired) Lee Van Arsdale described a "continuous haze" over the city of Mogadishu in 1993 from the open cooking fires.[35] Cars in developing cities tend not to have good environmental standards. So, pollution is a real threat to the health of a city, both the physical health of its residents and also in terms of what the people begin to expect or not expect from their government.

Spencer: During my research trip to Mumbai, India I took a tour of the Dharavi slum. I was fascinated to see that within the slum there were some limited services being provided and there was a modicum of the economy and commerce—even if many of the jobs were dirty and dangerous recycling jobs. It was very eye opening in terms of what not to take for granted in our society.

Norton: I had a similar experience in South Africa. Like you, I was personally conflicted as to whether we are turning people into a human zoo, but if you do not go then you will never be able to see for yourself what it is like and meet people who live in these environments. Seeing is important, and I think it is important to remember that, as you point out, there are people who live there, they raise their kids there, their families are strong, and there is commerce. Granted, it might not be something we would want our children to do. I would prefer my children not have to pick through the dump for bits and pieces of computers or engage in ripping asbestos out of old machinery. But life goes on, not to sound like Jeff Goldblum in *Jurassic Park*.[36]

Spencer: Yes, it is as if different cultural assumptions are programmed into us, and we do not even realize they are there until we get a cultural shock. Your taxonomy of feral cities is great, but how is it relevant to the United States military? Why should the military care about feral cities?

Norton: That is a great question and it is promising to see that the army and the Marine Corps have been examining megacities and the

[35] See chapter 8, "The Battle of Mogadishu," in this book.

[36] In the movie *Jurassic Park*, actor Jeff Goldblum playing Dr. Ian Malcolm says "life finds a way" to a scientist who is briefing him that they control the chromosomes of the dinosaurs to prevent them from breeding. *Jurassic Park*, directed by Steven Spielberg (Universal Studies, 1993).

challenges that they pose for the last decade or so.[37] The military should care because it will likely find itself operating in a feral city in the future, because war and humanitarian operations tend to occur in nations and cities that trend toward the red part of the spectrum. The military could conduct humanitarian operations in a feral city as it did in Somalia in 1993. Although the "Black Hawk Down" mission was not a humanitarian mission, Task Force Ranger was deployed to support the humanitarian operations.[38]

Often when a massive humanitarian event occurs, such as the 2021 earthquake in Haiti, the United States military takes part in it, because nongovernmental organizations, commercial organizations, and even other parts of the United States government lack the logistics, the manpower, the expertise, and the machinery to deal with that environment.[39] So, supporting humanitarian operations is one way the United States military could end up in a feral city and they do not want to make things worse and they want to protect their force, and that requires understanding.

One look at history or any of the case studies in this book, and it is clear that the military will likely find itself fighting in a feral city or megacity in the future, and the conflict or the fighting itself could turn a healthy or marginal city into a feral one. If you look at some army doctrine, how does it say to deal with a city?

One answer is to bypass it.[40] You contain it and you let things take their course. That strategy worked for a period of time in Fallujah, Iraq, in 2004, but eventually the military had to enter and fight within this feral city.[41] Fallujah was a large city, but it was not a megacity where the challenges are exponentially greater. So, the military may want to avoid

[37] See, for example, Todd South "The Army and Marine Corps are looking at what troops will need to fight in megacities, underground," *Navy Times*, January 9, 2019, https://www.navytimes.com/news/your-army/2019/01/09/the-army-and-marine-corps-are-looking-at-what-troops-will-need-to-fight-in-megacities-underground/.

[38] Center for Military History, *The United States Army in Somalia: 1992-1994* (Washington, DC: U.S. Army Center of Military History, no date), https://history.army.mil/brochures/somalia/somalia.htm.

[39] Office of Press Relations, "SOUTHCOM Successfully Completes Haiti Earthquake Operation as USAID Continues Robust Relief and Recover Effort," U.S. Southern Command, September 3, 2021, https://www.usaid.gov/news-information/press-releases/sep-3-2021-southcom-successfully-completes-haiti-earthquake-operation-usaid.

[40] Department of the Army, *Urban Operations*, ATP 3-06 (Washington, DC: Headquarters, Department of the Army, 2017), https://armypubs.army.mil/epubs/DR_pubs/DR_a/pdf/web/ARN6452_ATP%203-06%20FINAL%20WEB.pdf.

[41] See Chapter 9, "The Second Battle of Fallujah" in this book.

cities, and it is prudent to do so, but it must also maintain the capability to fight within them.

The terrain of a modern city is extraordinarily complex. The terrain is three-dimensional with buildings stretching dozens of floors into the sky and layers of sewers underground. There is an old rule of thumb that says that an attacking force should maintain a three-to-one advantage over a defending force, but U.S. Army doctrine also notes that "history suggests urban offensive missions require three to five times greater troop density than for similar missions in open terrain."[42] So, wars in cities eat up infantry, eat up armor, and eat up rotary wing aircraft unless the military sits outside the city and pounds it with artillery, rockets, and bombs. But then you have the problem of collateral damage and the noncombatants residents that could suffer, which raises real questions about proportionality and appropriate response. Are you destroying the city to save it?

The urban environment is extremely challenging, but one that the military cannot avoid. I think it is indicative that we have some form of urban training site on nearly every army and Marine base, albeit some are extremely small. Some countries boast of theirs. The Israelis have a pretty large one. The U.S. Marine Corps has a spectacular complex at Twentynine Palms called Range 220. The U.S. Army has a large one at Fort Irwin that has noise, smell, and sound emitters that evidently produces very realistic effects. Ultimately, we better be ready to operate in cities because we are going to have to do it.

Spencer: I have a whole podcast series dedicated to understanding and preparing for urban operations, so you do not have to convince me.[43] Some argue that the military should never enter a megacity based on the numerical opposition or the sheer complexity.[44] I will not drag you into this debate, but there are also many that realize the military must be ready to operate within a megacity when called. I attended a conference last year where General (retired) John Allen said that as a former combatant commander he wished that he would have prioritized understanding the

---

[42] According to U.S. Army doctrine, "History suggests urban offensive missions require three to five times greater troop density than for similar missions in open terrain." See, Department of the Army, *Urban Operations*, 2-3.

[43] See, Modern War Institute, *Urban Warfare Project Podcast*, https://mwi.usma.edu/category/podcasts/urban-warfare-project-podcasts/.

[44] See, for example, Gian Gentile, et al, *Reimagining the Character of Urban Operations for the U.S. Army: How the Past Can Inform the Present and Future* (Santa Monica, CA: RAND, 2017), 15, https://apps.dtic.mil/sti/pdfs/AD1055510.pdf.

cities that he viewed as high risk—those in the marginal or becoming feral category given they were the ones where his command would most likely find itself engaged. He said that he wished that he would have dedicated more of his intelligence resources, whether they be analysts or researchers, to understanding the cities on that short list a little more.[45] Using your taxonomy, you could provide a list of cities that are currently in the red or trending red. What are some of the cities at the top of that list?

Norton: I am always a little nervous answering that question because the answer gives offense to any nation-state that has one of its cities named. I would offer that one way to start is to look at the cities at the top of the list for most violent cities in the world. By the way, the top 50 include four American cities: St. Louis (#15), Baltimore (#23), Detroit (#46), and New Orleans (#50).[46] I am not suggesting that they are turning feral, but it is one way to come up with a list. Although you said you did not want to drag me into the debate about entering a megacity, I would offer to those who say we just will not go there, we may not have the option. So, I think General Allen's recommendations are excellent.

Spencer: In 2019, I co-authored an article and took a shot at listing the top 11 cities that the United States should be watching and studying.[47] So, I realize that rankings are somewhat arbitrary based on data limitations and how you weigh the variables, but you have to start somewhere. There are many ways to develop lists and study cities. One way would be to develop a research center within the military that only focuses on large urban areas.

Norton: There are civilian institutions that do that, and there are several annual conferences that discuss the health of cities.[48] But it is not a topic or function that most military officers think about.

Spencer: The Rockefeller Foundation's 100 Resilient Cities Foundation which is now defunct is an example of an organization that used to do

---

[45] John Allen, *Analysis of Urban Warfare Conference*, Military Operations Research Society, Quantico, VA, April 4, 2019.

[46] "Most Violent Cities in the World 2021," World Population Review, https://worldpopulationreview.com/world-city-rankings/most-violent-cities-in-the-world.

[47] John Spencer and John Amble, "Every City is Different. That's Why a One-Size-Fits-All Approach to Urban Operations Won't Work," Modern War Institute, May 30, 2019, https://mwi.usma.edu/every-city-different-thats-one-size-fits-approach-urban-operations-wont-work/.

[48] See, for example, *International Conference on Resilient Cities*, https://waset.org/resilient-cities-conference-in-june-2022-in-copenhagen.

that.[49] You could create city-specific foreign area officers that are somewhat analogous to the coastwatchers that the Allies used in the Pacific during World War II.[50] You could put someone in these large cities and have them study them for years. We need to expend resources and dedicate intellectual power to really understand cities.

Norton: It seems to me that it does not have to be a uniformed military officer. It could be a State Department official. It does not matter where the expertise resides as long as it really is expertise. Military officers tend to think in terms of military solutions, but I think if you look at the whole of government, there may be far better organizations that produce the kind of people that would be the city experts that you are talking about. But then the military would have to respect them as experts and actually listen to them, and the military is not always good at listening to outside advice. I am curious, when you did your ranking, what were your top three cities?

Spencer: We listed 11 cities, but we did not rank them. For each city, we explained why they were on the list using evaluation criteria that included: the city is important to United States national interest, the city is among those the United States military is least prepared for, the city has important links to the global economy, the city has uncontrolled or rising political instability, the city and surrounding areas have a significant presence of armed groups, the city is at risk to natural disasters, the city is experiencing or at risk of an internally recognized crisis, among other criteria.[51] The first three that we listed were Caracas, Venezuela; Mexico City, Mexico; and Sana'a, Yemen; but as I mentioned, the top 11 were not prioritized.

Norton: Sana'a is a good choice, and it would make my list for sure. Caracas would as well, and the tragedy of Venezuela extends beyond Caracas. It does not require much imagination to think of scenarios where the United States might find itself in Caracas. I do not mean to stoke President Nicolás Maduro's concerns about a United States intervention, but if Venezuela were to collapse and pose a threat to the region, I could imagine some level of United States involvement as part of a larger Organization of

[49] Laura Bliss, "The Rise, Fall, and Possible Rebirth of 100 Resilient Cities," Bloomberg CityLab, June 12, 2019, https://www.bloomberg.com/news/articles/2019-06-12/the-demise-of-rockefeller-s-100-resilient-cities.

[50] The Allies used coastwatchers to gather intelligence and provide early warning of Japanese activity during World War II. The coastwatchers were placed on many islands in the Pacific and included both officers and indigenous personnel. "The Coastwatchers 1941-1945," Anzac Portal, https://anzacportal.dva.gov.au/wars-and-missions/world-war-ii-1939-1945/resources/coastwatchers-1941-1945.

[51] Spencer and Amble, "Every City is Different."

American States humanitarian mission. The idea of putting Caracas back on its feet is daunting. It would require a wide variety of experts: experts in power, water, and petroleum (for Venezuela's state-owned oil and natural gas company). Those things are critical.

A few years ago, I participated in an exercise that involved fixing an occupied megacity. The exercise designers gave it their best effort, but it was clearly lacking in scale and scope in many areas. But I think Sana'a, because of war, and Caracas, because of the tragedy of the Chávez and Maduro regimes, are two cities that belong on your list. Having said that, I am certain there will be people who will write to me and tell me you are wrong, but those two are very defensible of being on any list of cities to be concerned with.

Spencer: To further defend my list, it was a list of cities that the United States needed to understand better, it was not a list of cities that the United States was most likely to find itself engaged in. I have my own scar tissue from the Iraq War. Being in Kirkuk, Iraq in 2003, I witnessed the city move from what was green or at worst yellow before the invasion to yellow or red very rapidly after the invasion. Governance, economy, services, and security dissolved, almost overnight. I had no clue as to how to understand a city and most of the military was in a similar situation. In 2003, the army simply failed to provide this knowledge through its education and training.

The officers with the greatest understanding of cities gained their knowledge not through the army's professional military education, but instead through their civilian education. General David Petraeus gained this knowledge through graduate studies at Princeton University. As the commander of the 101st Airborne Division in Mosul, he made it a priority to jumpstart the political process and the local economy. Less than two weeks after arriving in Mosul, he created the first popularly elected government in liberated Iraq with a gubernatorial election on May 5, 2003.[52] Likewise, he "opened the border to get trade going despite the fact that that decision likely exceeded his authority."[53] Petraeus was by no means an urban specialist, but he at least understood the need to get the government, economy, services, and security going; the need to get civil society engaged; and some understanding of how to do that.

---

52 Fred Kaplan, *The Insurgents: David Petraeus and the Plot to Change the American Way of War* (New York: Simon & Schuster, 2013), 72.

53 Liam Collins, "Military Innovation in War: The Criticality of the Senior Military Leader" (PhD diss., Princeton University, 2014), 142, http://arks.princeton.edu/ark:/88435/dsp01br86b3727.

Norton: In the early days of the Iraq War, an outstanding scholar named Phebe Marr was suddenly in high demand due to her expertise as a historian of modern Iraq. So, you are right, it does not matter where the expertise resides as long as you can access it. That expertise can come from very unusual sources, and the military should not be predisposed to negate that expertise just because it comes from an unusual source.

Spencer: To return to my friend, Max Brooks, I consider him an expert in several different areas, so it is not surprising to see him in high demand today because of the time he spent researching disaster preparedness, supply chain management, you name it. His fiction is amazing because it is based on academic and historical reality. Some people may laugh at me for turning to a fiction author for expertise, but I refuse to ignore useful advice just because it comes from an unusual source.

Norton: I would agree, and to quote Admiral Wiscup, many years ago, when talking about Max and his book *World War Z*, "I think he truly understands the potential for global crisis and just how unprepared we might be."[54] He agreed with you that Max's book was about what happens when you come across a threat or an event for which you are unprepared, one that is outside of your expertise, and one that your normal reactions may be counterproductive in a big way. I give Max great credit for coming up with the idea of writing a book in oral history format—coming at it with a Studs Turkel approach—and making it very readable.[55] But it also raises extraordinarily important questions once you cast it in the realm of: "What if this is something wide sweeping and we are not ready to respond to it when asked or required?" The pandemic is the first thing that comes to mind when you do that.

Spencer: I think this has been a great conversation and I think you provided a great taxonomy to classify cities and one that would seem to have clear utility for the military. From my experience in Iraq, I would agree that the government, economy, services, security, and civil society are all critical components when trying to determine the health of a city. Thank you.

---

54 U.S. Naval War College, "Max Brooks: World War Z."

55 Studs Terkel wrote the popular book *The Good War: An Oral History of World War II* using interviews as the primary text and format for the book. Studs Terkel, *The Good War: An Oral History of World War II* (New York: The New Press, 1997).

3

Global Cities

As discussed in this book's introduction, there are different taxonomies that can be used to describe or categorize cities. This chapter discusses the concept of the "global city," a new type of city that emerged in the 1980s, one that shifted the power, ownership, function, and the relationship or stature of cities with their host nations.[1] A global city "is a significant production point of specialized financial and producer services that make the globalized economy run."[2] A city today can be more important economically and potentially politically than the nation in which it resides.[3]

Dr. Saskia Sassen popularized the term global city in the 2000s and deliberately selected the term to distinguish it from other terms, such as "world city," which had been used in the past. For Sassen, it was important to select a new term because the structures of the contemporary period were fundamentally different from those that existed previously and simply using the term "world city" would not capture this difference. The world cities of the past might have been designed with hierarchical institutions and local, regional, and international connections, but they were not economically integrated at the global level or characterized by the level of complexity that exists today.

Several chapters in this book discuss the importance of understanding power and the flows of various elements within cities. Sassen takes a similar approach, focusing on the important role of financial power and the flow of information and capital that emanates from global cities.[4] Global cities

---

1 Aaron Renn, "What is a Global City?" *New Geography*, June 12, 2012, https://www.newgeography.com/content/003292-what-is-a-global-city.

2 Ibid.

3 Dana Varinsky, "Cities are becoming more powerful than countries," *Business Insider*, August 19, 2016, https://www.businessinsider.com/why-cities-are-more-powerful-than-countries-2016-8.

4 Daniel Little, "The global city – Saskia Sassen," Understanding Society (blog), September 15, 2013, https://understandingsociety.blogspot.com/2013/09/the-global-city-saskia-sassen.html.

are major nodes in an interconnected system of information and money. They create, capture, and distribute great wealth. In global cities, many of the critical flows that allow these cities to flourish are not bound to national boundaries or traditional systems of regulation.

Economic policies in the 1980s created the emergence of global cities. Multinational corporations existed prior to this time, dating back to the Dutch East India Company in the 17th century, but this is when they started exploding in terms of size, scale, scope, and power.[5] One byproduct of this globalization was the emergence of highly specialized service firms, also called the corporate services complex, that allowed businesses to expand globally. These specialized firms included financial, legal, accounting, and other special services that allowed businesses to expand and hopefully succeed in foreign cities. A second result has been an increase in the power wielded by these multinational corporations with many arguing that they have too much power. One examination found that there are 70 multinational corporations that have revenue that exceeds the gross domestic product of all but the top 30 nations.[6]

Global cities attract people and investment and become powerful in their own right. A city's "globalness" is often ranked on their business activity, human capital, information exchange, cultural interaction, and political engagement. Cities such as New York, Paris, London, and Tokyo often come out at the top of various lists.[7] To relate to the previous chapter, not surprisingly, most global cities, or at least those that are considered most global, often tend to fall on the healthy end of the feral spectrum. Hence, it may seem a little strange to include a discussion of global cities in a book relating to urban warfare. But it is important for military professionals to understand this phenomenon because the globalization of the economy impacts cities everywhere, not just global cities.

---

[5] Ben Phelan, "Dutch East India Company: The World's First Multinational," PBS, January 7, 2013, https://www.pbs.org/wgbh/roadshow/stories/articles/2013/1/7/dutch-east-india-company-worlds-first-multinational/#:~:text=The%20mighty%20Dutch%20East%20India,was%20the%20first%20multinational%20corporation.

[6] Milan Babic, Eelke Heemskerk, and Jan Fichtner, "Who is more powerful – states or corporations?" *The Conversation*, July 10, 2018, https://theconversation.com/who-is-more-powerful-states-or-corporations-99616.

[7] See, for example, *Global Cities: New Priorities for a New World: 2020 Global Cities Report* (Chicago, IL Kearney, 2020), https://www.kearney.com/documents/20152/60890709/Global+Cities%E2%80%94new+priorities+for+a+new+world.pdf/d908884f-b786-a49b-ffed-1c08735603fd?t=1608472416000; and Norio Yamato et al., *Global Power City Index* (Tokyo: The Mori Memorial Foundation, 2021), https://mori-m-foundation.or.jp/pdf/GPCI2021_summary.pdf.

In this chapter John Spencer interviews Dr. Saskia Sassen to discuss global cities and why they matter. Sassen is the Robert S. Lynd professor of sociology at Columbia University, and a member of its Committee on Global Thought, which she co-chaired until 2015. She has been called "the leading urban theorist of the global world."[8] She has authored eight books including *Cities in a World Economy* and *The Global City*. She is the editor or co-editor of three books, with her most recent being *Cities at War: Global Insecurity and Urban Resistance*.

John Spencer: Dr. Sassen, I would like to jump in by asking you why you think cities are important?

Saskia Sassen: Cities outlast empires and survive dictators. Cities are a space where those without power get to make a history. A place where they get to leave a trace. All cities have a very broad spectrum of people, from the very poor to the very rich. So that tells us something about the unique qualities that a city brings. If you go to a plantation or a big institution, you do not have the mix that the city offers.

Spencer: The military often struggles to define an urban area, sometimes even calling something as small as a dozen or so small structures a "military operations on urbanized terrain" site. I was reading one of your past interviews where you said a city is much more than just a bunch of buildings.[9] How do you draw the line as to what constitutes a city and what does not?

Sassen: I am very glad you picked up on that. If you just have a huge building with a vast number of people, that is not necessarily a city. A city is a space, a place where people without power get to make a history, get to make a little economy, get to innovate, and get to do many other things. At the other end of the spectrum, there are very wealthy people who could go anywhere, but they are also in the cities. Why? Because that is where they can make connections, not just with people who are in that city, but with people who are traveling into that city from other parts of the world. There are very few spaces where that kind of mixing happens.

A city is a mix of all types of people, all types of stores, all types of businesses, all types of religions, all types of everything. And that

[8] Manzoor Ali Isran, "An Administrative Blackhole," *The News International*, August 28, 2019, https://www.thenews.com.pk/print/518665-an-administrative-blackhole.

[9] Saskia Sassen and Richard Sennett, "Global City into Perspective," Philippe Chiamberetta Architecte, 2014, https://www.pca-stream.com/en/articles/global-city-into-perspective-4.

is special. That can also humanize a situation. For instance, the poor do not necessarily have to feel as they would on a plantation where they are completely exploited. The poor in the city can also make a local economy, and many poor neighborhoods have little economies. They may not be great, but they are there.

Spencer: You wrote the book, *The Global City*, which is obviously about global cities and popularized the term. What is a global city and how is it different from an international city?

Sassen: The concept of an international city came from the "world cities hypothesis." This hypothesis took earlier urban theory that tiered cities at the national level and suggested an international hierarchy of cites. International cities were ones that served as major sites of production or finances for the world economy. It viewed the cities as being in the highest tier of an international economy that operated largely within a nation-state and interstate framework.[10] But a globalized economy is fundamentally different from the interstate economy of the past. So, what was once considered an international city is probably functioning as a global city today. At the heart of it, a global city is a centralized location of specialized financial and producer services that make the globalized economy run independent of nation-states.[11]

Global cities started emerging in the 1980s after nations started changing their economic policies. Many countries were undergoing similar transformations at that time, but I will focus on the United States. The United States deregulated and internationalized its economy in the 1980s. As a result, more and more American firms started establishing operations in other parts of the world. This was not new, but it really took off in a much more significant way. It included all kinds of businesses, and not just large ones. Certainly New York and Los Angeles were becoming global cities, but so were smaller ones.

As the United States and other nations around the world deregulated their economies, it opened the United States' economy to actors from across the globe. They could now set up businesses or offices in New York, Los Angeles, or any other city. That was a big transformation. It also meant that American businesses could set up businesses all over the world. So, the 1980s

[10] William Robinson, "Saskia Sassen and the Sociology of Globalization: A Critical Appraisal," *Sociological Analysis* 3, no 1. (2009): 5-29, https://robinson.faculty.soc.ucsb.edu/Assets/pdf/Saskia%20Sassen.pdf.

[11] Renn, "What is a Global City?"

marked a time when a new type of economy emerged: an internationalized economy with global cities at the center of this new economy.

The United States had quite a few global cities in that early period, maybe six or seven and they were all large cities. Cities in smaller countries also became global, but maybe in just a single type of business zone. But in countries with large economies, like the United States, cities could go global on a whole set of economic vectors. It was not just one thing, as it was previously, it was many. This globalized the major cities. The economy of smaller cities, however, did not really change. Their residents continued living their traditional lives. But something was happening in the big cities in a very significant way.

Spencer: What are some positives and negatives that have resulted from the emergence of global cities?

Sassen: This was very exciting. It internationalized the major cities of the United States and other countries. Many, but not all, businesses and individuals benefited from globalization. It also enabled some actors to become extraordinarily powerful. Some were individuals, but most were firms. It was not the large manufacturing firms, such as automotive firms in Detroit, that benefited most from this transformation, though they benefited as well. It was, instead, the highly specialized knowledge bearers in the banking, financial, and legal fields that benefited the most. These intermediaries enabled a typical firm, one that had never gone global or was barely global, to go truly global. They set up all kinds of instrumentalities, as we say in business, that enabled certain kinds of firms to go global.

Going global means entering a number of countries, and it requires establishing specific types of economic arrangements within each. These arrangements varied by business and nation, and the individual firms required these knowledge brokers or highly specialized firms to help them expand overseas since they lacked the knowledge of how to do that. These highly specialized firms, or intermediaries, specialize in "accounting, legal, public relations, programing, telecommunications, and other such services."[12] These intermediaries enabled American and foreign firms to connect with one another and enabled American firms to expand into other nations.

The intermediaries became powerful because they possessed this knowledge and they never "lost." They always won. What I mean is that the firms that they try to mate—connect with other firms—often go

[12] Saskia Sassen, "The Global City: Introducing a Concept," *The Brown Journal of World Affairs* 11, no. 2 (2005), 28, https://www.jstor.org/stable/24590544.

broke. Many of those operations do not necessarily work out. If you are a car manufacturer, for example, you are taking a risk by going global. The intermediaries, however, enable that firm to go global and then they are done. If it does not work out it is sad, but the intermediaries still profit. The intermediary firms, whether in New York, San Francisco, Chicago, or wherever, do not lose. Goldman Sachs is an excellent example of an intermediary.

This was a significant transformation, but not one that the average person would have noticed. The average citizen did not understand that a foundational shift of power transferred to a very small group at the top of the pyramid, to use a metaphor. Financiers are exhibit number one, but they were not the only ones. Some brilliant law firms also came out on top. These really big winners started to emerge in the 1980s and this economic system remains largely in place today. I believe this remains a very problematic feature of our economies today. So, this is one of the clear downsides, or unintended consequences, of this globalization.

Spencer: One concern that you have been very public about is "land grabs" by major corporations.[13] As populations migrate to the cities, major corporations are purchasing the lands that the populations are leaving. I read an article in which you wrote that this is altering "the historic meaning of the city" and causing a re-development of many cities around the world.[14] How does that contribute to instability in major urban environments that could subsequently lead to conflict?

Sassen: This is not just happening in urban environments; it is happening across the world. There are firms from dozens of counties that have been very active in acquiring land in Africa, Asia, and Latin America. In the past decade, Chinese companies alone have purchased or leased land that is equal to the total land area of Lithuania or Sri Lanka.[15] Accessing land is extremely important and offers possibilities ranging from mining to water extraction. I could discuss any number of examples relating to land grabs, but I will discuss water extraction.

---

[13] Saskia Sassen, "Land Grabs Today: Feeding the Disassembling of National Territory," *Globalizations* 10, no. 1 (2013), 25-46, http://saskiasassen.com/PDFs/publications/SS%20LandGrabs.pdf.

[14] Saskia Sassen, "Who owns our cities – and why this urban takeover should concern us all," *The Guardian*, November 24, 2015, https://www.theguardian.com/cities/2015/nov/24/who-owns-our-cities-and-why-this-urban-takeover-should-concern-us-all.

[15] See, for example, Daishi Chiba, Shin Watanabe, and Yuichi Nitta, "Chinese companies corralling land around the world," *Nikkei Asia*, July 13, 2021, https://asia.nikkei.com/Spotlight/Datawatch/Chinese-companies-corralling-land-around-world.

The Swiss company, Nestlé, is the second largest food and beverage company in the world.[16] They need to extract water, that is part of their business. In addition to domestic brands such as Nestlé Pure Life, Ice Mountain, and Poland Spring, they also produce international brands such as Perrier, and S. Pellegrino. But in the process of extracting water, Nestlé leaves lands dry and local water undrinkable all over the planet, because they exhaust the water.[17] This is not just a social concern, it is a security concern given that some experts have argued that water scarcity is the "most understated global security risk."[18] This is but one example, but it is a major issue.

This sort of thing is happening in many different forms and on many different levels. I will provide another example. I am not sure if you would consider this a micro-land grab, but it is definitely a downside of globalization that negatively impacts many global cities. In recent years, there has been a trend by some of the more powerful firms to buy up low-income housing complexes in cities around the world. They actually made a documentary about this. They are not buying small houses; they are buying large housing towers that house hundreds of low-income workers and their families. These firms buy them up. They forcibly eject the residents to renovate and upgrade the building, but after the renovation, the former inhabitants are priced out of the market because it is no longer low-income housing. Beyond the issue of deception to the initial tenants, there is also the issue that there are now hundreds of people who cannot find affordable housing.[19] To return to your question, this can definitely contribute to instability.

Spencer: What is the name of the documentary that you mentioned?

Sassen: The documentary is called *Push* as in "push out."[20] The main person in the film is a special rapporteur on the right to housing for the

---

[16] Casey Laughman, "The 2021 Top 100 Food and Beverage Companies," *Food Engineering*, August 27, 2021, https://www.foodengineeringmag.com/articles/99594-the-2021-top-100-food-and-beverage-companies.

[17] See, for example, Tom Perkins, "The fight to stop Nestlé from taking America's water to sell in plastic bottles," *Guardian*, October 29, 2019, https://www.theguardian.com/environment/2019/oct/29/the-fight-over-water-how-nestle-dries-up-us-creeks-to-sell-water-in-plastic-bottles; and Alexandra Shimo, "While Nestlé extracts millions of litres from their land, residents have no drinking water," *The Guardian*, October 4, 2018, https://www.theguardian.com/global/2018/oct/04/ontario-six-nations-nestle-running-water.

[18] See, for example, David J. Stuckenberg and Anthony L. Contento, "Water Scarcity: The Most Understated Global Security Risk," *Harvard Law School National Security Journal*, May 18, 2018, https://harvardnsj.org/2018/05/water-scarcity-the-most-understated-global-security-risk/.

[19] Sassen, "Who owns our cities."

[20] *Push*, directed by Fredrik Gertten (WG Film, 2019).

United Nations. In that role, she sees this happening all over the world. To the common observer, it looks like progress: a giant decaying tower for low-income people is transformed into a remarkable housing complex two years later. It looks very positive. But in doing that, the prior renters, hundreds of them, often become homeless because there is now one less affordable housing complex and few companies are in the business of building low-income housing, because the profit margins are so much smaller. And this is happening all over the world. You asked about how global cities and how cities around the world experiencing power shifts may contribute to instability, and this is one way that can happen.

Spencer: Something that you just said just keyed me to something that I struggle with as an urban warfare researcher. It is about cities as systems. I know you have described cities as complex but incomplete systems.[21] Others have described cities as organisms.[22] Can you describe what you mean when you say a city is a complex but incomplete system?

Sassen: I think describing a city as a complex, open, and incomplete system is a better analogy than describing it as an organism. I think incompleteness is a very important characteristic that is left out when describing a city as an organism. Each city is very unique because each is complex yet incomplete. The city is in an incomplete condition because it is an open system. That creates great possibilities. You can innovate, you can build great things, you can work collectively together to grow food, you can gain a sense of community. There are almost infinite ways in which that condition of incompleteness can be made positive. But at the same time, that incompleteness can also result in a negative, such as the housing example that I just described. Too often, powerful actors abuse their power for their own benefit and to the detriment of the city. And it does not have to be someone with exceptional power, it could be a local entrepreneur who owns a bunch of the housing in a modest neighborhood. They do not have to be super rich to be extractive.

Spencer: I like your concept of incompleteness. I deployed to Kirkuk, Iraq in 2004 shortly after the United States invasion and to Baghdad in 2008. You could definitely describe Baghdad as being incomplete, even if it was not considered a global city. As you describe, this incompleteness created

---

21 Nicholas Barrett, "Saskia Sassen and the incomplete city," *EUI Life*, June 20, 2014, https://life.eui.eu/saskia-sassen-and-the-incomplete-city.html.

22 David Kilcullen, *Out of the Mountains: The Coming Age of the Urban Guerrilla* (New York: Hurst & Company, 2013), 41.

great possibility. Like many military officers operating in that post-invasion situation, I was looking for ways to try to improve the security, economy, government, or whatever. One of the things that we attempted to do at the very tactical level was to help local economies stand back up by using things like micro grants, using what the military called the commander's emergency response program.[23] But most of us had no way to know if our actions were going to have the intended effect because we really did not understand the city as a system. So, one part of your research that interests me is what you call "tactical urbanism." It really made me remember being at the tactical neighborhood level and trying to understand the economies, understand the neighborhoods, and understand the system. Can you explain what you mean by tactical urbanism?

Sassen: I can understand why a military mind might be attracted to the term. Tactical urbanism is "a city and citizen-led approach to groups or neighborhoods using short-term, low-cost, and scalable interventions intended to create long-term change."[24] It centers on new approaches to the form and function of cities such as prioritizing the design of cities for people as opposed to cars.[25] It often involves projects designed to provide a better quality of life to the disadvantaged in poor or modest neighborhoods. Left to the normal state, these communities are underserved, so it takes action, "tactical urbanism," to change them for the betterment of the people that actually live in them.[26]

Spencer: I have been studying urban warfare and cities for nearly a decade, but even during this relatively brief time, I have seen a shift in how we understand cities. I have read of your desire to create some new version of the Athens Charter, a seventy-year-old concept of urban planning developed by functionalist architects that guided the development of cities in the 20th century.[27] I know that has included work with the United Nations

[23] For more on the program, see, Department of the Army, *The Commander's Emergency Response Program*, ATP 1-06.2 (Washington, DC: Headquarters, Department of the Army, 2017), https://armypubs.army.mil/epubs/DR_pubs/DR_a/pdf/web/ARN7910_ATP%20 1-06x2%20FINAL%20WEB.pdf.

[24] "Tactical Urbanism: Creating Long-Term Change in Cities Through Short-Term Interventions," parCitypatory (blog), July 30, 2020, https://parcitypatory.org/2020/07/31/tactical-urbanism/.

[25] Ibid.

[26] Saskia Sassen, "Deep Inside the Global City," *e-flux*, January 2018, https://www.e-flux.com/architecture/urban-village/169799/deep-inside-the-global-city/.

[27] Rebecca Greenwald, "The Quito Papers: A New Manifesto for Urban Planning in the 21st Century," *Metropolis*, February 10, 2017, https://metropolismag.com/viewpoints/quito-papers-new-manifesto-urban-planning-21st-century/.

to develop a new approach to the planning, design, and relationship between form and function in the world's growing cities. Has there been a change as to how we view or understand cities?

Sassen: There probably has been a change, but I think part of the change in how we understand cities is due to the fact that cities are constantly evolving. I will use New York as an example. There was a time in the United States when people were leaving the cities. When I first moved to New York City in the 1970s, it was really quite degraded, and it was dangerous at night. Some people referred to this time in New York City as "The War Years."[28] But because New York City was so broke, it also meant that it was cheap. So, artists came to New York from all over the world because they could afford to live there. It was one of the most exciting periods for Manhattan when it came to art.

That of course changed as New York started to become a global city in the 1980s. It changed the economy of the city. The city became richer, so it could invest more in city services, such as the police. People started investing in the city again, and the upper classes started to return to the city. When you look at New York today, it has recovered, but as I have mentioned throughout our discussion, not everyone was a winner as a result of this transformation.

I do want to make one final point. I am concerned that many of our cities are getting too big. I think they are in deep trouble because many people who work there cannot afford to live there. Some have to travel 90 minutes or more just to get to a modest paying job. Cities should be designed in a way that the poorest workers do not find themselves in the situation where they get up at three or four in the morning just to get to their job and are just dead tired, and, as a result, do not perform well. I think the notion of the ever-expanding, endless city is not a good one.

Spencer: I have read a lot about the rapid rise of modern urbanization. The United Nations calls it the largest wave of urban growth, citing, of course, how half the world's seven billion inhabitants now live in urban

[28] "The War Years" was a period of dwindling prosperity in New York City in the 1960s and 1970s where many property owners turned to arson to burn down their property for the insurance payouts and firefighters were routinely attacked. Some also postulated that New York City was ungovernable. See, Celeste Neill, "The New York City Fire Department: A Timeline of the City's Firefighting History," HistoryHit.com, September 9, 2021, https://www.historyhit.com/new-york-fire-department-history/; and Nathan Glazer, "Is New York City Ungovernable?" *Commentary*, September 1961, https://www.commentary.org/articles/nathan-glazer-2/is-new-york-city-ungovernable/.

areas. By 2030, that number is expected to balloon to five billion.[29] You have argued that we need more cities rather than expanding existing cities until they become unmanageable.[30] As a military professional who studies urban operations, the idea of unmanageable growth concerns me, because it could cause the military to deploy to one of these cities to conduct humanitarian or combat operations.

For example, Lagos, Nigeria had less than a million residents in 1960 and now has more than 15 million.[31] Because of its rapid, unmanaged growth, more than 90 percent of the city's population does not live in a home with sewer connections and 80 percent do not have access to tap water.[32] Do you think this unmanageable growth could lead to political instability?

Sassen: Absolutely. I think there should be a way to measure a city, to see if it is growing in an unmanageable way. I think that would be of great value not just for military practitioners, but also for urban designers, community planners, and urban policy writers.

Spencer: I think that is a great note to end on. I completely agree with your comments about the expanding role of cities beyond their institutions and governments. They are communities, global communities of complex populations, systems, economies, and connections that we need to continually study. I also like your characterization of a city as an open and incomplete system; one that can change and even reinvent itself. That is why understanding these environments is so important to me, especially from a military perspective. Urban is the future and cities are increasingly global.

---

29 "Urbanization," United Nations Population Fund, https://www.unfpa.org/urbanization.

30 G. Sampath, "I think we need more cities: Saskia Sassen," *The Hindu*, February 5, 2017, https://www.thehindu.com/books/I-think-we-need-more-cities-Saskia-Sassen/article17194604.ece.

31 "Lagos Population 2022," World Population Review, https://worldpopulationreview.com/world-cities/lagos-population.

32 Jillian Du and Anjali Mahendra, "Too Many Cities Are Growing Out Rather Than Up. Three Reasons That's a Problem," The City Fix (blog), World Resources Institute, January 31, 2019, https://thecityfix.com/blog/many-cities-growing-rather-3-reasons-thats-problem-jillian-du-anjali-mahendra/.

4

Smart Cities

While the term "smart city" became popular around 2008, the concept dates back to the 1970s when the community analyses bureau in Los Angeles started to use data to shape its urban planning. While rather crude by today's standards, the city was cutting edge at the time using "computer databases, cluster analysis, and infrared aerial photography to gather data, produce reports on neighborhood demographists and housing quality, and help direct resource to tackle poverty."[1] Investment into smart cities started to accelerate in 2005 when Cisco, an American multinational technology corporation, dedicated $25 million over five years to research making cities more sustainable. Cisco subsequently created its own connected urban development program.[2]

Not to be left behind, IBM launched its smarter planet initiative in 2008 to investigate the application of "instrumentation, interconnectedness, and intelligence" to address some of the world's most pressing issues.[3] Sam Palmisano, IBM's chief executive officer, introduced the term "smarter cities" in a talk entitled *A Smarter Planet: The Next Leadership Agenda* on November 6, 2008.[4] The following year IBM launched a smarter cities program with the goal of helping cities run more efficiently.[5] In 2011, IBM trademarked the term "smarter cities."[6] That same year, Barcelona hosted the world's first Smart City Expo World Congress, which it has hosted every

---

1 Jose Montes, "A Historical View of Smart Cities: Definitions, Features and Tipping Points" (working paper, June 2020), https://papers.ssrn.com/sol3/papers.cfm?abstract_id=3637617.

2 Pete Swabey, "IBM, Cisco and the business of smart cities," *Information Age*, February 23, 2012, https://www.information-age.com/ibm-cisco-and-the-business-of-smart-cities-2087993/.

3 Ibid.

4 Herman van den Bosch, "Smart City: Smart Story?" *Smart City Hub*, November 29, 2017, https://smartcityhub.com/governance-economy/smart-city-smart-story/.

5 Swabey, "IBM, Cisco and the business of smart cities."

6 Van den Bosch, "Smart City: Smart Story?"

year since.[7] By 2013, 143 cities across the world self-designated themselves as smart cities.[8]

Yet more than a decade later, there is nothing close to a universally accepted definition for smart city. Some academic papers spend as many as six pages examining the definitions of smart cities yet fail to provide a synthesized definition of their own.[9] It should come as no surprise then, that industry generally does a better job of defining the term than the academic community.

One definition of smart city is a "city [that] uses information and communications technology to improve operational efficiency, share information with the public and provide a better quality of government service and citizen welfare."[10] This definition captures elements that are common to most definitions: (1) a city or community, that (2) uses information and computer technology, to improve (3a) government services and (3b) residents' quality of life.

In this chapter John Spencer interviews Dr. Sokwoo Rhee, who was the associate director of Cyber-Physical Systems Program at the National Institute of Standards and Technology (NIST) when the interview was conducted in 2020. In that position, Rhee was responsible for the smart city program within NIST.

John Spencer: Dr. Rhee, I would like to start by asking you to describe what you do at NIST and how you got into the business of smart cities?

Sokwoo Rhee:[11] NIST is a bureau within the United States Department of Commerce. Its mission is to "promote U.S. innovation and industrial competitiveness by advancing measurement science, standards, and technology in ways that enhance economic security and improve our quality of life."[12] NIST covers a broad range of topics, including physical

---

[7] "About Smart City Expo World Congress," Smart City Expo, https://www.smartcityexpo.com/the-event/#:~:text=Held%20in%20Barcelona%20since%202011,urban%20innovation%20across%20the%20globe.

[8] Vito Albino, Umberto Berardi, and Rosa M. Dangelico, "Smart Cities: Definitions, Dimensions, Performance, and Initiatives," *Journal of Urban Technology* 22, no. 1 (2015), 3-21, https://doi.org/10.1080/10630732.2014.942092.

[9] Ibid., 4-10.

[10] "What is a Smart City? – Definition and Examples," TWI, https://www.twi-global.com/technical-knowledge/faqs/what-is-a-smart-city.

[11] The views and opinions expressed in this chapter are the interviewees and do not necessarily reflect those of the National Institute of Standards and Technology, the Department of Commerce, or the U.S. Government.

[12] "About NIST," NIST, https://www.nist.gov/about-nist.

measurements, atomic clocks, and even, surprisingly, peanut butter compositions.

I run the smart city program at NIST. Cities and communities across the "globe are seeking to deploy advanced technologies such as Cyber-Physical Systems and the Internet of Things to improve the quality of life of their residents. Such a system involves cyber-networking devices and other supporting systems working with physical infrastructure. When applied to transportation, health care, utilities, and other sectors, these [Cyber-Physical Systems and the Internet of Things] could expand and improve services, promote economic growth, and enhance quality of life."[13]

As such, NIST "works to develop the foundations and measurement science for Cyber-Physical Systems and the Internet of Things. NIST considers smart cities and communities to be an opportunity for studying [Cyber-Physical Systems and the Internet of Things] at scale, across sectors, and their interaction between devices, applications, and humans. In conducting this research, NIST seeks to facilitate stakeholders' development of replicable, scalable and interoperable smart cities and communities."[14]

Spencer: What is a smart city?

Rhee: That question is both easy and difficult to answer. If you asked 100 people to define a smart city, you would probably get 95 different answers. I consider smart cities and smart communities to be ones that use smart technologies such as the Internet of Things and Cyber-Physical Systems to improve the quality of life of their residents. But this is merely my definition. Even at NIST we do not have a single definition for smart city, but this captures what we consider a smart city.

Spencer: I have heard you describe smart cities in terms of different layers.[15] Can you describe these layers?

Rhee: Smart cities are comprised of four layers. The bottom layer is composed of the physical systems, the things that you can touch: sensors,

---

[13] A Cyber-Physical System (CPS) is "an integrated, hybrid network of cyber and engineered physical elements that are co-designed and co-engineered to create adaptive and predictive systems." Examples of CPS are the Internet of Things (IoT), Smart Robots, and Autonomous Vehicles. See, Sokwoo Rhee and Geoff Mulligan, "SmartAmerica Challenge," PowerPoint presentation, no date, https://www.nist.gov/system/files/documents/smartgrid/SGTB-Rhee-Keynote.pdf. Text in quotes comes from "Smart Cities and Communities," NIST, https://www.nist.gov/el/cyber-physical-systems/smart-americaglobal-cities.

[14] Ibid.

[15] TRADOC G-2 OE Enterprise, "TRADOC Mad Scientist 2017 Georgetown: Smart Cities w/ Dr Rhee," YouTube video, August 8, 2017, https://www.youtube.com/watch?v=7svLfitt7FA.

actuators, thermostats, chips, or even large devices like vehicles and airplanes. The second layer is the communications layer; it connects the physical structures by Wi-Fi, cellular networks, or Bluetooth so that they can send and receive information. These two layers form the foundational layer. These are the layers most people think about when they think of the Internet of Things, but that is only half of the system.

The third layer is the data analytics layer. This layer collects data from the communications layer and from the physical systems and turns the ones and zeros into actionable, useful information. The fourth layer, which I believe to be the most important, is the human and services layer. This is where a human or some type of automated service has to decide what to do with that useful information. If no one does anything with data, then it is just ones and zeros. I have even heard people say, "data is new currency," because they feel it is so valuable.[16] That is not true. Data has no value without these other layers; it is just ones and zeros.

For example, a person within the command center of an emergency response system can look at all the data coming from disaster response areas and decide to send more resources to one region than other. That is where the decision is made to take an action, and this decision typically results in action implemented through the reverse order of the layers. The action starts at the human and service decision layer, then passes through the data analytics layer, then through the communications layer, with action in the physical systems layer. It is a feedback system with information flowing in both directions.

When you overlay this concept of four layers onto a city or a municipality, it becomes a smart city or smart community. From a technical perspective, that is probably the easiest way to understand the concept of a smart city. It really comes down to how to improve the quality of life of a community's residents using these smart technologies. Obviously different cities may need to adopt different approaches because each city is unique.

Spencer: I love your point about people thinking that data is the new currency in a technologically advanced society, but it is just ones and zeros if no one is analyzing or actioning the data. There is a ton of data out there,

---

[16] See, for example, Natarajan Chandrasekaran, "Is data the new currency?" World Economic Forum, August 14, 2015, https://www.weforum.org/agenda/2015/08/is-data-the-new-currency/.

such as cellular phone data, traffic data, and essential services data. Can you provide an example of how this data is being used to improve the lives of smart city residents?

Rhee: Obviously, there are many examples, but I will use the example of traffic congestion management. Most large cities experience traffic congestion problems, and there are many ways that they can deal with it. One way is to collect real-time data from the traffic signals or traffic sensors spread throughout the city and figure out how to dynamically adjust the traffic signals of the city. It is fairly easy if you have ten traffic intersections, but it is a fairly complex, multi-variable, time-variant problem if you have 200 plus intersections.

The physical systems collect the data and transfer this data where it is turned into actionable information—there are more vehicles on the south side of the city, cars are running faster on the north side of the city, there are more cars in this intersection, and so on—that a human or system can use to decide how to change the traffic signals and improve traffic flow. That is one example that demonstrates how collecting the data, transferring it, converting it into actionable information, and acting can improve residents' quality of life.

I will provide a second example from a project that I worked on in Virginia Beach to predict local flooding. Virginia Beach suffers from periodic flooding. Often the flooding is not caused by a particularly historic storm. It rains, and because of extreme localized weather and a low water table, one street may get flooded very quickly while another is just fine. This microscopic weather cannot be accurately predicted with weather models.

The city's solution was fairly simple: they deployed a number of flood sensors to detect water levels. These sensors transmit their data to the cloud where it is translated into useful information that an analytics engine uses to predict which parts of the city are going to be flooded. The pilot was fairly successful, and I believe they have benefited from the system. Those are but two examples of how sensors, data, information, and action based on that information, provides a real benefit to cities.

Spencer: You discussed the importance of deciding what to do with the data, but in this example, the city actually had to create the data. It had to place sensors at critical points throughout the city.

Rhee: Yes, cities have a lot of data, but often the existing data is inadequate. When analyzing the existing data for new applications or new purposes, you often find gaps. Thus, it is absolutely critical to invest in physical systems, which are more than simply sensors. Previously, many people overly focused on the physical layer. They thought, "just throw out tons of sensors and see how it goes" without really thinking through how to use that data to make decisions. In recent years, the pendulum has swung to the opposite end of the spectrum. Now there is too much focus on analyzing the existing data, as opposed to thinking about how to collect the data that is required to make the best decisions. It has to be balanced. You have to have the sensors to collect the right data and the systems to transfer, translate, and action that data.

Spencer: I want to back up a little bit. Who is in the business of saying: "This city is smart?" Or is it a self-designation system, where a city declares: "We are a smart city." Does NIST classify cities as smart?

Rhee: I get these questions all the time: "What is a smart city? Can you define a smart city? Can you classify a smart city? Can you measure a smart city?" The short answer is no one can. Every city is unique, so the notion of smartness varies from one city to the other. That is completely acceptable, and I encourage cities to think of smartness in the context of their own city. To answer your question, NIST does not classify a city as smart. However, that does not prevent NIST, or any organization for that matter, from developing tools and methodologies to help cities measure their progress on a smart spectrum—even if we do not have a formal spectrum. There have been efforts to develop consensus-based standards on what it means to be smart. It may start with the things that are easy to measure such as the number of sensors that are deployed, the number of Wi-Fi hotspots, or the number of smart traffic control systems.

The physical systems may be the easiest to measure but they are also the least important. What is important is whether those sensors are actually improving the residents' quality of life? Do the systems improve the quality of municipal services? Those are more difficult to measure. I realize that I am not providing an answer or a checklist that you can use to determine if a city is smart or not, but there are various indicators that research communities are developing. Yet at the same time, I do not expect there to be one single metric that can be used at any time in the near future. As I said, every city is unique, and each city must find the metrics to measure their progress that are relevant to it. Still, organizations like NIST and other research communities can provide tools to assist.

Spencer: I went to a website that you referenced in one of your talks about the Global City Teams Challenge.[17] The website listed over 200 smart cities that had opted into this smart city community. Is that correct?

Rhee: Yes, that is the program I am running at NIST.

Spencer: I was really surprised by the number and variety of cities listed as participating in the challenge and attending the annual meeting. They ranged from Syracuse, New York to Bharatpur, Nepal. I never would have thought that a city in Nepal would have opted into a conversation about improving the quality of life of its citizens through the Internet of Things.[18] Where are the world's smart cities? Are they everywhere?

Rhee: Yes, smart cities are everywhere. New York City, San Francisco, and Chicago invest large sums of money to install and deploy smart systems to improve the quality of life of their residents and the service quality of their municipality, as do other large cities across the United States and around the globe. But there are also smaller towns, with populations under 5,000, that implement smart projects, albeit on a much smaller scale. Cities and communities of any size can become a smart city or a smart community, as long as they implement the right projects. If the community is implementing technology that can help it gather data, translate that data into information that is then actioned to improve the quality of life of its residents, then it can be considered a smart community.

Spencer: As you said, smart cities are being developed around the world and you track some at NIST. Where are smart cities not being tracked or developed? I think that is relevant for the military should it find itself in a smart city for any range of military operations, from combat to humanitarian.

Rhee: That is an important question, especially for the military. As I mentioned earlier, NIST "develops and disseminates the standards that allow technology to work seamlessly and business to operate smoothly,"[19] but we believe the system works best when the standards are developed

[17] The website was, "Global City Teams Challenge," NIST, https://pages.nist.gov/GCTC/. The talk Spencer references was, TRADOC G-2 OE Enterprise, "TRADOC Mad Scientist 2017 Georgetown: Smart Cities w/Dr Rhee."

[18] Sokwoo Rhee and Martin Burns, *Global City Teams Challenge 2018 Kickoff and IES-City Framework Workshop: Smart and Secure Cities and Communities Challenge* (Portland, OR: National Institute of Standards and Technology, 2018), 15, https://nvlpubs.nist.gov/nistpubs/SpecialPublications/NIST.SP.1900-201.pdf.

[19] "Standards," NIST, https://www.nist.gov/standards.

collaboratively. The development process has to be industry-led and consensus-based. It has to have voluntary participation from the beginning. Standards are more likely to be effective and efficient if they are developed collaboratively as opposed to being developed arbitrarily and forced.

You mentioned NIST's Global City Teams Challenge. We launched the program "to encourage smart city and Internet of Things stakeholders to develop and deploy standard-based solutions that are replicable, scalable, interoperable, and that provide measurable benefits for cities and communities."[20] It is essentially an information sharing and collaboration program. We encourage any city, company, university, or non-profit to join and share their best practices so they can learn from each other and replicate successes and identify best practices among them. It is completely voluntary.

NIST is a United States government agency, so most of the cities participating in the program are in the United States, but 40 percent are foreign with approximately 25 percent in Europe and 15 percent elsewhere in the world. Of course, as you can imagine, there are cities and whole nations that choose not to participate in this program, but we have good participation from the United States, Europe, and some parts of Asia. China has little, if any, participation in the program. That does not mean there are no smart city programs in China. The way China is growing its smart cities is sometimes mind blowing. So, the list of the cities in the Global City Teams Challenge is completely voluntary and probably a good survey of the United States, Europe, and some parts of Asia.

Spencer: If a community has X amount of dollars to invest in a certain quality of life project, the Global City Teams Challenge is a forum where lessons can be shared so that a community can make a smart investment with its limited resources. I think it is great that there is a forum where information can be shared to improve people's quality of life.

Rhee: Absolutely. Every city has its own government and its own way of doing things, and because every city is different, cities are very much siloed in terms of developing smart solutions. Returning to the example of traffic congestion, nearly every large city in the world experiences traffic congestion. Nearly all of them have developed their own traffic congestion management system. Each developed a home-grown solution for their unique challenge, but I suspect all eventually evolved into fairly similar

[20] NIST, "Global City Teams Challenge."

systems. But each city made a significant investment to develop that solution, often learning by trial and error along the way. From a standards or emerging technology perspective, what happened was that each city "reinvented the wheel." Instead, they should have worked collaboratively and shared their experience to develop a solution at a cheaper cost. That is the point of the Global City Teams Challenge: to create more collaboration between cities and to help cities use their resources more efficiently.

As I mentioned, the program also includes companies. If you look at smart cities from an industry's perspective, no company wants to develop a solution for one single customer. They want to develop a solution that works for as many customers, or cities, as possible. If every city developed a custom solution, the industry would never grow or grow extremely slowly. It would fail to develop economies of scale that could attract investment and produce cheaper products that could then be sold to smaller communities with smaller budgets. Because cities experience similar problems, they can improve more efficiently through a collaborative process.

Once again, I will return to the traffic congestion problem. There is probably a baseline or foundational approach that can solve 70 percent of the problem. But every city is unique, so a custom approach is required to solve the final 30 percent. But learning the best practices to get the first 70 percent is better than developing everything from scratch. It is all about collaboration, sharing, and replication.

Within the Global City Teams Challenge are ten categories, or "SuperClusters" as we call them: transportation, public safety, utility, smart buildings, agriculture and rural, data and city platform, wireless, education, cybersecurity and privacy, and healthcare.[21] The traffic congestion example that we keep returning to falls within the transportation SuperCluster.

Spencer: I love this topic! It reminds me of a podcast that I recently listened to. A PhD candidate discussed how large municipalities attempt to make the right investments to address their cities' problems, but far too often, the investments do not pan out. City officials are trying to do the right thing, but there are so many different sources of information that cities are often unable to determine which information is appropriate for their city.

She discussed a number of failures, ranging from how one community misapplied crime prevention tactics that worked in another community to their own, to what we have been discussing, which is the investment

[21] Ibid.

in sensors, data analytics, and artificial intelligence to improve that city's or that community's problems. It seems to me that the Global City Teams Challenge could minimize the likelihood of bad investments.

Returning to the traffic congestion example, during a recent research trip to Mumbai, India, I could not help but to think that they could use some help with their traffic. I am not sure what investments the city has made, but the decision ultimately rests on the city government to invest in that bottom layer, the physical systems layer.

Rhee: You are absolutely right. Governments have their own reasons to invest or not invest in solving a particular problem. In some cases, governments may be hesitant to emplace systems due to security or privacy concerns. But there are often ways to solve problems without violating an individual's privacy.

Spencer: You mentioned that China has chosen not to participate in the Global City Teams Challenge. That is not surprising. I saw one of your presentations at another conference in which you said that some studies have shown that China has invested more than $80 billion into their smart cities, which, as you said earlier, is mind blowing.[22] I understand why they might not want to share that information but failing to collaborate could affect the betterment of human societies and communities in general.

Rhee: There are two reasons some cities are not part of the Global City Teams Challenge program. The first is that some cities and countries are reluctant to share their experience. The second is that they do not know about this program, maybe due to language and geographic distance.

China is an important player in the smart city landscape, but they have a very different approach from the United States. Many Asian countries use a top-down approach, meaning the central government pays for and directs the smart investment in its cities. The United States conducts a bottom-up approach, meaning that each city or community determines how it will conduct its smart investments.

There are pros and cons to each approach. Top-down is generally faster because it is typically an investment at scale for a particular program. Bottom-up is slower, it involves more trial and error, but it is more "democratic" in that municipalities get to invest in the projects that they need most as opposed to being directed by the central government. While

[22] "Cities of Tomorrow: Safety, Smarts, and Surveillance," CSIS, January 23, 2020, https://www.csis.org/events/cities-tomorrow-safety-smarts-and-surveillance.

slower, bottom-up often ends up more stable and more reliable, because it is not the cookie-cutter approach that top-down may be.

China, obviously, uses a very top-down approach. Their government investment in smart cities ranges between $80 billion to $130 billion. From my perspective, that level of investment is hard to even imagine, because the U.S. federal government's role in smart city development is mostly about encouraging collaboration and facilitating collaboration as opposed to directing the actual investment. Not surprisingly, with this level of investment, many cities in China are moving really, really fast, but because they are not part of our program, we do not have as much visibility on their smart cities as we do for cities who are part of the program.

Spencer: What are some of the common issues in the development of smart cities?

Rhee: The lack of collaboration that I already discussed is a big problem. It is not a technological problem; it is a structural problem: cities are simply not used to working together. All the various offices within a municipality report to the mayor's office or its equivalent. Offices with similar functions in different municipalities will not reach out to one another to collaborate unless there exists the political will of both mayors to do so. This results in siloed or custom-tailored solutions.

A second challenge is also structural: municipalities are not designed to take on the risk of investing in new technologies or systems; that is the role of industry. Generally, tax dollars are not spent in this way, but because of that, the deployment and trial and error of new solutions is slowed. It is also tough for industry to enter the field and recoup its investment because each municipality can be thought of as a single market if there is no sharing or collaboration. Thus, with nearly 90,000 different jurisdictions, there exist 90,000 different markets in the United States.[23] That is not a good thing to have from an industry perspective.

Spencer: That makes a lot of sense. I recently interviewed a local police chief and we discussed how his city has invested in technologies to help policing, but as you said, it was a political decision that ultimately fell on the mayor and the city council.[24] I would now like to discuss military

[23] In 2012, the U.S. Census Bureau reported 89,004 local governments in the United States. See, "Census Bureau Reports There Are 89,004 Local Governments in the United States," Census, August 30, 2012, https://www.census.gov/newsroom/releases/archives/governments/cb12-161.html.

[24] John Spencer, "Urban Policing 101," *Urban Warfare Project Podcast*, Modern War Institute, August 21, 2020, https://mwi.usma.edu/urban-policing-101/.

operations—combat, humanitarian, or other—in urban areas. What are some of the implications, concerns, advantages, disadvantages that smart communities pose to a military that is operating there?

Rhee: That is a very important question. It really comes down to infrastructure and I mean a new generation of infrastructure. Previously, infrastructure consisted of buildings, bridges, and roads. It was stationary, you could look at it on a map and figure out how to deploy around it. Now, infrastructure includes all the physical sensors that are communicating with one another. They are constantly collecting information, they may not be stationary, and they may also act if that data is turned into action. Suddenly you need a completely separate set of virtual maps to understand how the city, as a system, operates. This sounds pretty abstract, so I will provide an example.

In 1997, the Defense Advanced Research Projects Agency funded a project called Smart Dust. Smart Dust was designed to solve the information problem on the battlefield: the problem of not really understanding the operational environment. The project involved the development of millimeter sized sensors that would be dropped over a battlefield or quickly emplaced in an urban environment to provide information about the environment to a central command.[25]

Thus, before sending in any troops, the military would have a much greater understanding of what they would be dealing with. A smart city already has a network of sensors in place. If the deployed military can hack into the networks of these various systems, then a smart city can provide an amazing amount of information. On the other hand, if the network is controlled by an enemy, it provides the enemy a marked advantage.

With smart cities, the military must think very differently about operating in an urban environment. While it must still be concerned about the physical infrastructure, it must also consider the informational infrastructure. It requires more than a physical map for planning and understanding. The infrastructure no longer just sits there; some of it can collect data and be directed to act on that data. I am not a military expert, so I do not have a lot of answers, but I can imagine operating in smart cities will require more planning and understanding.

Spencer: I love that you describe the city as a system. When I provide presentations about urban warfare, I say there are two requirements: (1)

[25] Jessica Jones, "Dust in the Wind," *Government Technology*, July 27, 2010, https://www.govtech.com/health/dust-in-the-wid.html.

you must understand the city, and (2) you must understand how best to conduct military missions in a city. Talking to you, it only emphasizes the challenges of understanding the city. Each one of these sensor systems in the city has a vastly different array of connections to the information infrastructure. The challenge will be how to quickly gain what we call situational awareness of the vast information infrastructure that is there. How do you quickly learn what is there and how to use it?

Rhee: That is a million-dollar question. First, the military must understand the city, it has to have expertise in how smart cities operate. It involves understanding and expertise in artificial intelligence, the Internet of Things, Big Data, smart sensors, and vehicles and drones. This will become more and more important. Whoever gains the advantage of understanding the city and its smart aspects will have a significant edge.

Spencer: Absolutely. The recommendation that the military requires experts in cities has come up many times throughout the course of my research. I do not think that the military necessarily has to grow its own experts, but it must know where the experts reside so they can tap into them when needed. There is so much that the military does not understand about smart cities and that could hurt it in future operations. But I have seen a few wargames where the data analytic center, if a city had one, was considered key terrain—a location or area which affords a marked advantage to either combatant—because the information infrastructure aspect of the city is so important.[26] The city manager might also be considered a key individual because of his or her knowledge of the smart systems.

There is some limited understanding within the military, but without some kind of map of the informational infrastructure that you mention, planning and understanding will remain limited. The military relies on maps for planning. Beyond two dimensional maps, imagery can provide some understanding of a city in three dimensions. There are also maps to understand the subterranean environment and sometimes maps to understand the tribal, religious, or ethnic structure. I think all these various layers—two dimensional, three-dimensional, subterranean, human—are important, but I have never seen anything that resembles a virtual map for

[26] The United States military defines key terrain as, "any locality, or area, the seizure or retention of which affords a marked advantage to either combatant." See, Joint Chiefs of Staff, *Department of Defense Dictionary of Military and Associated Terms*, JP 1-02 (Washington, DC: Joint Chiefs of Staff , 2021), 125, https://www.jcs.mil/Portals/36/Documents/Doctrine/pubs/dictionary.pdf.

the informational infrastructure. I think that would be important for any type of military operation.

Rhee: I think you are right; the military does not need to have expertise on everything, but it needs to know where to find that expertise. Likewise, a smart city expert cannot generate a military plan because that is not his or her area of expertise. I think it requires collaboration—between the military and smart city experts—to develop a good operational plan.

Spencer: Definitely. Getting perspectives that the military may not, but should, consider, is one of the reasons that I love interviewing experts like you.

Rhee: Thank you, this was really fun. I really hope that when people discuss smart cities and smart communities what they think about first is improving peoples' quality of life. It is possible that the smart investments we are making in cities today will impact not only the current generation, but future generations as well. There is a long gain to this investment and that is how it should be viewed.

5

Megacities and the U.S. Army

Megacities is a term that denotes any city with a population exceeding ten million residents.[1] The first cities to reach megacity levels were New York City and Tokyo in the 1950s.[2] The United Nations reported there were 33 megacities in the world in 2018 and another ten projected by 2030.[3] The rise of megacities is a product of global urbanization and the benefits of living in urban areas as opposed to rural: increased job opportunities, access to centralized markets, and higher pay and individual wealth.[4]

But the rapid growth of some megacities around the world has exceeded their nations' ability to maintain security and political stability. Some scholars have begun to sound alarms about the ability for militaries to operate in a foreign megacity if called to do so for any mission—from disaster relief to major combat operations. In 2014, a research team of soldiers and civilians from the Chief of Staff of the Army's Strategic Studies Group published a report about the military's readiness to conduct operations in megacities. The team found the U.S. Army "ill prepared" to operate in megacities for many reasons to include the difficulty for the military to understand the scale, connectedness, and complexity of these large urban areas.[5]

---

1 Department of Economic and Social Affairs, Population Division, *The World's Cities in 2018: Data Booklet* (New York: United Nations, 2018), https://digitallibrary.un.org/record/3799524?ln=en.

2 "The History of Cities," National Geographic Society, https://www.nationalgeographic.org/article/history-cities/.

3 Department of Economic and Social Affairs, Population Division, *The World's Cities in 2018*.

4 "Urbanization and the Megacity," World Population History, https://worldpopulationhistory.org/urbanization-and-the-megacity/.

5 Marc Harris et al., *Megacities and the United States Army: Preparing for a Complex and Uncertain Future* (Arlington, VA: Office of the Chief of Staff of the Army, 2014), 3, https://api.army.mil/e2/c/downloads/351235.pdf.

The report offered several recommendations to better prepare the army for future operations in megacities. Two of the recommendations included building a new depth of understanding of megacities and educating leaders to appreciate the complexity of these environments so they can provide options to military and national security leaders.[6]

In this chapter, John Spencer interviews Colonel (retired) Patrick Mahaney. Mahaney is the co-founder and director of the National Center for Urban Operations. In 2014-2015, Colonel Mahaney served as the director of the Chief of Staff of the Army's Strategic Studies Group while the group continued researching the implications of megacities and urbanization for the U.S. Army.

John Spencer: Could you discuss your background and how you got drawn into or started to focus on urban operations?

Patrick Mahaney: I was literally born into "urban." I was born in Flatbush, Brooklyn, one of New York City's five boroughs, and raised in the southern part of New York City. After retiring from the army, I returned, where I am now working as the director of the non-profit National Center for Urban Operations at Fort Hamilton, here in Brooklyn. My father was a firefighter during "The War Years," a period of dwindling prosperity in New York City during the 1960s and 1970s where many property owners turned to arson to burn down their property for the insurance payouts and routinely attacked firefighters.[7] It was a time of tremendous turbulence. So as a child, I grew up in a world of firefighters, police officers, and military veterans. As a teenager, I could see some of the challenges posed with living and operating in a dense urban environment. The city back then, I sometimes say facetiously, was more like a zombie apocalypse than the sparkling gem that New York City later became.

I was fortunate enough to grow up around many people who were on the front lines at that time. Obviously, that was mainly from a first responder perspective, but when I graduated from New York University and entered the army, it was something I was very much drawn to. I started my military career as a military police officer, but I finished with 24 years in special forces. As a military police officer, the first place I wanted to be stationed

---

[6] Ibid.

[7] Celeste Neill, "The New York City Fire Department: A Timeline of the City's Firefighting History," HistoryHit.com, September 9, 2021, https://www.historyhit.com/new-york-fire-department-history/.

was Panama. In the late 1980s, Panama—and particularly Panama City, but Colón as well—was in a period of tremendous instability. Panama City was a dense urban environment and since I spoke Spanish and had grown up in New York City during "The War Years," I was drawn to Panama.

Had I not gone to Panama, I would have liked to have been stationed in either Berlin or Korea. This was prior to the dissolution of the Soviet Union when there was still an East and West Germany and an East and West Berlin. The U.S. Army had a brigade of soldiers, the Berlin Brigade, stationed there to deter Soviet aggression and defend the city from Soviet threat.[8] I also wanted to be stationed in Korea because the greater Seoul metro area fascinated me and still does today.

Spencer: I have a ton of questions that I want to ask, especially about the Berlin Brigade. Creating an urban-focused unit was a real novelty. I think it is a great concept to focus a unit on urban areas. So now that you are retired are you still doing urban things?

Mahaney: Yes. That is pretty much what I do. I do that wearing a "government hat" as a special government employee working with the under secretary of defense for research engineering on an innovation program called the National Security Innovation Network.[9] I also co-founded a nonprofit called the National Center for Urban Operations that is obviously focused on urban operations.[10] The big idea in all of that was seeing the need within the Department of Defense, and especially for the ground forces—the army, the Marine Corps, and special operations—to invest in the conceptual development, material development, training, and education of urban operations. The United States military has made tremendous strides since the Chief of Staff of the Army's Strategic Studies Group first started working on this in 2014, but it has also just scratched the surface. There is so much work that has yet to be done.

Dense urban terrain and megacities are part of the emerging complex operational environment that we find ourselves in today. I was drawn back into this urban operations world when I commanded the army's Asymmetric Warfare Group from 2011 to 2013. One of the group's missions was to focus on complex operational environments. I would argue that a

[8] "The Story of the Berlin Brigade," Army Heritage Center Foundation, https://www.armyheritage.org/soldier-stories-information/the-story-of-the-berlin-brigade/.

[9] The MD5 National Security Technology Accelerator was renamed the National Security Innovation Network (NSIN) in 2019. See, "MD5 Adopts New Name to Reflect Redefined Mission," NSIN, May 6, 2019, https://www.nsin.mil/news/2019-05-06-md5-adopts-new-name/.

[10] National Center for Urban Operations, https://ncuo.net/.

dense urban area, and certainly megacities, are among the most complex operational environments that exist. Therefore, any work involving dense urban terrain, and particularly megacities, by definition is going to be intricately interwoven with the issues of complexity and how we prepare our troops to operate in that type of environment.

Spencer: I know from working in the Pentagon that the army gave the Asymmetric Warfare Group some of its toughest problems and expected the group to rapidly develop material, doctrinal, or broader solutions to those problems. How did you, as the commander of the Asymmetric Warfare Group, define a complex operational environment and how did that inform the work you performed with megacities later?

Mahaney: First, let me take you quickly back to 2011. In April 2011, General Robert Cone took command of the U.S. Army's Training and Doctrine Command (TRADOC).[11] Lieutenant General David Perkins, who later became the commanding general of TRADOC, took command of the Combined Arms Center at Fort Leavenworth, Kansas in November 2011.[12] Then Major General H.R. McMaster took command of the Maneuver Center of Excellence at Fort Benning, Georgia in June in 2012.[13]

By the summer of 2012, we had these intellectual giants leading our army. We had been at war for a solid ten years and they started asking hard questions such as: Are we winning? What does winning look like? Are we learning the lessons that we need to learn from emerging operations that are increasingly being labeled as hybrid, asymmetric, irregular, or unconventional? I say all of this because these questions and this thinking spurred a tremendous amount of important and useful work.

It was General Cone who helped promote the term "complex operational environments" within the army's vernacular. The easiest way to describe the complexity is to focus on what the army calls operational variables: political, military, economic, social, information, infrastructure, physical environment, and time (PMESII-PT).[14] Too often, the military

[11] Pleasant Lindsey, "Gen. Robert W. Cone assumes command of TRADOC," Army.mil, April 29, 2011, https://www.army.mil/article/55767/gen_robert_w_cone_assumes_command_of_tradoc.
[12] Melissa Bower, "CAC welcomes new commander, CSM," Army.mil, November 23, 2011, https://www.army.mil/article/69846/cac_welcomes_new_commander_csm.
[13] Vince Little, "Fort Benning ushers in new commanding general," Army.mil, June 20, 2012, https://www.army.mil/article/82136/fort_benning_ushers_in_new_commanding_general.
[14] Department of the Army, *Operational Environment and Army Learning*, TC 7-102 (Washington, DC: Headquarters, Department of the Army, 2014), vii, https://armypubs.army.mil/epubs/DR_pubs/DR_a/pdf/web/tc7_102.pdf.

focuses almost exclusively on the military variable. There is a reason for that: warfighting is incredibly difficult and very dangerous. It is the essence of much of what the military does.

But in these complex environments, the other variables may be just as important, if not more important, than the military factors. The salient point is that you have to consider all relevant factors in a complex operational environment. If that is true, which the army believes since the army includes it in its doctrine, then you have to ask yourself, have we prepared ourselves to understand all the factors within complex operational environments like megacities?[15] What are the ways in which we prepare our troops to be able to deal with, for example, the political aspects of the operational environments they are going to enter? The economic aspects? The social aspects? Especially in megacities where the interactions of each variable to other variables can be extremely complex.

The U.S. Army addresses some of these aspects, to a degree, in its professional military education. For example, the U.S. Army War College provides education on the political aspects through lessons on international relations between nations. That is very important. But what happens when you are in an operational environment where those elements are still important but then you must also consider issues, for example, of sub-national governance? In Afghanistan and Iraq, we had to consider the local and provincial conditions. Military leaders often dealt with tribal groups and with what I used to refer to as "gray area players," formal and informal leaders who did not align with one side or the other but often had a powerful role in influencing the environment whether we realized it or not.

That brings up the question of whether we really understand power and everything relating to power as it is executed in terms of those operational variables: the power of money, the power of ideas, and the power that may reside at the local or neighborhood level that may, in fact, be more influential than what is being said in the capital city. That is complex. That is hard. I think that many of us would emphasize that our troops eventually became pretty good at understanding power during our recent wars, but they often had to learn it by experience, or on-the-job training, repeatedly. Our professional military education did not do a good job of preparing the force for these types of urban environments, despite the fact that we fought in that environment for nearly 20 years.

---

15 See, for example, Department of the Army, *Operational Environment and Army Learning*.

That reminds me of an analogy that General Cone used to use that involved a three-legged stool. He said knowledge is the seat of the stool, which is learned through training, education, and experience—the three legs. During the wars in Afghanistan and Iraq, we became really good at learning from experience. We had a tremendously experienced force that had figured out many tactical challenges on the battlefield.

But there were questions as to whether we had incorporated this knowledge into our training and our education. Or was our education and training suffering as we relied too heavily on experience, which was often purchased in blood and treasure. Generals Cone, Perkins, and McMaster believed that, indeed, we were relying too heavily on experience to educate the force. Returning to Cone's analogy, the stool was falling over because one leg was longer than the others. These leaders wanted to level the stool.

One of the things that struck me from my combat tours and is something I felt when I was growing up here in New York, is that some things must be experienced firsthand. But with the right education and training, you can shorten the learning process. For example, you do not want your first experience in a dense urban area or a large city to be in combat, as it was, too often, for many of our soldiers that found themselves in Baghdad, Mosul, or Fallujah.

I will offer a couple examples that demonstrate the importance of training and education. When the United States found itself fighting a counterinsurgency in Iraq, the army was completely unprepared since it had no doctrine, no training, and no education on counterinsurgency. Thus, it is no surprise that we struggled. However, the officers who are widely lauded at performing best in the early years of the conflict—officers like General David Petraeus, Lieutenant General H.R. McMaster, and General Peter Chiarelli—gained the knowledge that they required to understand this complex operating environment not through their professional military education, but instead, through their civilian graduate school studies. Others, lacking this education, could only learn through experience.[16]

Likewise, critical knowledge can be gained through training. The special operations forces involved in the battle of Mogadishu were well trained at operating in an urban environment, so when they found themselves involved in urban combat, their learning curve was much

[16] Liam Collins, "Military Innovation in War: The Criticality of the Senior Military Leader" (PhD diss., Princeton University, 2014), 116-7, http://arks.princeton.edu/ark:/88435/dsp01br86b3727.

shorter than, say, the Armed Forces of the Philippines in Marawi or the Canadians in Ortona who had no urban training prior to combat.[17] Their first experience with urban terrain came under fire!

It is possible, with the right training and education, to create immersive experiences so that learning results from an experiential experience that is not a combat experience. For example, to learn about the complexity of megacities and major urban areas, you have to go into big cities to get a sense of what they are like.

There are plenty of people, however, who take a contrary opinion, but I argue that you have to physically get into the particular environment, in this case a city but it could be any complex environment, to understand what makes it tick. In fact, a number of learning theories, including experiential learning theory, stress the importance that concrete experiences have on learning.[18] How do people—first responders, utility workers, or anyone for that matter—operate and survive in that environment? I would argue that you cannot really get that from a book. Books are useful. Theories are useful, very useful, in fact. But there is a component, when it comes to complex operational environments, that you have to experience.

What we should be striving for is to provide as much knowledge as possible prior to deploying soldiers into cities so that the experiential learning while operationally deployed is accelerated. Other professions may be able to afford the luxury of on-the-job training for learning to occur, but the lethality of combat makes that a risky proposition for the military. Thus, we should redouble our efforts with respect to the training and education aspects of understanding complex operational environments such as dense urban areas.

Spencer: One of the things that you mentioned that I keyed on was power, and it reminded me of a comment that General Stephen Townsend made at the TRADOC megacities conference in 2018.[19] Townsend had recently returned from Iraq where he helped the Iraqi security forces liberate the city of Mosul from the Islamic State of Iraq and al Sham. He said one thing that he missed when planning for the battle of Mosul was

[17] See chapter 8, "The Battle of Mogadishu;" chapter 14, "The Battle of Marawi;" and chapter 7, "The Battle of Ortona;" in this book.

[18] See, for example, David Kolb, *Experiential Learning: Experience as the Source of Learning and Development* (Upper Saddle River, NJ: Prentice Hall, 1984).

[19] Russell W. Glenn, *Where None Have Gone Before: Operational and Strategic Perspectives on Multi-Domain Operations in Megacities* (Fort Eustis, VA: G2, U.S. Army Training and Doctrine Command, 2018), Multi-Domain Operations in Megacities Proceedings.pdf.

understanding power. He said that if he could do it all over again, he would have tried to understand power a little better. He also said both the flows of a city and power variables should be incorporated into how we try to understand cities.[20]

I know that you are very familiar with the flows of a city. Understanding the flows of urban environments is complex: knowing what needs to come in—such as water, food, and medical supplies— and what must go out—such as trash or sewage. But power is something I have thought a lot about recently.

Like you were mentioning earlier, I am not sure if our current model, PMESII-PT, which is supposed to be used to understand any operational environment, gets down to the necessary level of detail to understand who the power brokers are in an urban environment. I wonder if we are going to continue to use the same methodologies, whether it is PMESII-PT; areas, structures, capabilities, organization, people, and events (ASCOPE);[21] or mission, enemy, terrain, troops, time, and civil considerations (METT-TC).[22] All these military frameworks and acronyms help us understand the operational environment, but are they adequate for operating in urban terrain?

Mahaney: That is a good question. In PMESII-PT, you could argue that the political, military, and economic variables are elements of power in and of themselves. But I would agree with you that power has to be emphasized more. Power is the unifying theme throughout the ages. You can have economic power that turns into political power. You can have military power that turns into political power. You can have political power that turns into military or economic power. The point is that power is fluid and fungible. As you correctly noted, understanding the very nature of the flow of power, as well as the other flows of a city, is critical. I cannot tell you how many times during my various operational and combat deployments that I had to tell people that anybody with power counts.

---

[20] TRADOC G-2 OE Enterprise, "TRADOC Mad Scientist 2018 Megacities: 1.06 Scenarios, Foundations & Convergence / GEN Townsend," YouTube video, April 12, 2018, https://www.youtube.com/watch?v=ARz0l_evGAE.

[21] Department of the Army, *Intelligence Preparation of the Battlefield*, ATP 2-01.3 (Washington, DC: Headquarters, Department of the Army, 2019), 3-6, https://armypubs.army.mil/epubs/DR_pubs/DR_a/ARN31379-ATP_2-01.3-001-WEB-4.pdf.

[22] Department of the Army, *Training the Command and Control Warfighting Function*, TC 6-0 (Washington, DC: Headquarters, Department of the Army, 2021), 1-11, https://armypubs.army.mil/epubs/DR_pubs/DR_a/ARN31776-TC_6-0-000-WEB-1.pdf.

So, the question is: Who has power? It may be someone with a lot of money. It may be someone with an important neighborhood or family position, for example, but they only count to the degree to which they have power, to which they can get things done. The person with the title is not necessarily the one with the power. Power or politics is all about getting things done. It is the art of the possible. It is the ability to affect your will on the situation or on someone else. I think it has important implications for how we, the military, look at the world and urban areas specifically.

To me, power and politics can emanate from political institutions; non-government organizations; and unelected tribal, community, or regional leaders. Sometimes these unelected leaders are called warlords or gang leaders, sometimes they are not, but they often have tremendous power and do not subscribe to the approaches that a diplomat would. They may not be "credentialed" by a nation-state, yet they have tremendous influence—they have power to compel and to persuade, both directly and indirectly.[23]

But if people simply default to raw power—the power of lethal force—they often forget the diplomatic or political element of power. Political power spans from the national to the local level. The military also cannot simply say this is a political thing, so we are going to pass it off to the diplomats to handle. That does not work when you are operating at the local level or if you are in a city. You cannot rely on some deus ex machina,[24] somebody to come in from off stage and just suddenly take care of that political stuff, allowing us to deal with the military stuff.

The military is going to have to understand the dynamics of power itself. As much as we do not want to get involved in that, it is almost inevitable. In every military operation that I have ever been involved with, I found myself having to be concerned with politics and power. So yes, I think our models miss some of the important variables needed to understand urban environments. Additionally, there is a rich field of academic study relating to power that the military needs to pay more attention to.

Spencer: I agree with everything you just said. In most environments, the military can impose some measure or form of power on the environment that we enter. We try to impose order and control; it is in our nature. But I think the complexity of the urban environment and all these different

---

[23] Patrick J. Mahaney Jr., "Observations for practitioners of Complex Operations," in *Complex Operations: NATO at war and on the margins of war*, ed. Christopher M. Schnaubelt (Rome: NATO Defense College, 2010), 46-68, https://www.files.ethz.ch/isn/118999/fp_14.pdf.

[24] Deus ex machina is defined as "an unexpected power or event saving a seemingly hopeless situation, especially as a contrived plot device in a play or novel," by Google's dictionary.

power structures are significantly overlooked. In Iraq I think we learned the difficulty. We learned by discovery what happened when you tried to impose your own power on a complex organism like a city, or if you attempted to destroy all existing power structures.[25] We may have thought we were the only remaining power structure, but we were not. That is what really hit me when General Townsend talked about understanding the power structure in megacities. I think we mistakenly believe that we can impose our will in a complex urban environment, like Baghdad, and fail to correct this belief with the appropriate education and training.

Mahaney: Absolutely. One of the first things I ask myself when entering any operational environment is what is really going on here? Who owns this neighborhood? There is usually an official story, but it may or may not be true. The central government may not know who controls this or that, or it may be deliberately overstating its level of control. The central government may appoint the provincial governors and district chiefs, but it is never monolithic. I simply think of my life as a youth in New York City. By growing up in New York's neighborhoods, especially during "The War Years," I learned that the powerbrokers did not necessarily operate from within the formal government structure.

In fact, there was an old saying that New York City was "ungovernable."[26] This term has also been used to describe some nations.[27] Sticking with the urban theme, people argued that the city was too big, too complex, too everything, to be governable.[28] Now we have proven that New York is governable. Today, New York is a really wonderful city. That does not mean it does not have any problems, but the present state is really quite remarkable, especially considering the low point of "The War Years." My point is that you have to recognize that it may in fact be the civic, religious, and cultural organizations, or the mafias and the criminal groups that are really in charge.

---

[25] In Iraq, Coalition Provisional Authority Order 1, the de-Ba'athification of Iraqi Society, and Coalition Provisional Authority Order 2, the Dissolution of Entities, essentially disbanded the entire Iraqi military, security, and intelligence infrastructure, and a vast majority of government officials since most had been members of the Ba'ath party.

[26] See, for example, Nathan Glazer, "Is New York City Ungovernable?" *Commentary*, September 1961, https://www.commentary.org/articles/nathan-glazer-2/is-new-york-city-ungovernable/.

[27] See, for example, Vicki Huddleston, "Opportunity in Somalia," Brookings, January 3, 2007, https://www.brookings.edu/opinions/opportunity-in-somalia/.

[28] Glazer, "Is New York City Ungovernable?"

Most people will establish a modus vivendi,[29] even with elements or individuals that they find odious, such as criminals or mafia elements. People and societies have to establish a rhythm of life and anybody who can provide stability, or some measure of order, often rules. Individuals living in these societies might not like it. They might find other parties odious, but they learn to live with it. And if not, they pick up and move. I experienced that as a kid here in New York, certainly.

Again, I always ask myself, "What is really going on here?" when entering a new operational environment. My concern is that far too often—by failing to understand the complexity of large urban areas—we allow our military and political leaders to be very naive. That is not a word that I often hear being used to describe our military leaders, but I think that in many cases the military is very naive when it comes to understanding how cities really work. The military may not want to work with the mafia, corrupt officials, or terrorist elements, but it must understand their influence; it cannot afford to be naive.

And that speaks to the very essence of power. So, the question becomes how do we make people less naive? I believe—returning to General Cone's example of the stool—the answer lies in training, education, and experience. You can at least educate people on some of the realities that they are likely to face. It can be accomplished by examining case studies, traveling to different cities, and by incorporating urban scenarios into training exercises to make people think through these very complex challenges.

Spencer: I had my own experience in Iraq. To help achieve stability during the surge, we had to supplement the coalition and Iraqi security forces with "Sons of Iraq," which were basically volunteer civilians—though many were paid a modest stipend—who served as local or neighborhood militias.[30] As a company commander in Baghdad, I had over one hundred of them and we suspected many of being low level criminals. It took us a long time to understand, at that basic level, how power and control was exercised. As we established a new Sons of Iraq security element, we eventually learned that this new source of power improved short-term stability, but it also created first, second, and third order effects.

---

29 Modus vivendi is defined as "an arrangement or agreement allowing conflicting parties to coexist peacefully," by Google's dictionary.

30 Greg Bruno, "Finding a Place for the 'Sons of Iraq'," Council on Foreign Relations, April 23, 2008, https://www.cfr.org/backgrounder/finding-place-sons-iraq.

By empowering one group, we were effectively delegitimizing another group that we often also sought to empower.

I think we learned a lot of lessons in Iraq through trial and error, and we made a lot of errors. Even on the economic front, we often seemed to be pumping money into areas without realizing that it was destabilizing the natural flows or existing statuses within the urban systems. Now that I have studied urban environments I can look back and go, "Man, I think we really messed up some things there." Other times, we figured it out eventually, but it took us far longer than it should have.

So, it is comforting to find someone, like you, that really understands the challenges that the urban environment presents to the military, but I am afraid you are somewhat of a unicorn in that regard. You were raised in a large megacity and had multiple operational experiences in urban environments, starting as a military police officer in Panama and later in multiple locations as a special forces officer. You also oversaw multiple intellectual organizations tasked with solving some very complex problems for the army. You have amazing insight into complex operational environments in general, and especially urban environments.

You discussed the need to incorporate operating in urban environments into training, education, and experience, but does the army even have the right personnel model to address the type of soldiers needed to be sent into these complex operational environments like megacities. Are we developing leaders who are capable of understanding these environments quickly enough? How is the army either creating more unicorns like you or educating its force more broadly?

Mahaney: There are some interesting aspects to that. Part of it comes down to talent management. Another part of it comes down to recruiting. Again, returning to the stool analogy, if you grew up in a city, you might have a leg up on understanding a complex urban environment because you grew up immersed in it. We have to have the right knowledge and understanding to know what questions to ask. For example, if I enter neighborhood X for the first time, I should want to know who "owns" or "runs" it. Probably the only time the military thinks this way is when a city is divided between various warring clans, as Mogadishu was for many years, but this thinking is relevant in all situations. People that have experience living or operating in a city are more likely to think this way, but for those who do not, they can learn through education and training.

The army must manage the talent of its people better; specifically, those who are best able to handle the challenges posed by complex operational environments. They tend to be creative thinkers and very adaptable, but often they tend not to fit as well as some of their peers into the somewhat rigid norms of the military because by their very nature, they tend to be intellectually curious and out-of-the-box type thinkers.

The military, like many organizations, is somewhat resistant to change.[31] Out-of-the-box thinkers are not always rewarded. Often it is those who conform to the organization who are rewarded. In fact, one U.S. Army War College study found that most successful officers score lower in "openness"—"the recurrent need to enlarge and examine experience"—than the general population, and officers who were selected for brigade command, which is basically a prerequisite for becoming a general, score lower than their peers within the army.[32] In other words, the army is producing leaders who are arguably the least adaptable at operating in complex operational environments.

Sometimes the people who do not fit neatly into our linear hierarchies are the best people at going into complex environments, showing their adaptability, resilience, and dare I say, anti-fragility. You want people who are able to not only survive in these environments but thrive in them. They love these environments and seek them out. You do not want to have a career penalty for people like that.

I think that the special forces, civil affairs, and psychological operations branches have done a pretty good job of attracting and promoting those kinds of people. But you also do not want people to stray too far outside the norm because then they become outliers that people do not want to be associated with. So, there is a balance, but I think that the conventional army has to get more comfortable with the discomfort of complex operational environments and recognize there is no school solution. You cannot simply reduce complex problems to a formulaic approach.

---

[31] Kurt Lang, "Military Organizations," in *Handbook on Organizations*, ed. James G. March (Chicago: Rand McNally & Co., 1965), 838.

[32] Robert R. McCrae and Paul T. Costa Jr., "Conceptions and Correlates of Openness to Experience," in *Handbook of Personality Psychology*, eds. Robert Horgan, John Johnson, and Stephen Briggs (New York: Academic Press, 1997), 826; and Stephen J. Gerras and Leonard Wong, "Changing Minds in the Army: Why it is so Difficult and What To Do About It" (Carlisle, PA: Strategic Studies Institute and U.S. Army War College Press, October 2013), 8, https://press.armywarcollege.edu/cgi/viewcontent.cgi?article=1514&context=monographs.

So, the army must improve its talent management. I am not alone in this recommendation as I know that over the last several years talent management issues have come up and it continues to gain the attention of senior military leaders. There is not going to be a sea change overnight. But what I love about the dense urban and the megacity aspects of this is that once you accept that you have to at least be prepared to operate in this sort of an environment, it forces you to take a hard look at who you are, how you are doing your training and education, and how you are managing your people. And likewise, it forces you to say, "We may need some skill sets, approaches, or certain people who appear to be non-conformist in some ways, and yet, as long as they remain within the professional norms, maybe they are the ones with the good ideas and the flexibility to actually get the job done on the ground."

Spencer: I think we could have a long conversation just on talent management within the military. I think you are right about the army doing a lot of work in trying to identify the talents of individual soldiers and having that information accessible. I often point to the battle of Mosul in 2017 as an example of how the army lacked this ability. As General Townsend and his staff were helping the Iraqis plan the battle, there was no way to quickly search army personnel records to see who had previously operated in Mosul. So, we had individuals with experience, but Townsend was unable to leverage it unless those individuals happened to be on his staff at the time.

We now have the Integrated Personnel and Pay System – Army which gives the army the ability to electronically record all the skills, experiences, and backgrounds of soldiers and search when needed for the special skills or backgrounds of a specific environment.[33] This system was only recently implemented, so it is not clear how well this new capability will improve the way the army works, but it is promising. Returning to the Mosul example, as an organization, the army had knowledge of Mosul, but no way to leverage that knowledge. So, the Integrated Personnel and Pay System can now do that, provided the right data is entered into the system. We cannot make everyone experts in everything, but we must be able to leverage the experts that we have.

Mahaney: Yes. When I was with the Chief of Staff of the Army's Strategic Studies Group back in 2014 and 2015, we knew that was a problem and had a group focused on fixing that problem. There are many people

[33] "What is IPPS-A?" IPPS-A, https://ipps-a.army.mâil/what-is-ipps-a/.

who have come to a similar conclusion about our talent management who had nothing to do with our work on urban operations. They looked at the simple inventories of skills and experiences of everybody coming into the army and determined they needed to be cataloged. It could be a language that they spoke prior to entering the army, but if they never took a defense language proficiency test when they entered, the army would be unaware. Now we have the ability to do that, but we have to wait and see how the organization uses the information that it collects.

Spencer: The information is only useful if someone looks at it. I think that advances in machine learning could help, having a system that knows all that information and can turn it into knowledge: a push system, not a pull system. The end user would not have to pull information on a person, the machine would push it to them at the right time, such as during the lead up to an urban operation. The army has more than one million soldiers.[34] We collect a ton of information, whether it is tactical, after actions reports, and much more, yet we struggle with taking that information and turning it into knowledge for the right leaders at the right moment in time. Some of the advances we are seeing in machine learning is hopefully getting us closer to being able to not just collect a bunch of data but also being capable of turning that data into useful information.[35]

Mahaney: I agree. I think that there is a human element with that as well. I fully concur with the potential for artificial intelligence and machine learning given the information technology revolution that is currently taking place. I think people will also self-select into the right jobs. For example, in the case of dense urban areas and megacities, there are people who are attracted to this kind of environment, whether it is because they have an urban background, or they are simply fascinated by the environment's complexity and challenges. They will self-select into positions that allow them to tackle these problems.

Some may say, "I really want to be stationed in the greater Seoul metro area, not just because it is Korea, but because it is a big city." There are some people, conversely, who are going to be more comfortable in the most rural of areas. Some may even say, "I love the desert, so I want to be stationed at the army's National Training Center at Fort Irwin, California."

[34] For fiscal year 2020, the regular force included 480,000 soldiers, the National Guard 336,000, and the Army Reserve 189,500. See Mark Cancian, "U.S. Military Forces in FY2020: Army," CSIS, October 15, 2019, https://www.csis.org/analysis/us-military-forces-fy-2020-army.

[35] See, for example, Shelly Kramer, "How Machine Learning is Changing Recruiting," Future of Work, June 26, 2017, https://fowmedia.com/machine-learning-changing-recruiting/.

With changes in the assignment process, the army is better at doing that now, but we will have to wait to see what type of impact it has on preparing leaders for complex operational environments like megacities and dense urban environments.

Spencer: I agree. It makes sense that if someone who grew up as a hunter and is a master at tracking, then you might want to consider assigning him or her to a scout unit. You could argue that we used to take advantage of those skillsets more so than we do. But it is very hard to do at scale. General Perkins would often remind people of the challenge of operating at scale when discussing potential changes to the army's personnel system. He would say, "Some people can train small organizations. Great. I have to hire 10,000 soldiers per month and give 500 soldiers a new job every day. I have to deal with talent management at scale."[36] It is a challenging problem, but I agree that capturing those types of factors is important.

So, this brings me to my final question. In New York City there is an urban unit, the New York National Guard's Joint Task Force - Empire Shield. The task force is made up of hundreds of individuals who are protecting New York City against a terrorist attack. If I am not mistaken, the unit is made up of volunteers from dense urban areas who wanted to be deployed into urban operations? Is that true and why can't we have more urban focused units within the U.S. Army?

Mahaney: Joint Task Force - Empire Shield is located at Fort Hamilton, a small army post at the foot of the Verrazano-Narrows Bridge in Brooklyn. It has about 750 troops and is treated as a brigade size element. The last time I checked it was 86 percent Army National Guard, 13 percent Air National Guard, and one percent New York Naval Militia. It is a remarkable unit that was established shortly after 9/11 and has been in existence ever since. While few people have ever heard of it, the unit has tremendous expertise in urban operations, though not necessarily urban combat.

Their mission is to "conduct Homeland Security operations, in the New York City Metro Area, in order to deter and prevent terrorist acts" and "on order, conduct Defense Support of Civil Authority operations in order to save lives and reduce human suffering."[37] If you visit New York, you will

[36] The quote comes from an engagement with General Perkins at West Point, NY in October 2015. To see one example of similar remarks, see Anthony O'Bryant, "Perkins to civilians: See the big picture, the big problems," Army.mil, November 13, 2014, https://www.army.mil/article/138110/perkins_to_civilians_see_the_big_picture_solve_the_big_problems.

[37] "Joint Task Force Empire Shield," no date, http://dmna.ny.gov/blog/resources/jtfes_info.pdf.

see them in Penn Station, Grand Central Station, the World Trade Center area, La Guardia and JFK airports, and at the bridges and tunnels.[38] They develop tactics, techniques, and procedures for their unique mission. I think one of your questions was why are we not learning from them? I would say that we are.

Empire Shield is not the only unit that trains on operating in urban environments. The New York Army National Guard's Weapons of Mass Destruction Civil Support Team (the 24th Civil Support Team) works closely with interagency partners so that it is prepared to execute its mission in New York's dense urban terrain. I have only worked with the 24th Civil Support Team, but I would assume that the other 56 Civil Support Teams (there is one in each state, U.S. territory, and Washington, DC; and two each in California, Florida, and New York) would also be well trained on operating in urban environments since that is the most likely location for a weapon of mass destruction attack.[39]

To return to your question, as to whether we should stand up other units that are focused on urban operations or should we instead try to provide knowledge to the force through training and education in such a way that you could at least begin to understand the complexities of a dense urban environment? My answer is that we need to do both. We absolutely need some sort of a training and education center for urban operations to develop the necessary expertise within the army, but the only way to develop a baseline of knowledge at scale requires a broader education and training focus across the force.

I believe an army urban training and education center should be located in New York City because it is the largest, densest city in our country and, very importantly, we have a network of incredibly collaborative partnerships with world-class organizations like the New York City Fire Department, the New York City Police Department, the Office of Emergency Management, Joint Task Force - Empire Shield, universities, and many others. This collaboration contributes to conceptual development, material development, training, and education. An urban center in New York City should also be supplemented by a broader training and education effort.

[38] Ibid.

[39] "Weapons of Mass Destruction (WMD) Civil Support Team (CST)," National Guard, December 2017, https://www.nationalguard.mil/Portals/31/Resources/Fact%20Sheets/Weapons%20of%20Mass%20Destruction%20Civil%20Support%20Team%20Fact%20Sheet%20(Dec.%202017).pdf.

One of my big ideas is to create a three-tiered training system. Level one, the most basic, would involve an exposure to urban terrain at large training bases like Fort Benning, Georgia and Fort Bragg, North Carolina. Soldiers would get a basic introduction to the cities around those bases such Columbus, Georgia or Fayetteville, North Carolina. Even if those cities are much less complex than large megacities, it is a good way to start, and we have to start somewhere.

Level two would include regional hubs, such as Atlanta, which is 115 miles from Fort Benning. Military bases would send their soldiers to the regional hubs to gain a greater understanding of the complexities than they could get at a level one location.

Level three would include megacities like New York and Los Angeles, with the point being that the center should be located in an actual megacity, and the bigger and denser the better. To compare a level two city to a level three city, Atlanta's population density is roughly 3,600 people per square mile, while New York City's is 28,500, nearly eight times higher and arguably orders of magnitude more complex.[40]

Indeed, Manhattan has a resident population density of approximately 70,000 people per square mile, which swells to around 170,000 per square mile during weekdays when commuters are present, not to mention buildings and neighborhoods with tenements and underground facilities dating back to the 19th century.[41] These are highly relevant features.

Recognizing the throughput limitations, as you noted earlier, we cannot expect to handle 10,000 soldiers every month, but you can train unit cadres. The cadres could be taught at the level three location and these cadres would then lead the education and training at the level two and one locations.

The model is not too different from the way certain skills are taught in special operations. It is also similar to how the army conducts its mountain and jungle training. The army sends soldiers to the mountains of Vermont to be trained by Vermont's National Guard mountaineering experts to serve as trainers when they return to their home bases to run the mountaineering

[40] Rich Exner, "Cleveland is nation's 27th most densely populated big city; Columbus 37th, New York 1st," Cleveland.com, January 30, 2019, https://www.cleveland.com/datacentral/2018/05/cleveland_is_nations_27th_most.html.

[41] "Manhattan Population 2022," World Population Review, https://worldpopulationreview.com/boroughs/manhattan-population.

training there. We have employed a similar train-the-trainer methodology for jungle training and that is exactly what we should do for urban training.

Just as we have mountain experts, we need to have urban experts who have been to the school and get an additional skill identifier in their personnel file that clearly identifies them as an urban expert. In the case of jungle training, the army used to have a patch that soldiers wore on their uniforms that identified them as a jungle expert. When they got rid of the patch, soldiers still had the special skill identifier in their personnel file. Yet despite the army's long history of operating in urban environments, as demonstrated by the case studies in this book, the army still has no schools or special identifiers for the urban environment. An urban center is required not only for training and education, but also for material and conceptual development.

To return to your question of what is required for the operational force, I will start first by discussing special forces. As a retired special forces officer, I always thought we did a pretty good job through the years at managing and maintaining certain specialties within each special forces company. Every special forces detachment was competent at general combat missions, but within each company, each detachment typically had a specialty: High Altitude Low Opening (HALO) parachute teams to conduct high altitude infiltrations, scuba teams for missions that required water infiltrations, mountain teams to operate in the most rugged natural terrain, and close quarters battle teams that were highly trained at fighting in more urban environments. A team rarely specialized in more than one specialty skill due to the time that it took for the team to maintain the skill at a collective level, yet the company, as a whole, possessed at least one team with each of these specialties.

This same approach could be applied to the conventional army with urban skills. I think that we have to have some engineer units that become experts in the dense urban operations. There is a huge difference between engineering in a desert, for example, and engineering in a dense urban environment. We should have similar specialization with infantry, armor, and other units, so that each company, battalion, or brigade, has some resident expertise in urban operations.

On a grander scale, just as we designated the 10th Infantry Division the 10th Mountain Division, there should be a unit—maybe a brigade or a battalion to start with—that is focused on the urban mission set. There really is no excuse for not investing more into the urban problem set right now.

Spencer: That makes a lot of sense. If you think about the large urban training sites available to the U.S. Army for training today, there are only a few. There are the unit-based sites on different installations like Joint Base Lewis-McChord, Washington; Fort Hood, Texas; and Fort Bragg, North Carolina. Most of these "urban sites" have around 20 to 50 buildings.[42] While 20-50 buildings hardly meets the definition of dense urban terrain, the sites do offer some training value. The biggest training facilities reside at our combat training centers. The Joint Readiness Training Center's largest site, Dara Lam, contains 27 single- to three-story buildings, an overpass, and traffic circle.[43] The U.S. Army's largest urban site, Razish, is at the National Training Center. It has around 600 buildings that include 16 three- to five-story buildings.[44]

I also agree with the idea of creating an urban unit. We could take a very small organization and have them focus on urban operations to help develop doctrine; develop new equipment; or develop new tactics, techniques, and procedures. We have some experimentation, but it has been separate from a larger organizational commitment. I think one of the reasons for our lack of focus is that there is no single unit or individual who is responsible for the urban problem set.

When the U.S. Joint Forces Command was dissolved in 2011, the urban operations office that sat within that command was also dissolved.[45] I get frustrated sometimes thinking about how little the U.S. military commits to urban operations. I know you are working hard to address many of the education and training aspects, and I have multiple efforts here at the Modern War Institute's Urban Warfare Project, but we are just two people, so there remains way too many gaps.[46] My fear is that the army will only care about the urban environment when it finds itself operating in it once again, and that will come at a significant cost. One thing that can be learned

---

[42] For example, Leschi Town at Fort Lewis, Washington contains around 50 buildings, https://virtualglobetrotting.com/map/leschi-town-complete-fake-city-used-for-urban-military-training/view/google/.

[43] John Spencer, "Attacking the City of Dara Lam," *Urban Warfare Project Podcast*, Modern War Institute, July 10, 2020, https://mwi.usma.edu/attacking-city-dara-lam/.

[44] John Spencer, "Attacking the City of Razish," *Urban Warfare Project Podcast*, Modern War Institute, May 29, 2020, https://mwi.usma.edu/attacking-city-razish/.

[45] Jason Ukman, "U.S. Joint Forces Command Formally dissolved," *Washington Post*, August 4, 2011, https://www.washingtonpost.com/blogs/checkpoint-washington/post/us-joint-forces-command-formally-dissolved/2011/08/04/gIQAQbzBuI_blog.html.

[46] See, "Urban Warfare Project," Modern War Institute, https://mwi.usma.edu/urban-warfare-project/.

from this book is that the army will find itself fighting in dense urban areas in the future.

Mahaney: Right. Clearly there is no way that an urban training site consisting of a few hundred buildings with a maximum height of seven stories, with no residents or working infrastructure, is going to replicate the complexity of a real city or provide soldiers with the experience required to operate in many urban areas around the world. That is why I strongly believe we have to expose soldiers, leaders, and commanders to the complexity of actual urban areas with a three-tiered training system that exposes soldiers to increasing levels of urban terrain.

I also agree with the urban unit specialty approach. You could start very small, maybe with just a company sized element. I think it could do and learn a lot. The army has done a lot with specialty companies like the 911th U.S. Army Technical Rescue Engineer Company, an urban search and rescue company, at Fort Belvoir; or the 511th Dive Company out at Fort Eustis. The engineers have a history of creating specialty companies, so the question is, will the engineers, infantry, or armor be willing to commit a company to urban operations? I think this should absolutely be implemented for a company. It is a way to rapidly spur doctrinal development; material development; and techniques, tactics, and procedures. A small investment could produce outsized gains. I think it is critical that we get the training and education piece right.

Spencer: I know discussing big changes in the army is extremely complex. Although I have argued for the army to create an urban unit, we also have to get the personnel system right as well.[47] We have to build the expertise. We must perform the training and education to develop the expertise. The army is so big that we have to train trainers and then send them back to inculcate the army.

Mahaney: Indeed. A parting thought. This urban material that we have been discussing is not just close quarters battle—training to enter and clear rooms and other confined spaces—it is much broader than that. All too often we default to a very simplistic understanding of urban combat, and return to what we know, such as discussing close quarters battle. But complexity is the key here. Megacities and dense urban environments are complex and demand deliberate approaches to understanding and operating inside them.

[47] John Spencer, "It's Time to Create a Megacities Unit," Modern War Institute, January 31, 2017, https://mwi.usma.edu/time-create-megacities-combat-unit/.

6

Beneath the City

Underground warfare has been a part of warfare for time immemorial. While the underground is not restricted to urban area—the mountains of Afghanistan and Mali for example—the subterranean environment is much more common in the urban environment as demonstrated by the use of tunnels in Mosul, Aleppo, and Gaza City. This marks a shift from the historical uses of tunnels when they were typically used by and against combatants as an anti-personnel tactic. Only the Vietnam War stands out in this regard, as Vietcong tunnels served both fighters—as ammunition caches, traps, and bunkers—and civilians—as hospitals and living compounds.

The use of tunnels in cities enables fighters to conceal entry and exit points and move around undetected. Generally, underground networks inside cities make use of the existing infrastructure—basements, parking lots, sewage systems—which are then enhanced and connected thanks to additional tunnels dug for military purposes. These "urban warfare tunnels" threaten civilians living in the vicinity above such tunnels and endanger civilian infrastructure: dwellings may collapse due to poor engineering work, water aquifers may be damaged or contaminated, and anti-tunnel operations in retaliation for tunnel-digging creates further risks.

In this chapter John Spencer interviews Dr. Daphné Richemond-Barak, who is an assistant professor in the Lauder School of Government, Diplomacy, and Strategy at Reichman University, where she heads the international program in government. She also serves as senior researcher and head of the international law desk at the International Institute for Counter-Terrorism, adjunct scholar at the Modern War Institute at West Point, and senior fellow at the Institute of Law and Land Warfare.

Richemond-Barak is one of the leading scholars of subterranean or underground warfare, author of the book *Underground Warfare,* and founder of the International Working Group on Subterranean Warfare. Her work on underground warfare is recognized around the world as groundbreaking and is a call to arms that more work must be done in this critical area: beneath the streets.

John Spencer: I would like to start by asking you to define underground warfare and ask if there is a difference between the terms "underground warfare" and "subterranean warfare," or are they interchangeable?

Daphné Richemond-Barak: Underground warfare refers to the use of spaces that are truly below ground in times of war. The terms underground warfare, tunnel warfare, and subterranean warfare are largely interchangeable. The United States military divides the subterranean environment into three categories: (1) tunnels, caves, and natural cavities that can be rudimentary or sophisticated; (2) urban subsurface systems such as basements, parking garages, subways and sewers that can be substructures or civil works; and (3) underground facilities for military purposes such as silos, bunkers, and military bases that can be shallow or deep.[1] I think that is a pretty good survey of what is considered underground warfare. My book addresses the first two categories.

The third category is definitely an area of concern, but not one that I research. I will note that I do not consider trenches or trench warfare as subterranean even if the trenches have occasional underground bunkers, because while the trenches are both partly above and partly below ground, your head is essentially breathing the regular air. Though deeply buried facilities have largely driven the advancement of anti-tunnel technology, they are used primarily by states to hide military arsenals and ensure the continuity of command in times of crisis. As such they are a unique category of underground facility, and I would not include them in the definition of "underground warfare" in a narrow sense.

Spencer: That is a great description. Although there is relatively little scholarly or professional work dedicated to studying underground warfare, it has been around for a long time. Underground warfare dates at least as far back as Roman times.[2] There is a long history of underground warfare, including the Battle of Jericho, tunnels under the castles during medieval

[1] Department of the Army, *Subterranean Operations,* ATP 3-21.51 (Washington, DC: Headquarters, Department of the Army, 2019), 1-9, https://armypubs.army.mil/epubs/DR_pubs/DR_a/pdf/web/ARN19656_ATP%203-21x51%20%20FINAL%20WEB.pdf.

[2] Daphné Richemond-Barak, *Underground Warfare* (London: Oxford University Press, 2018), 4.

times, and in the Vietnam War.[3] There was an entire book written just about the tunnels and underground warfare in Vietnam.[4] So what is new?

Richemond-Barak: *The Tunnels of Cu Chi* is one of my favorite books, but it only addressed the use of tunnels in Vietnam, it did not compare the use of the tactic in Vietnam with the use of the tactic in other conflicts or terrains. All major works on underground warfare focus on one specific conflict or theater. My book examines underground warfare across many conflicts, assesses how it has evolved, and anticipates future trends.

Before we get there, however, I think it is important to consider how warfare has changed in recent years. Militaries continually seek more technological sophistication. In recent decades, we see less interstate war—the current war in Ukraine being an exception—and more wars being fought between the technologically advanced militaries of nation-states and less technologically sophisticated groups like al Qaeda, the Islamic State of Iraq and al Sham (ISIS), and Hamas.[5] This asymmetric pattern is not new, but it has become more common.

As states improve their ability to find and attack their enemies through advanced intelligence, surveillance, and reconnaissance assets, going underground is a natural response for less technologically sophisticated opponents. Thus, as states become more performant and technologically sophisticated, it becomes impossible for their enemies to maintain symmetry and they look for asymmetric ways to neutralize the military advantages that states have over them. The use of the underground terrain is an equalizer between two highly unequal opponents. This explains the spread of the tactic and the constant adaptation and innovation witnessed in recent years, particularly in Syria.[6]

But just as the digging technology has improved, so has the counter-tunneling technology. I think it raises some interesting questions here. Why is the use of tunnels on the rise? Why does the appeal continue as states improve their counter-tunneling technologies?

Spencer: I agree. You can point to the use of tunnels in warfare across time, especially in the American Civil War, World War I, and the Vietnam

[3] Ibid.

[4] Tom Mangold and John Penycate, *The Tunnels of Cu Chi: A Harrowing Account of America's Tunnel Rats in the Underground Battlefields of Vietnam* (New York: Berkley Books, 1985).

[5] Thomas S. Szayna et al., *What Are the Trends in Armed Conflicts, and What Do They Mean for U.S. Defense Policy?* (Santa Monica, CA: RAND, 2017), https://www.rand.org/pubs/research_reports/RR1904.html.

[6] Richemond-Barak, *Underground Warfare*, 30.

War, but their frequency has intensified for many reasons. I also believe that improved intelligence, surveillance, and reconnaissance assets combined with improved aerial strike capability have pushed enemies underground. We have to understand how the character of warfare is changing. Underground warfare may not be new, but it has definitely increased in recent years.

Richemond-Barak: That is correct. Underground warfare has not returned, it has always been there. We discern elements of underground warfare in nearly every conflict this century. Despite technological advancements, the use of underground terrain continues to be popular. We have seen an innovation and diffusion of the tactics. So, there is something new and concerning happening.

Spencer: I am interested in the convergence of urban warfare and underground warfare. Warfare continues to be pushed further into the urban space and further underground. As an underground warfare specialist, how do you view the intersection of underground and urban warfare?

Richemond-Barak: Tunnel warfare follows the more general trends affecting war. As battles are increasingly being fought in urban areas, underground warfare too moves to urban areas. Tunnels were used in the mountains of Mali in the 2012 war and in two successive wars in Afghanistan.[7] Today tunnels are dug in or in close proximity to civilian populated areas in Egypt, Israel, Libya, Iraq, and Syria, among other places.

The intersection of underground and urban warfare is extremely complex. Each environment is complex in its own right, but when you put subterranean and urban together, the challenges are magnified. Militaries already face tough dilemmas where civilians are caught in the middle of fighting, but it is a totally different ballgame when underground complexity is added to that mix. The fog of war increases; it is difficult to keep track of the enemy's moves; and operating in and around tunnels requires unique or entirely new equipment, training, and techniques. There is definitely a convergence of urban and underground warfare, and this convergence is particularly concerning because it puts civilians and soldiers in harm's way.

Spencer: In my research on urban warfare, I have found that many urban areas, especially large ones, have a labyrinth of underground structures such as subway tunnels, transportation hubs, infrastructure

[7] Ibid., 14-15, 19.

tunnels, or sewer tunnels. As cities grow, they have to expand vertically—both above and below ground—because of the population density. Older cities have a labyrinth of underground structures that have been built over time, and newer cities often have this underground infrastructure planned from the onset.

Richemond-Barak: Pre-existing tunnels and subterranean structures—subways, transportation, sewage, parking lots, commercial centers—can pose a challenge for military operations. For an enemy who wishes to use them, there is no digging involved, all they need to do is occupy them. Although pre-existing, a military may still have to worry about detecting and mapping these underground networks, because they may not be well mapped, or their maps may have been lost—those things tend to happen in cities ravaged by war. So, while different from the cross-border tunnels between Israel and Gaza or between Pakistan and India, they still pose many challenges: tunnels dug in or near civilian populated areas create a direct and often intentional danger to the civilian population, they will likely be used by civilians (to move around or seek protection from the hostilities), and they might even be used to launch attacks against civilians.

Spencer: In your book, you discuss the challenges posed by the intersection of underground and urban terrain. Can you describe some of these challenges?

Richemond-Barak: There are many challenges, and I will discuss a few of them. First, consider the challenge of destroying or neutralizing a tunnel. An "urban tunnel," as I like to call them, may connect to a civilian dwelling with an opening located under a kitchen sink. If a commander is concerned about collateral damage and civilian casualties, his or her options are limited when it comes to destroying such a tunnel. You cannot deploy your most powerful weapon to blow up the entire tunnel because the overpressure is likely to destroy the house with the tunnel under its sink. Maybe instead of destroying the entire tunnel, you will neutralize segments that you believe to be the most problematic. Your job, as you are trying to counter a threat, is going to be much more complicated than if you operate in a mountainous area where the tunnels are inhabited and used only by fighters.

The law of war, the body of law that regulates warfare, is going to place constraints on the methods that can be employed to neutralize or destroy an urban subterranean threat. Destroying a tunnel might damage water pipes, sewage pipes, or gas or electric lines that run underground and that would have to be taken into consideration. Another challenge is that fighters may bring their families into the tunnel to discourage their opponent from bombing the tunnel. Having civilians in the tunnel can, in some situations, render a strike unlawful or compel the military to select lesser firepower than it would otherwise choose, which puts the military at greater risk and would be less likely to fully eliminate the threat.

A more general point about the challenge of operating where the underground meets urban is that the consequences of a simple operation that involves the underground cannot be predicted with certainty. In war nothing can be predicted with certainty, but when tunnels are involved, it is even more difficult. The fog of war increases. I call this the "unknown factor."

Commanders will not know the number of stories, arteries, or precise route of the tunnel. They will not know to what extent such a tunnel network connects with civilian infrastructure or how much ammunition is stored in there. They may be unable to ascertain the presence of civilians inside the tunnel or get accurate intelligence about what goes on inside the tunnel. The network of interconnected tunnels can be very difficult to pinpoint and the collateral damage that will likely result from the strike is very difficult to predict. While the risks cannot be completely mitigated, a military commander must weigh them carefully and consider the unique features of the terrain.

I will provide an example that demonstrates these challenges. In May 2021, Hamas launched a rocket toward a civilian area in Jerusalem. In response, Israel "focused on a network of tunnels Hamas used to transfer weapons and fighters undetached" and subsequently responded with an airstrike consisting of 11 precision guided missiles against the tunnel that lay under Al Wahda Street.[8] Drone footage that filmed the strike showed what appeared to be a perfect strike: 11 craters evenly spaced along the 200-meter stretch of road. But unexpectedly, the explosions weakened the foundation of a neighboring building, causing it to collapse and kill 22 of the 38 residents inside. The tunnel was a lawful target, yet this example demonstrates the challenges when fighting at the intersection of the urban and subterranean environments.[9]

---

8 Patrick Kingsley, Iyad Abuheweila, and Evan Hill, "Dreams in Rubble: An Israeli Airstrike and the 22 Lives Lost," *New York Times*, June 17, 2021, https://www.nytimes.com/2021/06/17/world/middleeast/gaza-israel-airstrike-tunnel.html.

9 Ibid.

Spencer: As you are describing it, I get the overwhelming sensation that the military should never want to engage in subterranean warfare, and it should just avoid or bypass it altogether. Of course, I know that is not possible if that is where the enemy chooses to fight. Just as it is not possible to avoid urban areas if that is where the enemy is located. I agree that the subterranean and urban environments are both extremely challenging on their own, but together they are even worse. Thus, it is not surprising to see any foe, technologically advanced or not, leverage this environment for their benefit. The underground can offer significant advantages, by negating an opponent's advantages, or simply as a means of protection, stealth, or surprise.

I have studied the tactical challenges and the soldier psychology of standing in front of a tunnel and deciding whether or not to enter it. Through that research, I have discovered that some militaries have a culture of not entering tunnels while others, like the U.S. Army, have a doctrine of entering and clearing a tunnel in a similar manner that it would enter and clear a building.[10] The underground environment has all the challenges of the above ground environment with the additional challenges of how to see, breath, navigate, and communicate. It is such a different and challenging environment. I find all of that fascinating, but before I met you, I had never considered the legal, political, and strategic implications of subterranean warfare. Can you discuss some of the legal gaps that you discuss in your book?

Richemond-Barak: I will start by discussing whether it is even lawful to use tunnels in war. The purpose of a tunnel is often to conceal, mislead, and surprise the enemy. As we know, you can wage war but there are limits to what you can do, and you cannot resort to perfidy, meaning that belligerents cannot trick the enemy by, say boobytrapping wounded or dead bodies or children's toys.[11] So we could ask the question: Is it even lawful to use a tunnel or is that going too far in misleading the enemy? So that is an interesting legal question.

---

[10] Ralph Cohen et al., *From Cast Lead to Protective Edge: Lessons from Israel's Wars in Gaza* (Santa Monica, CA: RAND, 2017), 100, https://www.rand.org/content/dam/rand/pubs/research_reports/RR1800/RR1888/RAND_RR1888.pdf; and Department of the Army, *Subterranean Operations*, B-9.

[11] Additional Protocol I to the 1949 Geneva Conventions, Article 37(2), https://www.ohchr.org/en/instruments-mechanisms/instruments/protocol-additional-geneva-conventions-12-august-1949-and#:~:text=Article%2037%20%2D%20Prohibition%20of%20perfidy,-1.&text=(d)%20The%20feigning%20of%20protected,of%20war%20are%20not%20prohibited.

As I was examining the question of the legality of tunnel use, I encountered this same question with respect to submarines. When submarines began to emerge in the early twentieth century, the legal community questioned their legality in war. You would probably be surprised to discover that the British wanted to ban submarines altogether, while the Germans disagreed. Ultimately, submarines were accepted as a lawful practice in war and the law accepted this kind of invisibility, which is very close to deception, but still acceptable in war under certain conditions.[12]

Currently, there are no laws of war dedicated specifically to the use of tunnels in war. I do not think that tunnels are unlawful, but the law does prescribe limits to their use. For example, certain types of tunnels must be regarded as military objectives by nature, just like a military base or a weapons or ammunition factory or depot. This is the case for example with tunnels that serve as supply routes. By virtue of their location—underground—tunnels provide a tactical advantage and are therefore a military objective. These are ways in which the law of war can be interpreted to overcome the potential legal uncertainty. In most situations, underground structures and tunnels constitute legitimate targets in war.

In my book I also examine the kind of actions that a state can take to defend itself against cross-border underground intrusions, think of Israel, India, or Egypt. These are just a taste of the legal challenges alongside all the operational challenges that we have discussed.

Spencer: I love the submarine analogy. As an army guy, I view the underground as an extension of the ground, so I had never considered submarine warfare its equivalent in the sea.

Richemond-Barak: I think underground warfare should be regarded as a subset of land warfare, but perhaps understanding this analogy is easier. Because the law is silent on tunnels, one way to make sense of them is by making analogies with submarines, cyber, or other types of warfare. By way of legal interpretation and analogies, experts can, little by little, shape a legal framework that is more cognizant of the unique nature of this terrain.

Spencer: If you think about naval warfare in general, the realm of military forces for surface warfare versus undersea are completely different. There could be just a few hundred meters between the surface level of the sea and the deep water, but they are worlds apart. One simple difference is that you cannot breathe without assistance underwater. I hope these

[12] Richemond-Barak, *Underground Warfare*, 174-5.

analogies will help people realize that underground warfare is different, it is not simply surface warfare that is fought underground. I think case studies are a great way to illustrate the differences, and that is another reason that I love your book.

I did my own research on U.S. Army experiences dealing with underground warfare which of course led me to study Vietnam since it is one of the more recent wars where our enemy routinely leveraged the subterranean environment. The army developed a number of techniques to deal with the tunnels: dropping a grenade down a tunnel to close it, using a generator to pump teargas into a tunnel, dumping teargas powder into a tunnel, or using a variety of explosives to collapse and destroy them.[13] The army incorporated these techniques into its training and education by producing training videos demonstrating how to deal with tunnels.[14]

As I researched, I realized that most of these counter-tunnels tools, weapons, and practices had real political ramifications. The U.S. Congress, for example, debated the use of teargas in tunnels extensively because the enemy also used the tunnels as hospitals.[15] But as you said, there was no way for the military to know what was down there without actually sending a human into a tunnel, which was a very risky proposition. In this subterranean environment, the law of war, to include proportionality and discrimination, gets very challenging.

Richemond-Barak: Yes, it gets very challenging, but I do not think we need new laws to handle the challenges that the underground raises. All we need to do is interpret the existing law in a way that makes sense given the operational reality. For example, after the 2014 Gaza War between Israel and Hamas, the United Nations published a report that assessed the legality of the actions taken by Israel.[16] That was probably the war where tunnels featured most prevalently, at least in modern times.

---

[13] "Historical Vignette 062 - How Army Engineers Cleared Viet Cong Tunnels," U.S. Army Corps of Engineers, https://www.usace.army.mil/About/History/Historical-Vignettes/Military-Construction-Combat/062-Viet-Cong-Tunnels/.

[14] Nuclear Vault, "Tunnel Destruction (1969)," YouTube video, August 28, 2011 (video produced by the Department of Defense in 1969), https://www.youtube.com/watch?v=R1wsbwdifc4.

[15] Thomas Dethlefs, "Tear Gas and the Politics of Lethality: Emerging from the Haze," *Yale Historical Review* 2, no. 3 (2013), 83-118, https://historicalreview.yale.edu/sites/default/files/yhr_spring_2013_web.pdf.

[16] Human Rights Council, *Report of the detailed findings of the independent commission of inquiry established pursuant to Human Rights Council resolution S-21/1* (New York: United Nations, 2015), https://www.ohchr.org/en/hrbodies/hrc/coigazaconflict/pages/reportcoigaza.aspx.

The United Nations commission of inquiry analyzed all of Israel's actions during this conflict but failed to address how Hamas' tunnels affected the operational environment and possibly the application or interpretation of the law. The commission of inquiry took the view that the tunnels were incidental and unrelated to the legal assessment of the operation. They failed to understand that the presence and use of tunnels affects virtually every aspect of the fight—from the tactical to the strategic—and that this impact must be acknowledged at every step of the analysis.

If someone—a military lawyer, a judge, or a United Nations expert or prosecutor—does not understand how significantly tunnels change the operational environment, then they are not in a position to make this analysis in the first place. I show in my book how big the gap is in terms of understanding the underground environment, not just among lawyers but also among decision makers and military commanders. This is something I have been struggling with and that I am working to change.

Spencer: I know the following question sometimes arises: Do we need new laws or new conventions? It is good to know that your answer is we do not need new laws, we just need to interpret the existing laws. You mentioned the 2014 Gaza War. I know the challenges of underground warfare, like urban warfare, has a lot to do with context. What type of city? What type of mission? What type of operational or political constraints? How is the underground being used by combatants? Is the geology of the area conducive for tunneling? What technology or expertise is required for tunneling? But in Israel, all these questions exist in separate contexts. The tunnels in the north are very different from the tunnels in the south. Can you discuss some of Israel's experiences with tunnels?

Richemond-Barak: Yes, for sure. Due to its many wars, Israel has often been kind of a lab for modern war that spotlights new trends, new methods, and new challenges. I think that was the case in the 2014 Gaza War as well. Israel knew that there were tunnels inside the Gaza Strip and inside Lebanon for more than a decade. Israel understood the challenge that these tunnels presented for troops when they entered the Gaza Strip or Lebanon, but what was new was the use of cross-border tunnels from one state or territory into Israel itself. Israel woke up to this strategic challenge in 2014 and since then it has been a steep learning curve. But it is important to understand that cross-border subterranean threats are of concern to many other states, not just Israel. I mentioned India and Egypt earlier, but you can also add Turkey and, albeit in a different context, the United States.

States tend to be very concerned about underground threats because they are hard to detect, they are almost invisible, and they often catch states by surprise. The psychological impact of such an invisible threat is immense. Following in the footsteps of Israel, many states have invested time and resources into developing new and tailored-made technology, tactics, techniques, and procedures to deal with tunnels of various types.

It took Israel some time—some people said too much time—to take full measure of the security risk posed by underground threats.[17] Years after the 2014 Gaza War, tunnels remain a threat to Israel. While the tunnels in Gaza and Lebanon are very different, they serve similar purposes. The Hamas tunnels, dug from the Gaza Strip, are less sophisticated. Hamas creates passageways by digging in relatively soft soil. The Hezbollah tunnels, dug from Lebanon, are more sophisticated. The ground is hard rock, so Hezbollah creates tunnels by excavating ground. It takes a significant amount of time and effort to create these tunnels, so that is a testament to their appeal. At the Gaza border and at the Lebanese border, the tunnels need to be carefully monitored by Israel as they threaten the security of the state and its citizens.

In December 2018, Israel initiated Operation Northern Shield to locate and destroy Hezbollah tunnels that crossed into Israel. It marked the first time that Israel acknowledged the existence of cross-border tunnels emanating from Lebanon. Rather than uncovering and neutralizing the tunnels during a war, Israel initiated the operation while there was no war or active hostilities. Israel stationed soldiers along the border where they suspected the tunnels were located and then attempted to detect them. During the five-week operation, they found six cross-border tunnels.[18]

These impressive and threatening cross-border tunnels were part of Hezbollah's underground invasion strategy. The violent Shia group could have taken control of territory, cut off main road accesses to the north, and gained access to central Israel in no time as part of a surprise operation. That, of course, threatened the lives of Israeli civilians on a grand scale. One important lesson learned from Operation Northern Shield is that the terrain does not act as a deterrent against tunnel tactics. No terrain is too

[17] See, for example, Cohen et al., *From Cast Lead to Protective Edge*, 160.

[18] "End of Operation Northern Shield to expose and neutralize Hezbollah tunnels on Israel's northern border," Meir Amit Intelligence and Terrorism Information Center, January 15, 2019, https://www.terrorism-info.org.il/app/uploads/2019/01/E_010_19.pdf.

difficult, so do not rule out the presence of tunnels based on the geological conditions alone.

This operation will be taught and studied because it was the first time that an anti-tunnel operation was undertaken outside active hostilities, and it demonstrated the long and painful work that detecting tunnels entails despite advanced tunnel-detection technology.

Spencer: I have been inside one of the Hezbollah tunnels. When standing in the tunnel, dug through solid rock, it was clear that a lot of time and effort went into digging it. Tunneling used to be a specialty of miners, and many militaries had sappers—soldiers that were assigned tasks like digging—that specialized in mining. Digging a tunnel is much different than digging a foxhole or a trench. Digging these tunnels requires expertise and sometimes technology. Earlier, you discussed the diffusion of information. Can you discuss how a non-state actor, like Hezbollah, was able to figure out how to dig tunnels in solid rock?

Richemond-Barak: The resemblance between the Hezbollah and the North Korean tunnels already tells us about diffusion. The tactic—how to dig tunnels and how to use them—has spread from North Korea to Hezbollah and Hamas. But I would like to turn to Syria to illustrate the point about diffusion. In 2014, in the earlier stages of the ongoing Syrian civil war, the rebels wanted to attack President Assad's forces and they identified the Carlton Hotel in Aleppo as a potential target. So, they dug a very basic tunnel. The tunnel was 107 meters long and took 33 days to dig.[19] They used the tunnel to remotely detonate explosives that they had placed under the hotel to kill 40 of Assad's forces.[20] It was a very successful operation, which went viral and attracted a lot of attention.[21]

[19] Martin Chulov, "Aleppo's most wanted man – the rebel leader behind tunnel bombs," *The Guardian,* May 20, 2014, https://www.theguardian.com/world/2014/may/20/aleppos-most-wanted-man-rebel-leader-tunnel-bombs.

[20] "Syrian troops hit in Aleppo tunnel bombing," *Al Jazeera,* May 31, 2014, https://www.aljazeera.com/news/2014/5/31/syrian-troops-hit-in-aleppo-tunnel-bombing.

[21] See, for example, the following that has more than 300,000 views: *The Telegraph,* "Syria: footage emerges of Aleppo Carlton Hotel bomb blast," YouTube video, May 8, 2014, https://www.youtube.com/watch?v=zJbeo5WE8tc.

The rebels employed a tactic that dates back to at least World War I and applied it in a contemporary context.[22] During an interview, the mastermind behind the operation was asked how he came up with the idea, and he proudly responded that he got the idea from his "brothers in Gaza."[23] So it is clear that even if Hamas is not talking directly to rebels in Syria and Hezbollah is not talking directly to North Korea—though they may be—that this information is diffusing through interpersonal relationships, imitation, social media, and other means. The use of a tactic reminiscent of World War I shows the amount of innovation, adaptation, and creativity that goes with the use of this military tactic. Understanding this process makes it possible to anticipate future trends.

A foreign fighter in Syria who has been exposed to underground warfare could take this knowledge and experience back to Canada, Europe, or to the Russia-Ukraine theater, for example, and employ it there. Thus, it is important to consider diffusion patterns to anticipate where we may see tunnels next and how they will be used. What are we likely to see in five or ten years and where are we likely to see it?

It seems unlikely that underground warfare is going to go away; on the contrary, continued innovation is more likely. One of the things that we are likely to see, for example, is a conflation of maritime and underground warfare: tunnels that are dug at a distance from the sea to enable an "invisible" entry or exit into the water or a surprise infiltration into an adjacent state or territory. We must constantly ask ourselves, "What is next?" Looking at the diffusion patterns is helpful in that respect.

Spencer: It makes me question what Russia learned from its experiences in Syria. We know that they rotated many of their senior officers through Syria as a proving ground to expose them to everything from robotics to drones to, I would assume, the extensive use of the underground as both an offensive and defensive capability.[24]

---

[22] In World War I's Western Front tunneling was seen at large scale by both sides as they dug under the deadly no man's land between the static trenches that characterized much of the war. Specialist miners dug tunnels to place mines beneath enemy positions. At the Battle of Messines on June 7, 1917, 19 of 22 tunnels dug by the Allies detonated an estimated 1 million pounds of explosive under German positions. See, Dave Ross, "WWI's Battle of Messines: How Allies Used Massive Explosives and Tunneling to Win," History.com, May 3, 2021, https://www.history.com/news/battle-messines-world-war-i-tunnels.

[23] "Gazan terrorists share their tunnel expertise with Syrian rebels," *Jerusalem Post*, May 25, 2015, https://www.jpost.com/middle-east/gazan-terrorists-share-their-tunnel-expertise-with-syrian-rebels-404016.

[24] John Spencer, "Russia's Urban Warfare Tech," *Urban Warfare Project Podcast*, Modern War Institute, September 18, 2020, https://mwi.usma.edu/russias-urban-warfare-tech/.

Richemond-Barak: I tend not to classify tunnels in terms of offensive and defensive capabilities and more as being used for different purposes, because ultimately, a tunnel is a tunnel is a tunnel. A tunnel can be used for all sorts of different purposes. The United States has its own border-tunnel problem, but rather than being used by terrorists or militants, the tunnels along its Mexican border are used to smuggle drugs and illegal immigrants.[25]

A tunnel can be used offensively or defensively and can be repurposed at any time provided it has not been destroyed. It can be inherited by different factions and used for different purposes. The tunnels between Gaza and Egypt were initially used to smuggle goods. Later those same tunnels were used to smuggle weapons. Eventually they were used to kidnap an Israeli soldier.[26] So, the same tunnel can be used for many purposes.

Does Russia understand this? Absolutely. It has learned this from the Syrian theater where the subterranean environment was exploited by all sides. Given how the tactic has diffused thus far, I would not be surprised to see Russian forces use the underground in future conflicts.

Spencer: Studying the use of the underground in warfare is a fascinating topic, but as you mention, the underground also has utility outside the context of war. This brings me to my final topic. We both agree that underground warfare is important, yet at the same time, understudied, under-researched, and under-prioritized by most militaries. What do you recommend that we and others do about it?

Richemond-Barak: I have been researching this topic since 2013 and I agree that it is definitely under-researched and under-prioritized, but the scholarly community and practitioners have made significant progress in recent years. I think to some extent that ISIS acted as a trigger. Escalatory rhetoric between the United States and North Korea likely contributed to this as well.[27] In the past decade, the U.S. Army developed and published the *Subterranean Warfare Handbook* in 2013,[28] the *Subterranean Operations Quick*

[25] "US-Mexico border: 'Longest ever' smuggling tunnel discovered," BBC, January 30, 2020, https://www.bbc.com/news/world-us-canada-51304861.

[26] "Israel finds long tunnel from Gaza into its territory," *France 24*, December 13, 2013, https://www.france24.com/en/20131013-israel-tunnel-gaza-hamas-underground-attack.

[27] Anne Gearan and Simon Denyer, "Trump escalates a war of words with North Korea, calling leader Kim a 'madman'," *Washington Post*, September 22, 2017, https://www.washingtonpost.com/world/asia_pacific/china-tightens-banking-screws-on-north-korea-but-denies-trumps-talk-of-total-ban/2017/09/22/0533a168-9f08-11e7-b2a7-bc70b6f98089_story.html.

[28] Michael Beaton, "Interactive Notes from the Underground," Army.mil, July 15, 2019, https://www.army.mil/article/224511/interactive_notes_from_the_underground.

Reference Card in 2018,[29] and the *Subterranean Operations* army techniques publication in 2019.[30]

Hence, the field is attracting more attention. The International Working Group on Subterranean Warfare, which is a community of people interested in sharing information and raising awareness on subterranean warfare, was created in 2019 and it has brought experts together to discuss psychological, legal, tactical, strategic, historical, and geological aspects of the tactic.[31] Just the fact that this community exists and has been institutionalized under the umbrella of the working group is a step forward.

The Defense Advanced Research Projects Agency (DARPA) established the subterranean challenge in 2018, to "seek novel approaches to rapidly map, navigate, and search underground environments during time-sensitive combat operations or disaster response scenarios."[32] To incentivize innovative approaches, DARPA offered a total of $5 million in prize money for various categories within the challenge. Another DARPA effort is the underminer program which was designed to "demonstrate the feasibility of rapidly constructing tactical tunnel networks that enable secure, responsive resupply."[33] It embodies an important effort by the United States to re-own the subterranean environment from non-state actors that have successfully exploited it to their benefit in recent conflicts.

If I could make one call for action it would be for states to build a strategy that includes detecting, mapping, neutralizing, preventing, monitoring, and cooperating in the subterranean environment. These are the key steps to a successful subterranean strategy. There used to be almost no information sharing between states on the form underground threats may take, the challenges they raise, and the range and efficiency of methods to counter them. This is slowly changing: I cannot overstate the importance of cooperation and collaboration in this field.

It is also important to note that militaries have also stepped up and are now training soldiers, not just those in special units, for the underground

[29] Asymmetric Warfare Group, *Maneuver Company-Team Subterranean (SBT) Assault and Clearing Operations Planning, Preparing, and Execution,* GTA 03-11-003 (Fort Eustis, VA: Training and Doctrine Command, 2019), https://armypubs.army.mil/ProductMaps/PubForm/Details.aspx?PUB_ID=1006485.

[30] Department of the Army, *Subterranean Operations.*

[31] See, The International Working Group on Subterranean Warfare website at https://www.runi.ac.il/en/schools/government/research/subtwg/pages/home.aspx.

[32] "Subterranean Challenge," Subtchallenge.com, https://www.subtchallenge.com/index.html#about.

[33] Andrew Nuss, "Underminer," DARPA, https://www.darpa.mil/program/underminer.

environment. Militaries are also building underground systems into their training sites.[34]

I fear that as the United States shifts its focus to "great power competition" it may divert some of its effort away from the subterranean environment, which would be extremely unfortunate given the progress made in recent years. Too often, the focus of great power competition turns to technology, but we cannot forget that warfare is not just about using or fighting advanced technological capability. Underground warfare shows that low-tech warfare can and does create real challenges for states. In addition, both states and non-states are likely to use the subterranean environment if recent history is any indicator. Hopefully the progress that has been made in subterranean warfare in recent years will not be lost with a shift toward great power competition.

I will add that when a military unit finds itself in a complex operational environment that includes a subterranean environment, it should not hesitate to bring in outside experts if it lacks the resident expertise, just as it would pull in a cyber expert whenever cyber expertise is required. It is a field or domain where expertise could make a tangible impact. Things that units take for granted, such as communications, navigation, or breathing are often difficult underground. An expert versed into the unique features of subterranean warfare could help anticipate and overcome any difficulty that arises.

Also, deployed militaries should learn to secure their own critical infrastructure, such as a base, from all directions. They employ protective measures against air attack, such as bunkers or air defenses; ground attack, such as walls, fences, guard posts, and roving patrols; but often fail to consider the threat from underground. Recent trends suggest that ignoring the subterranean environment is a risky proposition, for both civilian and military infrastructure.

Spencer: I one hundred percent agree with you, and you can count on me to contribute to that effort. As you mentioned multiple times, the subterranean environment is a part of war and although its use by asymmetric actors has recently expanded, the subterranean environment is not likely something that would disappear if we found ourselves facing

[34] See, for example, Todd South, "Here's how the Marine Corps is dealing with the looming underground combat threat," *Marine Corps Times*, June 3, 2019, https://www.marinecorpstimes.com/news/your-marine-corps/2019/06/03/heres-how-the-marine-corps-is-dealing-with-the-looming-underground-combat-threat/.

a near-peer or peer enemy. We may want to avoid subterranean warfare, just as we want to avoid urban warfare, but I think you and I agree that it will be impossible to avoid. If you imagine potential future wars around the world, it will surely involve some underground warfare. Any final thoughts?

Richemond-Barak: I have no doubt that you will continue to play a central role in raising awareness for underground warfare. We both know that the underground maintains its appeal despite technological advances. I would even say that high-tech warfare has exacerbated the low-tech appeal of the underground. Looking at future trends, underground warfare will continue to converge with urban warfare. Naval warfare, about which we spoke earlier, will likely be another focal point of convergence with the underground going forward.

Ultimately, underground threats are unsettling and the risk they pose increases significantly in urban environments. States have countered them using violent means—such as B52 bombers in Vietnam, the "mother of all bombs" in Afghanistan, or flamethrowers in Syria—often without a clear strategy in mind, and without a real understanding of the terrain. But underground threats are not existential, and the challenge can be overcome by (1) detection and mapping, (2) neutralization and/or destruction, (3) prevention and monitoring, and (4) cooperation. These form the basic components of a successful underground warfare strategy. Anti-tunnel technology cannot, by itself, provide the solution to subterranean threats.

PART II

OPERATIONAL CASE STUDIES

7

The Battle of Ortona

The 1st Canadian Infantry Division was part of the British 8th Army for almost the entirety of the Italian campaign of World War II. The division landed in Sicily on July 10, 1943, and conducted 28 days of grinding combat from the southern through the central part of the island before it was pitched into the invasion of the Italian mainland on September 3, 1943. As part of the 8th Army, it continued to fight up the eastern side of Italy before it was pulled out of the line for several weeks to conduct smaller actions as the Allies advanced through a series of challenging defensive lines.

In December 1943 the division was once again put into the front lines just a few miles southwest of the small seaside town of Ortona as the far eastern unit of the 8th Army. The Allies were approaching what they thought to be the German's most challenging defensive line of all: the Gustav Line. The line stretched from the Tyrrhenian Sea to the Adriatic Sea, anchored south of Rome in the west to Ortona in the east.

In this chapter, John Spencer interviews Major Jayson Geroux, an infantry officer with The Royal Canadian Regiment of the Canadian Armed Forces. Geroux is an urban warfare instructor, urban warfare historian, and scholar who has done extensive research on the 1943 Battle of Ortona. The battle, sometimes called "Little Stalingrad" or the "Italian Stalingrad," took place in the town from December 20-27, 1943. The fighting pitted the 2nd Canadian Infantry Brigade against two battalions of German paratroopers from the 3rd and 4th Parachute Regiments. It was the first prolonged urban battle of the Italian campaign.

Geroux explains the strategic context of the battle, before discussing the unique terrain considerations, the tactics each side employed, and the details of the fighting that took place over the 8-day battle. He also highlights lessons from the battle that inform the way we should think about urban warfare and the potential for peer conflict in dense urban terrain today.

John Spencer: I would like to start by asking you to provide the readers with a brief summary of your background and how you became an urban warfare expert.

Jayson Geroux:[1] I have been very fortunate throughout my career to have been in the right place at the right time to have been with units that wanted to conduct a number of urban operations training events. My first opportunity came in 2001 when I was assigned to the 3rd Battalion of The Royal Canadian Regiment. After returning from a deployment to Bosnia-Herzegovina, we conducted a year-long urban operations training package at the Canadian Forces Base Petawawa, Ontario. I was then assigned to the Argyll and Sutherland Highlanders of Canada, an army reserve unit (the equivalent of the United States' National Guard), based at the John W. Foote Armoury in Hamilton, Ontario. We were very fortunate to have a company commander, now Lieutenant Colonel Lawrence Hatfield, who was a very big supporter of urban operations. He believed wars in the present and future would likely be fought in urban areas. So, we spent three years in this reserve unit conducting urban operations training, largely at the individual to platoon skill level.

Several years later, I was posted to the Tactics School at the Canadian Forces Base Gagetown in New Brunswick, Canada, as the directing staff for one of our career courses for army officers: the Army Tactical Operations Course. When I was a student in this course a few years prior, the course had included a module on urban operations, but it had since been removed. So, as the officer in charge of the course, I endeavored to put urban operations back into the training curriculum. I started small, inserting a two-hour general introduction of Canadian urban operations doctrine into the course. But thankfully, commandants and course officers of the other courses, including the Army Tactical Operations Course, the Combat Team Commanders Course, and the Infantry Dismounted Company Commanders Course, also believed that future wars were likely to involve urban combat, so they asked me to develop urban operations training modules for their courses. These modules focused on the planning, tactics, and sustainment at the company, battalion, brigade, and division levels.

After developing the modules, I soon found myself planning and executing operations training exercises which I have now been doing for the past six years. During that time, I also earned a master's degree after

[1] The views and the opinions expressed within this chapter do not necessarily reflect the Canadian Department of National Defence or the Canadian Armed Forces.

writing a thesis on the Battle of Ortona. While the battle may be unfamiliar to many Americans, it is one of the most well-known urban battles within Canadian military history. Very fortunately I have participated in, planned, executed, written about, and taught urban operations and urban warfare history for much of the past twenty years and very intensively over the past six years.

Spencer: That is amazing. I would call you a unicorn, since you have spent decades studying urban operations and urban warfare history as a student, practitioner, trainer, and teacher. So, without delay, I would like to turn the conversation over to you, to discuss the 1943 Battle of Ortona.

Geroux: I always like to begin with the enemy situation and will start at the strategic level and work down to the tactical level. The Apennine mountains and its deep valleys, located at the center of the Italian "boot," creates naturally restrictive terrain throughout Italy. The Germans supplemented these valleys with manmade defenses to create challenging obstacles to chew away at the Allies as they moved up the boot. Using these valleys, the Germans created a series of defensive lines to delay and weaken the Allies so that the Germans could stop the Allies advance at the Gustav Line.

Holding the Allies south of the Gustav Line would prevent the Allies from taking Rome. It would have been a huge psychological and physical victory for the Allies and at the same time a huge psychological and physical defeat for the Germans. So, the Germans built the Gustav Line as an attempt to halt the Allied advance north and to protect Rome. The line stretched across the country from the Tyrrhenian Sea on Italy's west coast to the town of Ortona along the Adriatic Sea on Italy's east coast (see figure 7.1).

Originally the British 8th Army intelligence stated that the Gustav Line was built on the Arielli River just north of Ortona. They thought this because throughout the campaigns in Sicily and Italy the Germans defended from urban areas for only 24-48 hours before withdrawing back to these natural valleys.[2] Thus, they believed the Germans would fight in a similar manner in Ortona: defending from Ortona for 24-48 hours before falling back to the main defensive line at the Arielli River.

Turning to the operational level, the Germans anchored the eastern end of the Gustav Line at Ortona, which was unknown to the Canadians. It was

[2] Reginald H. Roy, *The Seaforth Highlanders of Canada, 1919-1965* (Vancouver: Evergreen Press, 1969), 258.

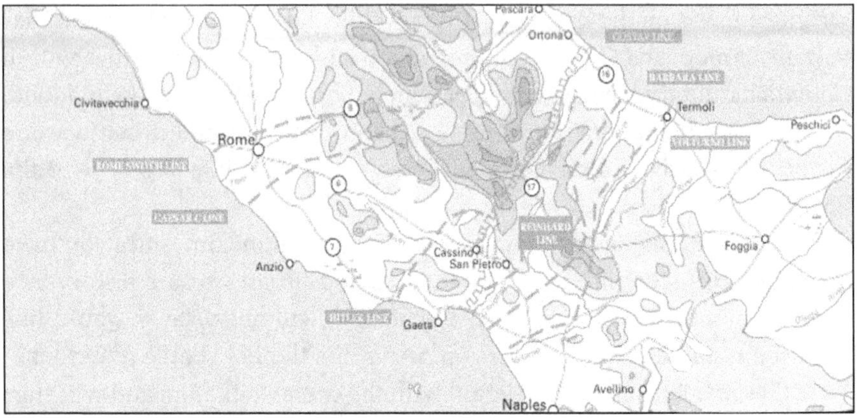

Figure 7.1. The Gustav Line and other defensive lines, 1943.
Source: Map courtesy of the United States Military Academy Department of History, Digital History Center.

vitally important for the Germans to hold Ortona, because if they failed, the Allies could easily advance north to Pescara, and then use the roads that run west from Pescara and cut through the mountain passes to enter Rome through this back door.[3] So, for many months, at the tactical level, regular German *Wehrmacht* units and eventually the 2nd Battalion of the 3rd Regiment of the 1st Parachute Division occupied Ortona and the areas surrounding it and built a defense within and outside of the town.

The intent of the German's concept of operations was to halt the Canadian advance with a carefully planned defensive stand at Ortona, building the defense from September to December 1943. Their scheme of maneuver was to establish a perimeter force at the southern edge of town, suck the Canadians into town with a disruption force battle through the middle of the town, and then have their main defensive area in the heart of town. They would put up limited defenses from the forward positions to delay the Canadians' movement and then fall back to the heart of town where they had established their main battle zones, using the five main *piazze* (town squares) as kill zones to mass fires on the Canadians.[4]

Fires refers to "the use of weapon systems or other actions to create specific lethal or nonlethal effects on a target."[5] Most often when the military

[3] Jayson Geroux, "The Urban Battle of Ortona," (master's thesis, University of Brunswick, 2020), 31-41.

[4] Ibid., 32-3 and 39-40.

[5] Joint Chiefs of Staff, *DoD Dictionary of Military and Associated Terms*, JP 1-02 (Washington, DC: Joint Chiefs of Staff, 2021), 86.

uses the term fires, they are referring to indirect fire, such as artillery. In this case, the Germans were attempting to mass both direct (such as rifle and machine gun fire) and indirect fire on the advancing Canadians.

Within Ortona, the kill zones were located in four piazze on the northern end of town—Piazza Municipali, Piazza San Tomasso, Piazza Plebiscita, and Piazza San Francesco—and the Germans were determined to halt the Canadian's advance along this line (see figure 7.2). A fifth *piazza*, Piazza Vittoria, was located in the southern side of the town. The Germans would occupy the buildings surrounding the squares with small arms and anti-tank guns, affording them excellent cover and concealment from the Canadians who would be exposed in the open squares. The Germans would keep their reserve in the railroad tunnel on the northern end of town.[6]

As part of the defensive plan, the Germans destroyed buildings to create rubble obstacles, attempting to block the side streets and force the Canadians up the main highway and into the open squares. Highway 16, which cuts through Ortona, served as the main supply route for the Canadians. The Germans left the highway clear, but rubbled the side roads, to canalize the Canadians into their kill zones. As a result, prior to the battle even starting, the Germans had destroyed a great deal of the town.[7]

Turning to the Canadians, the 1st Canadian Infantry Division was the most experienced division within the Canadian Army at the time. The unit had been involved in heavy fighting in Sicily in July and August, had participated in the invasion of Italy in September, and was subsequently taken out of the main fight to conduct small actions, recover, and refit. In December the division was thrown back into the fight just south of Ortona at the battle of the Moro River. The division fought at "the Gully," a very deep natural U-shaped ravine southwest of Ortona. It was a particularly challenging fight, and the 1st and 3rd Canadian Infantry Brigades took a real pounding.[8] Historians sometimes criticize Major General Christopher Vokes, the division commander, for his handling of the division during the Moro River Campaign and the Gully fighting.[9]

After finally seizing the Gully, General Vokes directed Brigadier General Bertram Hoffmeister ("Hoffy"), the commander of the 2nd

---

6 Geroux, "The Urban Battle of Ortona," 32-3 and 39-40.

7 Ibid., 32-5 and 194.

8 Terry Copp, "Clearing the Gully," *Legion*, January 1, 2007, https://legionmagazine.com/en/2007/01/clearing-the-gully/.

9 See, for example, Brereton Greenhous, "'Would It Not Have Been Better to Bypass Ortona Completely...?' A Canadian Christmas, 1943," *Canadian Defence Quarterly* 18, no. 5 (1989): 51-55.

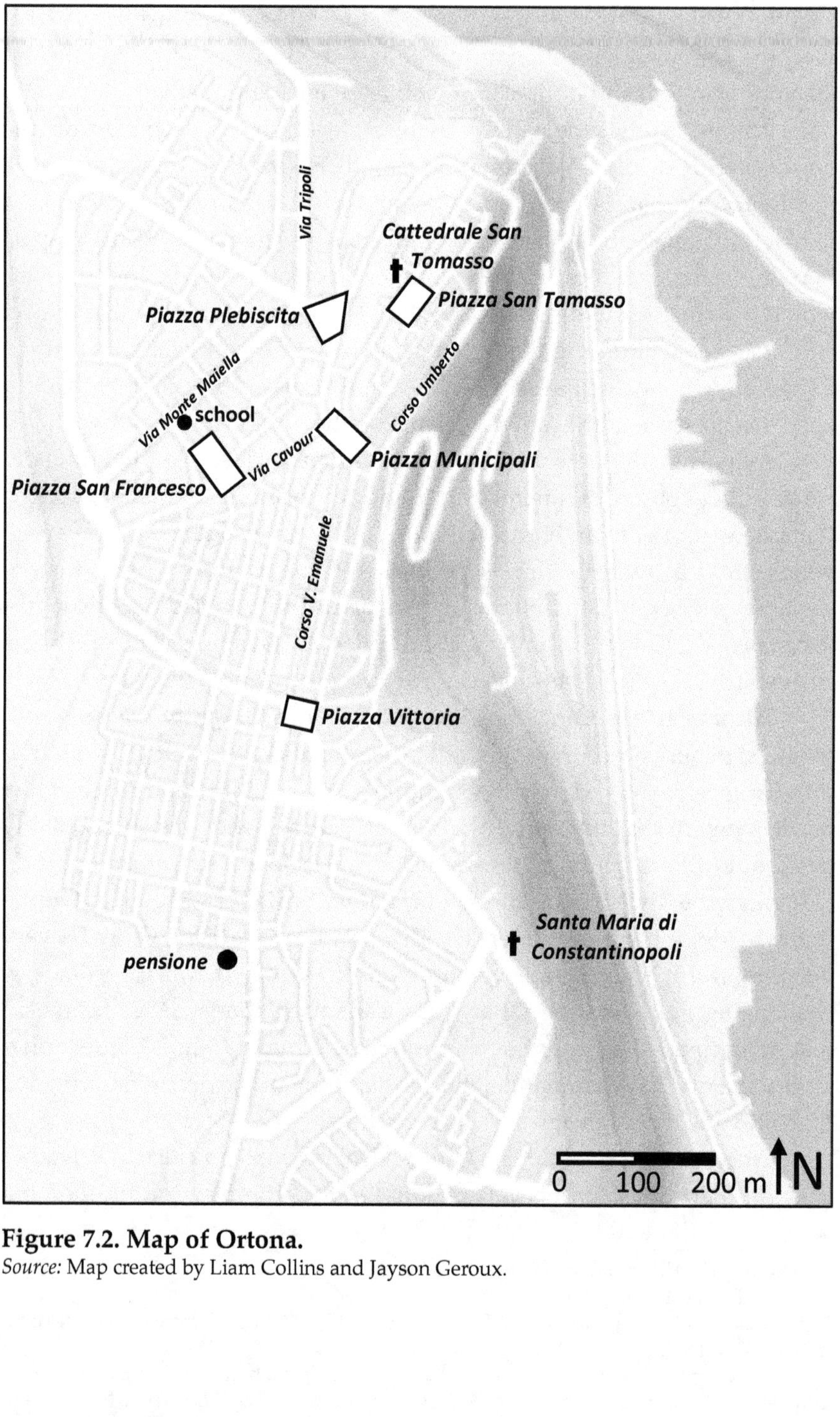

Figure 7.2. Map of Ortona.
Source: Map created by Liam Collins and Jayson Geroux.

Canadian Infantry Brigade, to enter Ortona.[10] Vokes told him there was only one battalion within Ortona and that he expected the Germans to remain in the town for no more than 24-48 hours. Vokes thought this because the Allies believed that the Gustav Line was located just north of Ortona, in the Arielli River valley.[11]

Vokes wanted Hoffy to go into Ortona and drive the Germans out so that the division could continue its advance to the Arielli River where they believed the Gustav Line to be located.[12] Hoffy's 2nd Brigade was comprised of the Loyal Edmonton Regiment (the "Loyal Eddies"), the Seaforth Highlanders (the "Seaforths"), and the Princess Patricia's Canadian Light Infantry (the "Patricias"). The brigade also included the Three Rivers Regiment (an armor unit), the 4th Field Company of the Royal Canadian Engineers, and the Royal Canadian Artillery's 90th Anti-Tank Battery in support.

Before discussing the battle, it is important to consider the geographic factors that matter for this particular battle. Ortona sits on a plateau, as do many Italian and Sicilian cities. Building on a plateau serves multiple purposes. First, plateaus are easier to defend because an enemy must fight upward to attack the city. A second reason is that the population can live near the crops, which are grown in the river valleys, yet avoid floods by living on the higher ground. Ortona's plateau is roughly 1,500 meters north to south, and about 500 meters east to west.[13] The town's south was comprised of many large buildings that were fairly close together and the town's north was comprised of tightly packed buildings that shared common walls.

The town's urban and physical geography was a defender's dream. A large ravine, 30-50 meters in depth, protected the town's west, and 50-75 meter high cliffs along the Adriatic Sea protected the town's east.[14] So, Ortona could not be flanked from the west or east. The only approach into the city for the Canadians would be from the south, so they would have to

---

[10] Roy, *The Seaforth Highlanders*, 257.

[11] J.A. Porter, "Operations of 1 Cdn Inf Div and 1 Cdn Armd Bde in Italy, 25 Nov 43 - 4 Jan 44," Canadian Military Headquarters reports 1940-1948, issue #165 (Ottawa: Department of National Defence, Directorate of History and Heritage, 1946), 84.

[12] Ibid., 257-278.

[13] To obtain these measurements Geroux referenced map 4 from Mark Zuehlke, *Ortona: Canada's Epic World War II Battle* (Vancouver: Douglas & McIntyre, 1999).

[14] Geroux, "The Urban Battle," 42-3.

fight uphill just to reach the town and then continue the fight through its difficult urban environment.

With any battle, but especially in urban battles, it is important to consider civilians since they often impact the tactical fight. There were about 10,000 *Ortonesi* civilians living in Ortona before the Germans occupied the town and started their defensive preparations. When the Germans arrived, they directed the *Ortonesi* to leave the city and told them that any fighting age males that remained would be shot or forced to assist the Germans with the defensive preparations. About 90 percent of the town's population departed, but approximately one thousand remained.[15]

The reasons civilians stay in active combat zones vary, but often they feel they do not have a choice. It may be their only residence, it contains all their personal property, and their livelihood may be there, so despite being a bad option, they may still feel it is the best way to protect their family. Unfortunately, many civilians were killed during the battle of Ortona from artillery, armor and, as will be discussed later, the incredible amount of violence that was exercised by both sides during this battle. Those that remained often hunkered down in their basements or the railroad tunnel at the northern end of the city.[16]

The battle kicked off on December 20. Hoffy directed Lieutenant Colonel Jim Jefferson, the commanding officer of the Loyal Eddies, to lead the brigade's attack into the city. As mentioned earlier, the Canadians expected a limited delaying action within Ortona, so Hoffy sent only a single battalion into the town. Hoffy directed the Seaforths to protect the brigade's eastern flank, and the Patricias to remain as the brigade's reserve and guard the brigade's western flank.[17] Hoffy did, however, assign a troop of tanks (what the American Army would call a platoon of tanks) from the Three Rivers Regiment to support the Loyal Eddies and a troop of tanks to support the Seaforths. Hoffy had engineers from the 4th Field Company co-located with his brigade as well. By the end of the first day, the brigade had consolidated its positions on the southern fringe of the town.[18]

---

[15] *The Canadians at War 1939/45, vol. 2* (Markham: Reader's Digest Association, 1969), 368; Saverio Di Tullio, *1943: The Road to Ortona* (Ottawa: Legas Publishing, 1998), 74; and Zachary D. Cavasin, *Hai visto i Canadesi?: A study of the Social Interactions between Canadian Soldiers and Italian Civilians before, during, and after the Battle of Ortona* (Ottawa: University of Ottawa 2010), 17.

[16] Cavasin, *Hai visto i Canadesi?* 48-9.

[17] Seaforth Highlanders of Canada, *Active Service Force War Diaries 1943/01-31 December*, RG-24-C-3, vol. 15256, Library and Archives Canada.

[18] Porter, "Operations."

As the battle continued into the second day, the Loyal Eddies continued to be surprised by the amount of fire they received. B Company, on the eastern side of the highway, experienced no challenges but Major Jim Stone—"Big Jim Stone" as he was sometimes called because he was very tall and stocky—the commander of D Company, was having trouble advancing up the western side of highway 16 into the town. After a second failed attempt, one of his platoon commanders, James Dougan, took some troops and infiltrated into a *pensione* behind the German defenses on the western side of the road. Dougan killed the Germans on the outer perimeter along highway 16, and D Company made their initial entry into the town.[19] Meanwhile, the Seaforths met heavy resistance as they tried to advance into town on the eastern flank.

After gaining an initial foothold in Ortona, the Loyal Eddies first observed the rubble piles blocking the streets, often 12-15 feet high and sown with booby traps and landmines.[20] They used the tanks from the Three Rivers Regiment to fire at the rubble piles in an attempt to reduce the obstacles and make them passable. At that point, Major Tiger Welsh, of the Royal Canadian Artillery 90th Anti-Tank Battery, remembered that they were fighting a German parachute division so it would not have tanks. So rather than having his six pounder anti-tank guns sit idle, he brought them forward to help reduce the rubble obstacles.[21]

Spencer: I think the Germans intentionally rubbling buildings as obstacles is fascinating. It is mentioned in U.S. military doctrine, but I have never heard of anybody actually rubbling buildings to support a defense.[22] The fact that Welsh brought anti-tank weapons forward to blow the rubble off the street is equally impressive. You have me hooked!

Geroux: On the afternoon of December 21, Lieutenant Colonel Jefferson ordered two of his companies up highway 16. One company on the left side and the other company on the right side, to thrust and clear

---

19 Zuehlke, *Ortona*, 253-254.

20 Roy, *The Seaforth Highlanders*, 263; Department of National Defence, *Unique Operations – Urban*, B-GL-322-007/FP-001 (Kingston, Ontario: Army Publishing Office, 2006), 6; Porter, "Operations," 8; and Zuehlke, *Ortona*, 258.

21 Lee Windsor et al., *Loyal Gunners: 3rd Field Artillery Regiment (The Loyal Company) and the History of New Brunswick's Artillery, 1893 to 2012* (Waterloo, Ontario: Wilfrid Laurier University Press, 2016), 309; and Daniel Dancocks, *The D-Day Dodgers: The Canadians in Italy, 1943-1945* (Toronto: McClelland and Stewart, 1991), 178.

22 Department of the Army, *Combined Arms Operations in Urban Terrain*, ATTP 3-06.11 (Washington, DC: Headquarters, Department of the Army, 2011), 6-9, https://armypubs.army.mil/ProductMaps/PubForm/Details.aspx?PUB_ID=102031.

all the houses up to *Piazza Vittoria* (see figure 7.3). The Seaforths on the eastern side found themselves engaged in a very heavy firefight inside a very large church, the Church of *Santa Maria di Constantinopoli*. It took most of the day for the Loyal Eddies to reach *Piazza Vittoria* and for the Seaforths to clear the church, and only after receiving tanks, anti-tank guns, and engineers to assist. As evening approached, the Canadians halted their advance and set up hasty defensive positions to prevent German patrols from infiltrating their lines.[23]

The following day, December 22, the Loyal Eddies started their attack on *Piazza Vittoria*. From their support position, the tanks engaged the German defensive positions, while the infantry assault force with their engineers started clearing the buildings around the *piazza* itself. Major Syd Thomson, the Seaforths' commander, moved his entire battalion into the *Santa Maria di Constantinopoli* church and remained in place. While that might seem strange, it is important to remember that up until this point, the Germans typically fought for 24-48 hours and then fled the city, so after two days of fighting, Hoffy assumed the Germans would be withdrawing from the city to the Arielli River. Thus, the Seaforths had not yet been committed to the battle. By the evening, the Canadians once again had only a single battalion—the Loyal Eddies—in the fight along with their tank troop, engineers, and anti-tank gun attachments.

After clearing *Piazza Vittoria,* Lieutenant Colonel Jefferson wanted to continue north along highway 16, which is named *Corso Vittorio Emanuele* in that part of the city, so he told Big Jim Stone that he needed him to continue to press north. But after two days of heavy urban fighting where his company had suffered many casualties, Stone had had enough of building clearing and instead asked Jefferson if he could bypass the buildings, rather than deliberately clearing each, and advance to *Piazza Municipali* as quickly as possible. Jefferson supported Stone's request for a surprise charge.[24]

Stone's plan was to have the tanks drive up the *Corso Vittorio Emanuele* with their horns blaring, with the engineers leading the column to clear any landmines while the infantry provided security to the sides, to prevent anti-tank gunners from engaging the tanks from buildings' second floors. Surprisingly, all went according to plan until they reached *Piazza Municipali* where they encountered an enormous rubble pile which blocked their route.

---

[23] Geroux, "The Urban Battle," 61-73.

[24] Ibid., 79-80.

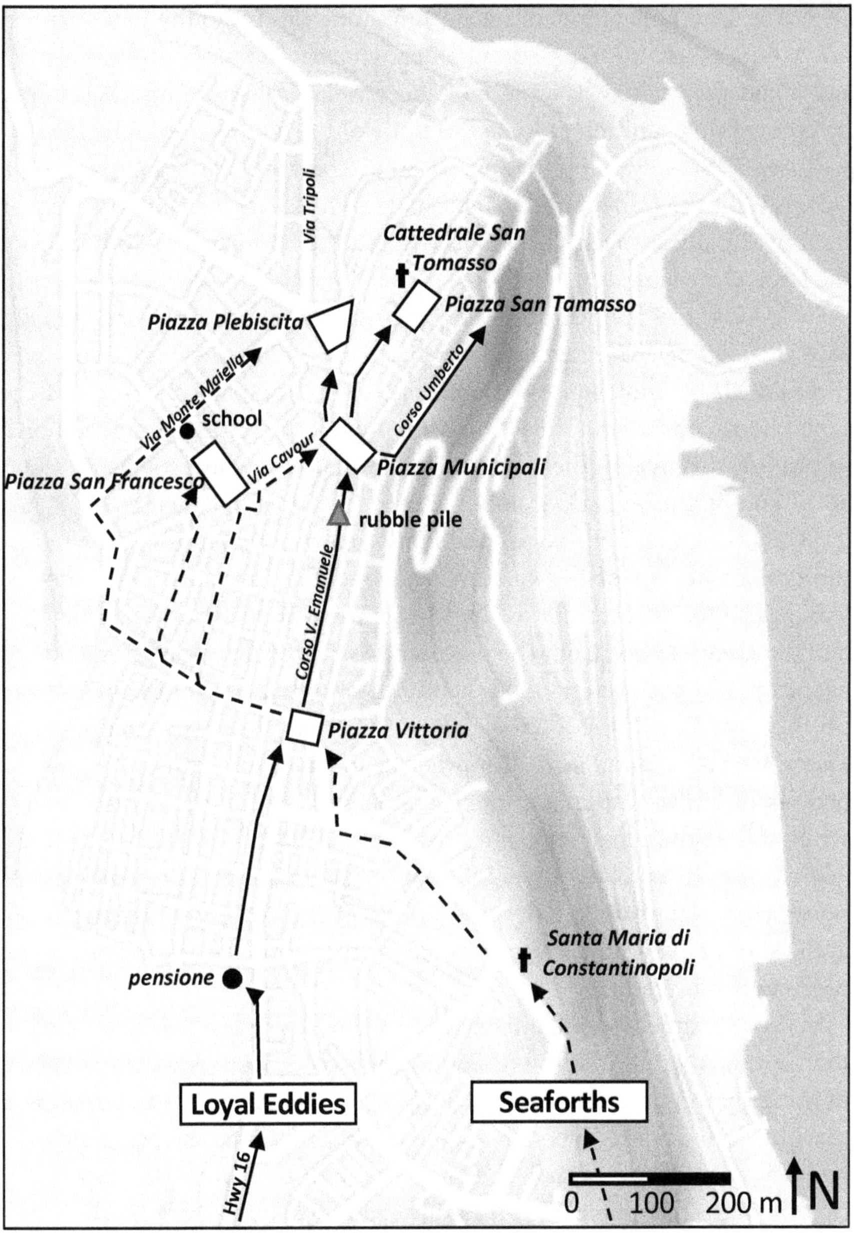

Figure 7.3. Map of the Battle of Ortona, December 1943.
Source: Map created by Liam Collins and Jayson Geroux. The routes of the Canadian advance on the map comes from map 4 of Zuehlke, *Ortona*, xx.

The lead tank stopped, so Big Jim Stone ran up to the tank, banged on it, and asked the tank commander why he had stopped. The tanker explained that there was a piece of sheet metal in front of him that might be concealing a landmine, so he did not want to continue until he knew it was clear.[25] Stone was pretty angry at that point and started cussing up a storm. With the Canadian advance stalled, the Germans started firing on the Loyal Eddies, so the Canadians entered the buildings on both sides of the street near the *piazza*.

Meanwhile the Seaforths had moved out of the cathedral and up to *Piazza Vittoria*. The Loyal Eddies had had the tougher fight, so Hoffy directed Major Thomson to provide one of his companies to support the Loyal Eddies' advance. Thus, Thomson moved the company moved north to protect the Loyal Eddies left flank. And that was where the battle stood at the end of the day on December 22, the third day of the battle.

That evening both Vokes and Hoffy realized that Ortona was in fact the eastern anchor of the Gustav Line, and the Germans would not be leaving. Hoffy realized that he had to commit more resources to the fight and directed Thomson, who he promoted to lieutenant colonel,[26] to take his Seaforths and pass behind the Loyal Eddies so that he could fight his way up the west side of the town to *Piazza San Francesco* before continuing to *Piazza Municipali* and *Piazza Plebiscita*. He attached a tank troop, engineers, and six pounder anti-tank guns to the Seaforths, to support their attack, while still holding the Patricias in reserve.[27]

Meanwhile, Vokes ordered the 1st Canadian Infantry Brigade to move north, west of the ravine, to bypass and isolate the town from the north. So, at the start of December 23, the Canadians went from two battalions in the battle to two brigades.

By this time, the Canadians had reached the main defensive area of the Germans within the town. The Loyal Eddies were engaged in heavy fighting in-and-around *Piazza Municipali*. The Germans had at least five machine gun posts, snipers, marksmen, and other infantrymen inside the buildings around the *piazza*. The tanks and anti-tank guns started the day clearing the large rubble pile that Stone's tanks encountered when they had

[25] *Fighting in Built Up Areas 1939-1945, Part 2, Ortona* (Britain: Ministry of Defence, 1991) minutes 12:14 to 12:56; Zuehlke, *Ortona*, 263-265.
[26] Seaforth Highlanders, *Active Service Force War Diaries*.
[27] Zuehlke, *Ortona*, 266; Douglas E. Delaney, *The Soldiers' General: Bert Hoffmeister at War* (Vancouver: UBC Press, 2005), 101-2.

reached the *piazza*.[28] Then the infantry, engineers, tanks, and anti-tank guns went to work to destroy German positions in-and-around the *piazza*.

While fighting to clear the buildings surrounding and north of the *piazza*, Captain Bill Longhurst, the commander of the Loyal Eddies' A Company, made a smart tactical move. Until that point of the battle, the Canadians had always entered buildings using a door on the ground floor and then fought their way up when clearing each building. That, of course, gave the Germans the advantage and they built their defenses inside the building with this in mind. Unsurprisingly, the Canadians suffered many casualties as a result.[29]

After taking a building at *Piazza Municipali*, Longhurst called up his engineers and infantry pioneers to blow a hole in the wall on the top floor of the building, so they could enter the neighboring building from the top and fight their way down.[30] So, the engineers and pioneers blew a hole through the wall with some quickly rigged explosives. The infantry immediately followed by throwing grenades and firing through the hole to clear the room and then entered and cleared the building from the top-down. This completely surprised the Germans.[31] Of course, that technique only worked when buildings shared common walls, which they did in Ortona's north. The technique would not have worked in the southern part of the town where there was space between the buildings.

The technique of blowing holes in walls to avoid fighting on the streets is called mouseholing.[32] Like many things in battle, it was not new, but it was new for the Canadians. British doctrine at the time also included something they called the "vertical technique." Instead of a hole through a wall, the vertical technique called for making a hole in a roof or ceiling, to surprise the enemy, and fight from the top floors down.[33] But until that point, there had been little urban fighting in Italy, so the Allied units

[28] G.R. Stevens, *A City at War: The Loyal Edmonton Regiment* (Edmonton: Edmonton Regiment Associates/Charters Publishing Company, 1964), 274; and Department of National Defence, *Unique Operations – Urban*, 6.

[29] Zuehlke, *Ortona*, 285-286; Porter, "Operations," 89.

[30] In the Canadian Army, infantry pioneers are soldiers that have been trained with basic engineer skills. Stephen J. Thorne, "Assault pioneers make a comeback in Canada," *Legion*, November 14, 2019, https://legionmagazine.com/en/2019/11/assault-pioneers-make-a-comeback-in-canada/.

[31] Zuehlke, *Ortona*, 285-286; Porter, "Operations," 89.

[32] "Mouse Holing at Ortona," Canadian Military Engineer Association, https://cmea-agmc.ca/heritage-moment/mouse-holing-ortona.

[33] Ministry of Defence, *Fighting in Built Up Areas*, minutes 21:50 to 21:54; and Dancocks, *The D-Day Dodgers*, 177.

involved in the Italy campaign had little experience with mouseholing. Longhurst simply thought up the technique on the spot. It seemed like an easy way to take away the German's advantage.

Spencer: It is the same technique that was used in the battle of Marawi nearly 74 years later.[34]

Geroux: While the Loyal Eddies were fighting through *Piazza Municipali,* Hoffy co-located three tanks from the Three Rivers Regiment and two 17-pounder anti-tank guns to support their advance. He placed the weapons 1,500 meters southeast of Ortona and had them blast away at the German defenses just northeast of the Loyal Eddies. They continued this bombardment for the remainder of the battle.

On December 24, the Seaforths passed behind the Loyal Eddies and started their advance towards the west side of the town. But the Seaforths, very much like the Loyal Eddies, met stubborn German resistance. *Piazza San Francesco* was also part of that main defensive area, so the Seaforths had a tough time breaking into the *piazza,* and even had to defend against a German counterattack. Also, as Hoffy had committed the Seaforths with its additional tanks and anti-tank guns, the Germans responded in kind by reinforcing the 2nd Battalion of the 3rd Regiment with the 2nd Battalion of the 4th Regiment on this day. So now it was two battalions fighting against two battalions in this urban fight.

Urban fighting is particularly challenging and deadly and the battle of Ortona was no different. As an example, on Christmas Eve, an infantry section (the equivalent of an infantry squad in the United States Army) from the Seaforths seized a school at *Piazza San Francesco.* Surprisingly, the Germans had put up only token resistance before withdrawing. A few minutes later it was clear why when the building imploded. The Germans had wired the building with explosives, sucked the Canadians in, and then blew the building. Only a single soldier, Private Gordon Currie-Smith, survived and it took his fellow Seaforths three days to discover him and dig him out. That is but one example that demonstrates the brutality of Ortona's urban fight.[35]

On December 25, there was no Christmas truce; instead, the fighting continued. But the quartermasters for both the Loyal Eddies and the Seaforths did their best to brighten the day with a Christmas dinner. The

---

[34] See Chapter 14, "The Battle of Marawi," in this book.

[35] Seaforth Highlanders, *Active Service Force War Diaries;* Roy, *The Seaforth Highlanders,* 268; Zuehlke, *Ortona,* 309-310; and Department of National Defence, *Unique Operations – Urban,* 6.

Loyal Eddies' quartermaster was able to provide some pork chops, some other food and, of course, the most important thing, cigarettes and beer, to the troops on the front lines. The Loyal Eddies were able to pull back, maybe 50 to 100 meters off the line, quickly eat some cold pork chops and drink a beer before returning to the fight at *Piazza Municipali.*[36]

The Seaforths celebrated their Christmas dinner with a little more style. Their quartermaster, Captain Borden Cameron, asked Lieutenant Colonel Thomson if they could do something to boost the troops' spirits after days of fighting in that urban hell and requested that they serve Christmas dinner in the *Santa Maria di Constantinopoli* church, which Thomson enthusiastically supported. So, Cameron and his quartermaster staff scoured the countryside and literally begged, borrowed, and "borrowed" a smorgasbord of soup, roast pork, vegetables, mashed potatoes, gravy, pudding, mince pies, chocolate, nuts, fresh fruit and, again the most important thing, cigarettes and beer. Thomson pulled out one company at a time to have that fantastic Christmas meal, and after a two-hour reprieve, each company returned to the fight at *Piazza San Francesco.*[37]

Spencer: That is pretty cool, but it makes me think about the Canadians lines of communication (supply routes). Clearly, they must have applied combat resources to keep those lines secure to be able to resupply those frontline troops.

Geroux: By that point in the battle, the Germans had withdrawn to the northern side of Ortona, so they were not able to interdict the Canadians' lines of communication with direct fire weapons, but they did drop artillery rounds on the southern end of the town. But after five days of urban fighting, the Canadians were very much urban veterans, so they knew to get inside buildings to avoid being killed by artillery.

December 25 was also an interesting day for the Germans because they realized that they were going to lose Ortona and implemented a deception plan to cover their withdrawal.[38] By that time, the battle was making international news. The Canadian Broadcasting Corporation's Matthew Halton had been reporting on the battle for the past five days and Ortona's "courtyard of hell" was now being called "Little Stalingrad" in reference to one of the most well-known and deadly urban battles in

[36] Loyal Edmonton Regiment, *Active Service Force War Diaries 1943/09-1944/06*, RG-24-C-3, vol. 15114, December 28, 1943, Library and Archives Canada; and Zuehlke, *Ortona*, 325.

[37] Roy, *The Seaforth Highlanders*, 268; and Seaforth Highlanders, *Active Service Force War Diaries.*

[38] Geroux, "The Urban Battle," 131-133.

history.[39] So for reasons of pride, the Germans felt that they had to hold onto the town as a result. However, as the Canadians steadily increased their use of force to seize the city, using the tanks and anti-tank guns to clear the city and engineers employing creative improvised explosive devices to destroy buildings, the Germans realized that they could no longer hold out.[40]

To deceive the Canadians into believing that the Germans planned to stay at all costs, the Germans increased their level of violence in the city. But the deception plan could have just as easily been called a destruction plan because the Germans started planting explosive devices throughout the northern end of the town to rubble the buildings into the streets, and created obstacles with anything they could find, like tram cars and other vehicles, to make it as hard as possible for the Canadians to advance.[41]

On December 26, the Loyal Eddies finally fought their way through *Piazza Municipali*. As they continued their advance to the northern end of the town, *Corso Vittorio Emanuele* (highway 16) split into three directions. *Via Cavour* split off to the west and continued to *Piazza San Francesco*; the main road, *Via Tripoli*, continued north to *Piazza Plebiscita*; and a third road, *Corso Umberto*, split to the northeast and continued to *Piazza San Tomasso* (see figure 7.3). So, the Loyal Eddies found themselves split in three directions, sending one company to each *piazza*.[42] The fighting remained extremely bitter, and the Loyal Eddies were happy that the three tanks and the two 17-pounder anti-tank guns located 1,500 meters away continued to blast away at the German positions.[43]

The Loyal Eddies, at *Piazza Plebiscita* and *Piazza San Tomasso*, continued to fight the stubborn Germans. At *Piazza Plebiscita*, the Germans had cleverly dismantled two *Pak* 38 anti-tank guns, carried them up to the second floor of a building, and then reassembled the guns. As the tanks moved into the square, the Germans destroyed three of them with the anti-tank guns. So, the Loyal Eddies had to regroup and reattack *Piazza Plebiscita* with more tanks, more infantry, more engineers, and more anti-tank guns

---

39 See, for example, Matthew Halton, "Under fire in Ortona," CBC Radio News, December 24, 1943, https://www.cbc.ca/archives/entry/attacking-ortona.

40 Geroux, "The Urban Battle," 131-133.

41 Carl Bayerlein, "Parachute Engineers in Combat, Ortona 1943: A German Perspective," in *Canadian Military History* 8, no. 4 (1999), 47-50.

42 Three Rivers Regiment, *Active Service Force War Diaries, 1943/01-31 December*, Library and Archives Canada.

43 Ibid.

to destroy the German anti-tank guns before they could continue their advance through the town.[44]

Meanwhile, on Ortona's western side, the Seaforths, now with eight tanks in support, fought their way through *Piazza San Francesco*, blasting away at the German positions within the buildings in and around the square. They then continued their advance westwards up the *Via Monte Maiella* towards the Loyal Eddies at *Piazza Plebiscita*.[45] As the battle waged on, the Canadians continued to throw more tanks, engineers, and anti-tank guns into the fight, so the level of violence continued to increase.

December 27 turned out to be the battle's last day. The Canadians intercepted a message referring to "Operation Ortona," with details about a counterattack.[46] The Canadians, falling for the German deception, prepared even more forces for the battle. With the Loyal Eddies and Seaforths exhausted, Hoffy readied the Patricias, which had been held in reserve, for the fight. He planned to have the Patricias and their supporting Three Rivers Regiment tanks conduct a forward passage of lines through the Loyal Eddies and the Seaforths to deliver the knockout blow to the Germans.[47]

At *Piazza San Tomasso* the Germans again demonstrated the bitterness of urban combat. The previous evening, the Germans had let a Loyal Eddie platoon into a building near the *Cattedrale San Tomasso*, where the platoon remained overnight. In the morning, the Germans blew the building and Lieutenant "Bunny" Allen and his platoon were buried underneath the building's rubble. Like the school at *Piazza San Francesco*, only a single soldier, Lance Corporal Roy Boyd, survived and was recovered three days later.[48]

That enraged the Canadians. Captain Longhurst, who had developed the mouseholing technique days earlier, decided to exact revenge by doing the exact thing to the Germans. So, he infiltrated into a building that was being held by the Germans and had engineers wire the building. After pulling out, the engineers detonated the charges which destroyed the

---

[44] Geroux, "The Urban Battle," 135-7.

[45] Ibid.

[46] Porter, "Operations," 90; and Headquarters 2 Canadian Infantry Brigade, *Active Service Force War Diaries, December 1943*, Library and Archives Canada; Zuehlke, *Ortona*, 345-346

[47] Headquarters 2 Canadian Infantry Brigade, *Active Service Force War Diaries.*

[48] Seaforth Highlanders, *Active Service Force War Diaries*; Roy, *The Seaforth Highlanders*, 268; Zuehlke, *Ortona*, 309-310; and Department of National Defence, *Unique Operations – Urban*, Dancocks, *The D-Day Dodgers*, 153 and 179.

building and killed the German platoon inside. Longhurst successfully applied that tactic a second time before the battle ended.[49]

Meanwhile, Lieutenant Colonel Thomson employed the same tactic on the west side of town after meeting stubborn German resistance from a mill on *Via Monte Maiella*. Some of Thomson's Seaforths infiltrated into the mill's basement and the engineers and pioneers installed a number of 4- to 25-pound explosive charges before they pulled out and detonated the charges. The blast brought down the building and killed the German soldiers inside, allowing Thomson to continue his advance.[50] By the end of the day, the Seaforths and the Loyal Eddies had linked up at *Piazza Plebiscita* in Ortona's north.[51]

On the morning of December 28, two *Ortonesi* civilians who remained in the town approached the Seaforths to tell them the Germans had withdrawn from the town. As the Patricias readied for their forward passage of lines, the Canadians sent out patrols and confirmed that the Germans had indeed withdrawn. Thus, the battle of Ortona came to an end after eight days of brutal fighting.[52]

While the battle of Ortona may have been over, the campaign was not, so the Patricias conducted their forward passage of lines and continued their advance north towards the Arielli River.[53] Meanwhile, Hoffy, anticipating that the Germans would bomb the town since there were no longer German soldiers there, brought forward some anti-air guns to protect the city.[54] His intuition proved correct as three German Focke-Wolfe fighter-bombers arrived to attack the city, but the heavy anti-aircraft fire from the guns prevented the planes from inflicting any casualties from their bombs and strafing run.[55]

I will now turn to discuss the post-battle events. In Italy, the 8th Army directed that Ortona be turned into a rest and recreation area, so

---

[49] Stevens, *A City at War*, 278; Loyal Edmonton Regiment, *Active Service Force War Diaries*; and Zuehlke, *Ortona*, 343.

[50] Seaforth Highlanders, *Active Service Force War Diaries*; Three Rivers Regiment, *Active Service Force War Diaries*; and Headquarters 2 Canadian Infantry Brigade, *Active Service Force War Diaries*.

[51] Zuehlke, *Ortona*, 347-348.

[52] Ibid., 349-50.

[53] Geroux, "The Urban Battle," 159-160.

[54] Seaforth Highlanders, *Active Service Force War Diaries*; Three Rivers Regiment, *Active Service Force War Diaries*; and Headquarters 2 Canadian Infantry Brigade, *Active Service Force War Diaries*.

[55] Headquarters 2 Canadian Infantry Brigade, *Active Service Force War Diaries*; and Loyal Edmonton Regiment, *Active Service Force War Diaries*.

the Canadians spent several weeks removing all of the improved explosive devices and booby traps that the Germans had planted throughout the town.[56] They also helped the *Ortonesi* rebuild the town. Together, the Canadians and Italians removed the rubble, de-mined the town, inspected the buildings to make sure they would not collapse, cleaned the houses, and cleaned the polluted water wells.[57]

The Canadians had bath and shower parades and film nights.[58] The Canadians provided water, food, shelter, livestock, and medical aid to the civilians.[59] The military contracted a number of services from the local community, thereby helping to jumpstart Ortona's economy.[60]

It was not all, of course, good. Military police officers were stationed inside the town to ensure that no looting would occur, but, unfortunately, some Canadian soldiers did "find" souvenirs that they mailed home. And of course, when it comes to soldiers on rest and relaxation after stressful fighting, there was some drunkenness, but the military police kept a grip on it.[61]

Over the next several months, from January to April of 1944, the Canadians, even though they had destroyed almost all of the town in their fight against the Germans, redeemed themselves to the *Ortonesi* by rebuilding the town. Eventually, they established very strong, friendly relationships with the *Ortonesi*, to the point where the *Ortonesi* were inviting Canadian soldiers in for home-cooked meals, giving haircuts, and doing all kinds of nice things like that.[62]

In conclusion, the Germans certainly had the upper hand at the start of the battle. The city was defended by the German 1st Parachute Division, a division full of professional, determined warriors. They had weeks to prepare the town's defense, and had the advantage of the town's physical geography, which forced the Canadians to fight up the plateau from a single direction. The houses acted as natural fortified strongpoints and the urban geography within the city was very challenging, which made it a defender's dream. The Germans were able to use a railway tunnel at the north end of the town to house their reserve and protect it from artillery

56 Cavasin, *Hai visto i Canadesi?* 43; and Roy, *The Seaforth Highlanders*, 272.
57 Stevens, *A City at War*, 282; and Cavasin, *Hai visto i Canadesi?* 53-4 and 60-5.
58 Stevens, *A City at War*, 282
59 Cavasin, *Hai visto i Canadesi?* 82-85.
60 Ibid., 109-113.
61 Ibid., 72-82.
62 Cavasin, *Hai visto i Canadesi?* 98-9.

fire. They also had an element of surprise since the Canadians expected only token resistance since they believed the Gustav Line was located to the north of the town. So, the Germans had many advantages in this particular battle.

Yet, the Canadians were able to prevail. One reason was soldier improvisation. The Royal Canadian Artillery used the anti-tank guns to blow away the rubble piles and to destroy German positions around the squares. Likewise, the engineers developed a number of cunning techniques, such as rigging up Teller anti-tank mines that the Germans had left behind with simple fuses to blow mouseholes in walls. They also built 30-pound bombs using Nobel 808 plastic explosive to level buildings.[63] They used their tanks' armor-piercing rounds to punch holes into walls protecting German machine gun positions and then followed with high-explosive rounds to kill the Germans inside them.[64]

They employed excellent combined arms techniques with the infantry, tanks, engineers, and artillery all working together. This was not just an infantry fight. The infantry and engineers often led the tanks and protected them by clearing buildings to their flanks, but the tanks also protected the infantry and engineers by destroying barricaded German positions. The engineers, co-located with the infantry, used their explosives to destroy buildings and create mouseholes, which allowed the infantry to seize the advantage in the close tactical fight. Finally, the artillery's anti-tank guns were extremely effective at leveling rubble piles and blasting away barricaded German positions around the squares. The mouseholing technique allowed the Canadians to clear buildings from the top-down.

And finally, the Canadians prevailed through sheer determination and pure and utter rage. The Canadians fought in an "urban hell" and took many casualties, but rather than getting deflated, they became more and more angry and used more and more violence to destroy the Germans throughout the town. So, the Canadians' determination, rage, and overwhelming violence allowed them to achieve victory in this battle.

As a result of the battle, the Edmontons suffered 109 members wounded and 63 killed; the Seaforths suffered 62 wounded and 41 killed; and the Three Rivers suffered 20 wounded and 4 killed. German numbers are challenging to verify as the paratrooper division was reported to

[63] A.J. Kerry and W.A. McDill, *The History of The Corps of Royal Canadian Engineers* (Ottawa: Thorn Press, 1966), 170.
[64] Windsor et al., *Loyal Gunners*, 310; and Stevens, *A City at War*, 276.

have lost 68 killed, 159 wounded, 205 missing, and 23 sick for a total of 455 casualties, but the Canadians counted over 100 unburied German dead in the buildings and streets, confirming that many of the missing Germans were killed.[65]

You could say that Ortona was a small urban battle, there certainly were larger urban battles during World War II. Stalingrad involved entire divisions if not armies and resulted in an estimated two million casualties. The American Battle of Aachen, in October 1944, involved two battalions from the United States' 1st Infantry Division. The Battle of Manila and the Battle of Berlin also involved many more troops than Ortona. Yet Ortona was no less horrific for the Canadians and Germans that were there and the lessons that can be learned from Ortona are similar to those that can be learned from studying any of these other urban battles. In fact, after the battle, some of Hoffy's soldiers toured Allied lines to share the lessons they learned in Ortona so that other troops would not have to learn these same lessons under fire.

Spencer: I find this battle amazing. I often hear people argue that you should never fight in a city; instead, you should avoid or bypass it. But Ortona shows this is not always possible. Ortona was the eastern anchor of the Gustav Line, so it was strategically and operationally important for the Germans to hold the city and hold the line, so the Canadians had no choice but to enter the city and conduct a brutal, urban fight. And when examining the tactics, there are so many similarities with other battles, such as the battle of Marawi 74 years later where we see house-borne improvised explosive devices, mouseholing, and the creative use of high explosive concrete-penetrating weapons. It really is amazing that an urban battle from World War II has so many lessons that remain applicable today.

[65] Geroux, "The Urban Battle," 169-170.

8

The Battle of Mogadishu

In January 1991, President Siad Barre fled Somalia, ending his decades-long and tumultuous reign. The country immediately lapsed into clan warfare and the situation only worsened as a drought brought famine to hundreds of thousands of the nation's poor.[1] With the worsening plight, the United Nations approved Resolution 751 on April 24, 1992, which authorized humanitarian operations and established the United Nations Operations in Somalia (UNISOM).[2] Shortly thereafter, a small group of peacekeepers deployed in an attempt to stem the worsening crisis.

UNISOM provided some limited relief, but many of the humanitarian supplies were hijacked by clan militias.[3] In response to the deteriorating situation, the United States launched Operation Provide Relief on August 14, 1992.[4] The United States supported the relief effort by airlifting supplies from Kenya to remote Somalian airfields to bypass the congested ports and clan armies.[5] When this limited support still failed to stop the starvation, the United Nations passed Security Council Resolution 794 on December 3, 1992, and President George H.W. Bush announced Operation Restore Hope on December 4, 1992.[6]

For Operation Restore Hope, the United States deployed a sizable military formation under the direction of a unified task force. The task

---

[1] Center for Military History, *The United States Army in Somalia: 1992-1994* (Washington, DC: U.S. Army Center of Military History, no date), 6, https://history.army.mil/brochures/somalia/somalia.htm.

[2] United Nations Security Council Resolution 751, April 24, 1992, http://unscr.com/en/resolutions/751.

[3] Center for Military History, *The U.S. Army in Somalia*, 8.

[4] Walter S. Poole, *The Effort to Save Somalia: August 1992 – March 1994* (Washington, DC: Joint Chiefs of Staff, 2005), 9, https://www.jcs.mil/Portals/36/Documents/History/Monographs/Somalia.pdf.

[5] Center for Military History, *The U.S. Army in Somalia*, 8.

[6] United Nations Security Council Resolution 794, December 3, 1992, http://unscr.com/en/resolutions/794; and Poole, *The Effort to Save Somalia*, 23.

force included 16,800 Marines from the I Marine Expeditionary Force; 10,200 soldiers from the army's 10th Mountain Division; 1,800 marines from the offshore Marine Expeditionary Unit; special forces, civil affairs, and psychological operations forces; and coalition forces from France, Italy, Belgium, Morocco, Australia, Pakistan, Malaysia, and Canada. In total, 38,000 soldiers from 23 different countries supported Operation Restore Hope.[7]

Within Somalia, clan leader General Muhammed Farah Aidid, who headed the Somalia National Alliance, feuded with other clan leaders for control of Mogadishu and the country. With the deployment of troops and the declaration of Restore Hope, United States Ambassador Robert Oakley was able to broker a ceasefire with the clans, allowing more than 40,000 tons of grain to be offloaded and distributed in December alone. The mission proceeded fairly well with few incidents of violence through March.[8]

Yet on March 26, 1993, the United Nations passed Security Council Resolution 814 which broadened its mandate considerably from a Chapter VI peacekeeping role to a Chapter VII peace enforcement role.[9] In response, Aidid's Somalia National Alliance forces ambushed Pakistani soldiers, killing 24 and wounding another 44.[10] This prompted the United Nations to authorize Security Council Resolution 837, allowing for more aggressive actions toward Aidid and asking member states to contribute additional troops and equipment.[11] On June 7, UNISOM issued an arrest warrant for Aidid. As tensions between Aidid's forces and UNISOM continued to escalate, Secretary of Defense Les Aspin authorized the deployment of Task Force Ranger on August 23 to capture Aidid and his lieutenants as part of Operation Gothic Serpent.[12]

Task Force Ranger consisted of special operations ground forces, special operations helicopters, U.S. Air Force special tactics personnel, and Navy SEALs. In August and September, the task force launched six missions into Mogadishu to capture Aidid associates.[13] On October 3, Task

[7] Ibid., 24; and Center for Military History, *The U.S. Army in Somalia*, 10.

[8] Center for Military History, *The U.S. Army in Somalia*, 9-13.

[9] United Nations Security Council Resolution 814, March 26, 1993, http://unscr.com/en/resolutions/814.

[10] Center for Military History, *The U.S. Army in Somalia*, 16.

[11] United Nations Security Council Resolution 814, June 6, 1993, http://unscr.com/en/resolutions/837.

[12] Center for Military History, *The U.S. Army in Somalia*, 17.

[13] Ibid., 18.

Force Ranger launched its seventh mission, which is more colloquially known as *Black Hawk Down*, referring to the book by Mark Bowden.[14]

The mission was a tactical success—the task force captured two of Aidid's senior lieutenants.[15] But when a Black Hawk helicopter was shot down, the mission quickly changed to recover the bodies of the two dead pilots trapped in the wreckage and what should have been a 20-minute mission lasted nearly 15 hours. In the end, Task Force Ranger lost 16 soldiers and another 57 wounded, the 10th Mountain Division lost two and 22 wounded, the Malaysians lost two and seven wounded, and the Pakistanis suffered two wounded. By contrast, it is estimated that Somalis suffered between 500 and 1,500 casualties.[16]

In this chapter, Liam Collins interviews three veterans of the battle of Mogadishu. The panel was conducted at the United States Military Academy at West Point on October 3, 2018, 25 years to the day after the battle. Joining him were Colonel (retired) Lee Van Arsdale, who was a lieutenant colonel within the Task Force headquarters during the operation; Sergeant Major (retired) Kyle Lamb who was part of the assault force; and Colonel (retired) Larry Perino who was a platoon leader with the 75th Ranger Regiment.

All three were awarded Silver Stars for their actions in Mogadishu on October 3-4.[17] Van Arsdale was also awarded the Purple Heart for injuries he sustained during the operation. The interview was conducted in front of a cadet audience, so the interviewees often spoke directly to the audience, addressing them as "you," and some of the audience laughter is indicated in the text as "[laughter]" to provide context to the interviewees' comments.

Liam Collins: To put the ground operation in context, I think it is useful to start with you, Lee, since you were in the headquarters. What was the Task Force Ranger mission?

Lee Van Arsdale: It was a very simple mission: kill or capture Muhammed Farah Aidid, a clan leader and former general in the Somali Army. Why did we want to do this? The Somali central government had

[14] Mark Bowden, *Black Hawk Down: A Story of Modern War* (New York: Penguin Books, 1999).
[15] Todd South, "The Battle of Mogadishu 25 years later: How the fateful fight changed combat operations," *Military Times*, October 2, 2018, https://www.armytimes.com/news/your-army/2018/10/02/the-battle-of-mogadishu-25-years-later-how-the-fateful-fight-changed-combat-operations/.
[16] Center for Military History, *The U.S. Army in Somalia*, 23.
[17] "U.S. Army Silver Star Recipients," Department of Defense, https://valor.defense.gov/Recipients/Army-Silver-Star/.

recently collapsed, so they were in a state of civil war. It was a clan-based society with each clan possessing its own militia. Aidid's militia had massacred a group of Pakistani peacekeepers in June 1993, and it fell to the United States to do something about this.[18] So, we activated Task Force Ranger in August to kill or capture the leader of this militia.[19] The elements that made up Task Force Ranger—the Rangers, an elite special operations unit, and the 160th Special Operations Aviation Regiment—had been training and rehearsing for the mission following the June attack on the Pakistanis, so we were ready.

Collins: Larry, you were a platoon leader in the Ranger Regiment, and Kyle, you were part of the elite special operations unit. You were from two different units, but you trained together for the mission. Can you describe that a little?

Larry Perino: Within the Rangers, we had no idea that deploying to Somalia was even a thought. We were in the middle of a fairly routine large joint readiness exercise at Fort Bliss, Texas in June. In the middle of training, several days into this large training exercise, we were pulled aside and told nothing more than to pack our bags. The next day, we loaded onto a C-5 Galaxy cargo plane, and we found ourselves at Fort Bragg, North Carolina a few hours later. At that time, we were finally told that we were preparing for a possible deployment to Somalia. We knew Somalia was in the news, but never thought we might actually deploy there. Things moved pretty quickly, but also very smoothly. We spent the next couple weeks training with Kyle and his special operations unit.

Kyle Lamb: My unit always maintained a high readiness standard, so I felt we were already trained and ready to go, so once the Rangers arrived, we saw it as our job to take the Rangers under our wing and show them some of the tricks to the trade because going into an urban environment, kicking in doors, rolling up the right bad guys, shooting other bad guys that need to be shot, and not shooting civilians who should not be shot, was our specialty. The Rangers were experts at seizing airfields, and they could teach us something there, but when it came to the urban environment, that was our specialty.

The Rangers and my unit had a great working relationship. The team that I was on, A Team, worked closely with a Ranger platoon led

[18] "26 U.N. Troops Reported Dead in Somalia Combat," *New York Times*, June 6, 1993, https://www.nytimes.com/1993/06/06/world/26-un-troops-reported-dead-in-somalia-combat.html.
[19] Center for Military History, *The U.S. Army in Somalia*, 10-18.

by Lieutenant Tom DiTomasso, and we got them spun up for the mission. After arriving in Mogadishu, we continued to train when we could. We did not stop training just because we were deployed to combat. We continued to do physical training and firearms training, and we took the Rangers to train with us. And since my unit had no privates, we had special operations staff sergeants and sergeants first class pulling kitchen patrol duty with Ranger privates, so that was kind of fun.

Collins: You arrived in Somalia in August 1993 and conducted a few missions prior to the mission on October 3. Can you describe how those missions went?

Lamb: We conducted six missions before the mission on October 3. The very first mission was kind of a "let's go spank something" because the Somalis hit us with mortars pretty heavily one night. Major General William Garrison, the task force commander, said, "We have a target, and we are going to go hit this target." That was my first real combat raid. I had been with the 5th Special Forces Group during Operation Desert Storm, but that was not "storming the castle" type of stuff. So, you could not have driven a stickpin up my ass with a sledgehammer on that first mission we went out to. [laughter]

The cadets have never heard that saying before, I guess. [laughter]

Van Arsdale: Why are they laughing?

Lamb: You guys will see what I am talking about! The first time you get called out to do a mission and you are looking at the face of the enemy, you will realize that it is not the same as being on a rifle range. You are going to have feelings or emotions that you just have to fight through to continue with your mission. You cannot stop. If you are scared, worried, hungry, whatever, you just cannot stop. You have to continue the mission, and our mission was a little nerve-wracking.

On my first mission, I kicked in a door after "shotgunning" it.[20] The shotgun rounds failed to breach the locking mechanism, so when I attempted to kick the door in, instead of the door flying open, I kicked so hard that I found myself lodged in the door all the way up to my crotch. The next several missions went better because I started to understand the flow of combat and I was not as scared. By the mission on October 3, we were all pretty comfortable since we had already performed six similar missions.

[20] "Shotgunning" a door refers to using a shotgun to breach a door.

Perino: I would agree, but I would also add that we were not overconfident. I think we were confident in our ability to handle anything we might face. Or maybe we were just blissfully ignorant of what was to come. When you find yourself in the heat of combat for the first time, I would just say that thing with the stickpin, you know, it is almost like running in sand. And as a young leader you are going to have to will yourself to keep pushing forward. And that stress or fear hits people differently.

Lamb: In my opinion it is probably tougher if you are just worried about yourself as opposed to when you are leading and you have to worry about your subordinates. So, for you future leaders, you are lucky because when you experience combat for the first time, hopefully you will be thinking more about the soldiers that you are leading than yourself. But as a special operations "private," a staff sergeant, I had nothing to worry about except keeping my gun zeroed, doing physical training, and being technically and tactically proficient. Lee and Larry had to worry about that bigger picture, and I think you need to keep in mind that if you ever find yourself in combat, do not worry about yourself.

The mission is more important than you and if you do not like that, then you are in the wrong business. It is about doing a mission for the United States of America—I am going to start singing the national anthem here in a minute—you have to have that perspective every time you go out. You have young men and women that you are leading, and they are looking to you for guidance and direction in the combat situation. You want to be someone that they look to and say, "I respect Larry Perino because we spent the night of October 3 in Mogadishu together and he never lost his composure." It was not very romantic, but we spent the night out there in the battlefield together, and he was very calm in that crisis and that helped to keep all of his guys calm as well.

Collins: Going into October 3, after doing six previous operations, did you feel like you were short of any assets or was there kind of a consensus, "We have what we need over here"?

Perino: We had all we were allowed to have if that answers your question.

Van Arsdale: When we did our pre-mission training from June through August, we always had AC-130 gunships[21] as part of our package. We just assumed that they would deploy with us, but when we arrived in

[21] For more on AC-130s, see "AC-130H/U Gunship," Military.com, https://www.military.com/equipment/ac-130h-u-gunship.

Somalia, they were not there. To this day I do not know who decided that they would not be part of the package. That was kind of a major "aw crap" moment. We were expecting to have a weapons platform with its vision and fires up in the sky above, and when we arrived it was not there.

Perino: And, Lee, during training, we also had an additional Ranger platoon, but we lacked that in Somalia, so we lost a reserve that we thought we would have.

Van Arsdale: Yeah.

Perino: So, we lacked a very large reserve and a key fire support platform. And that had huge impacts when the, you know, "oh crap" happened.

Collins: When you lost the Ranger platoon, that reserve, what was the plan then? Who pulled the reserve or served as the quick reaction force?

Lamb: The 10th Mountain Division had to provide the quick reaction force, but they were not organic to our task force, and they were not even collocated with us. So, we had to rely on an external element that we had never trained or worked with to serve as our quick reaction force. Keep in mind, a quick reaction force is required when things have taken a turn for the worse, so it is not the time to bring a new force into the fight as we had to do on October 3. But our commander, General Garrison—who was one of the greatest generals I have ever known—looked at what we had and said we had what we needed to conduct the mission. So once our boss said that, that was it, we were going to do it.

We would have loved to have had those other assets, but you rarely get everything you want. There are certain nights out on the battlefield when you are not going to get everything you want because some other target has higher priority, or as was the case in Somali, it might be a political decision that limits what you are able to bring into the country. It may just be for one night, but there will be nights like that. Unfortunately for us, we lacked those critical assets for the entire deployment, but we had to deal with it.

Collins: I rarely had everything I wanted in Afghanistan and Iraq. I often had to rely on conventional units that I had never worked with to provide my quick reaction force, so that is probably more the norm than the exception. Returning to October 3, you received intelligence that caused you to launch the operation. Can you describe, from the headquarters or operator perspective, the plan for the mission.

Van Arsdale: Many mistakenly believe that our mission was to capture Aidid, but our mission that day was to capture two of Aidid's key lieutenants. Since our arrival in July, we had never known Aidid's location, so we decided to take out his infrastructure. We captured his key financier, and we took out a radio station that he used to broadcast propaganda and send instructions to his troops. After six weeks we finally had what we considered to be actionable intelligence on two of his key lieutenants, so of course we were going to launch. Yet many journalists wrote that we failed the mission because we did not capture Aidid, but that was never the mission that day.[22]

Collins: Can you describe the infiltration plan?

Lamb: Let me first walk you through how the alert worked for us. So, when we got notified that we were about to launch for a mission we were out running or—

Perino: We were playing volleyball.

Lamb: Volleyball?! Who the heck plays volleyball!

Collins: *Top Gun*! We have all seen it.

Lamb: Yeah, *Top Gun*, you know, whatever. The radio call came in that we had a mission, so we all ran back. I ran back and got my gear on and headed out to the aircraft to get the fast ropes rigged up and everything ready to go while the leaders ran to the tactical operations center to receive a quick mission brief. The co-pilots started the aircraft while the crew chiefs loaded the miniguns. The team leaders and the pilots arrived with the plan that had been printed out on paper by a mimeograph machine.[23] Does anybody know what a mimeograph machine is?

Van Arsdale: Any history majors here?

Lamb: Because you are like 400 years old! Maybe not quite that old, but you know what I mean. Did you serve with General Robert E. Lee when he was here? Sorry, I have to give you a hard time. A mimeograph machine was the predecessor to the Xerox copier, and it would give you these really coarse, purple copies from a hand-cranked machine. Did that one crank?

Van Arsdale: Yes, it was a hand-crank machine.

---

[22] See, for example, Michael Gordon and John Cushman Jr., "Mission in Somalia; After Supporting Hunt for Aidid, U.S. is Blaming U.N. for Loses," *New York Times*, October 18, 1993, https://www.nytimes.com/1993/10/18/world/mission-in-somalia-after-supporting-hunt-for-aidid-us-is-blaming-un-for-losses.html.

[23] A mimeograph "is a duplicating machine that uses a stencil consisting of a coated fiber sheet through which ink is pressed." See, "mimeograph," *Britannica*, https://www.britannica.com/technology/mimeograph.

Lamb: So, you would hand-crank the machine and these purple print copies would pop out with our mission orders; that is what we had. But we really did not need it. We knew that we would infiltrate on a helicopter, land or slide down a fast rope, breach a gate, and do our thing. So, my team leader showed up and put on his headset, which was like a suction cup hearing piece, and said that when we get on the ground there will be a big gate and I would have to breach it. Everyone on the team carried breaching charges, but I was the team breacher, so I had some extra-large demolition charges. That was really the extent of our planning. How long did that take? I do not recall clearly, but I think it was maybe ten or fifteen minutes.

Perino: We did not need a lot of time to plan because we were so well-trained for this type of mission. Regardless, it was a time sensitive operation so taking additional time to plan actually made the mission more likely to fail by allowing the high value targets time to flee. We had just received the intelligence, so we had to act on it very quickly. In training we had built a series of mission templates based on potential scenarios. The target might be in a single-story or multi-story structure or in a vehicle or convoy moving on the ground.

We built generic templates for the assault force and the Ranger security force and then tailored them for each situation. In training, we rehearsed them over and over again. There was no Google Maps in 1993 that we could use for planning. Typically, we had some really awful overhead photos that we used to draw up the plan—essentially modifying the template based on the terrain in the photo. It was a lot like drawing up a football play, except we were using a piece of blotchy paper to circle the target and draw little stick figure helicopters for each "chalk." So, the sketch would include the infiltration of the assault force on their MH-6 Little Bird helicopters[24] and the infiltration location and blocking position for each of the Ranger chalks on their Black Hawk helicopters.

Van Arsdale: We had different levels of planning. When Kyle says we made the plan in 10-15 minutes, it was built onto an existing plan. As Larry said, we had created the templates, had trained on each dozens or hundreds of times, and because, as Larry said, time is of the essence, we had to launch shortly after receiving the intelligence. We would take the template and modify it for the mission: A Team you are in this MH-6 Little Bird and your primary breach point is right here, so you will land here, and

[24] Little Bird helicopters are small MH-6 McDonnell Douglas helicopters used by the U.S. Army's 160th Special Operations Aviation Regiment.

your alternate breach point is here. So that is what we drew up on butcher paper and then transferred by hand to a piece of typing paper so we could run it through the mimeograph machine.

As the team breacher, all Kyle really needed to know was where his primary and alternate breach points were located and any intelligence on them; for example, there is a big gate here, or a door there, so he will need some breaching charges, his shotgun, or whatever. And then we took a step back from the target to figure out where the Rangers needed to go to isolate the target. Each Black Hawk helicopter had a dozen or so Rangers to provide the external security, so we just drew it up and figured out the infiltration location for each aircraft. That was how planning went. It took less than five minutes because most of the planning had already been done. All we had to do was modify our existing template based on the specific target and then launch the force.

Collins: Yes, if you are highly trained, it is just a battle drill. Before turning to the assault, can you describe what Mogadishu was like? How wide were the streets? How big were the buildings?

Van Arsdale: Remember, there was a civil war, and the population of the city grew as a result of the war. In 1993, the metro area of Somalia was estimated to have more than one million people.[25] Many of the internally displaced persons that came to the city built and lived in lean-to shelters. We had overhead photos that had been taken before the crisis that showed a pretty normal grid pattern for the city, but when we were there, nearly every building had a lean-to built on it. Many of the lean-tos had lean-tos built on them, so many streets were too narrow for vehicles to drive down. As a result, most of the traffic was foot traffic or donkey carts.

The photos looked nothing like the city as it existed for us. There was no electrical grid in Mogadishu at the time, so we did not have to worry about flying into electrical lines because there were none. There were open cooking fires which produced a continuous haze over the city, open sewage, livestock everywhere, and people walking around with nothing to do but go out and walk around. And many of those people were well armed because there was a raging civil war. It was almost a medieval setting there. So, it was a very difficult environment.

[25] In 1993, the population was estimated to be 1,126,000. See, "Mogadishu, Somalia Metro Area Population 1950-2022," Macrotrends, https://www.macrotrends.net/cities/22477/mogadishu/population.

Perino: Also, because it was a clannish society with a civil war, the city was segmented. We quickly learned which clan controlled which area and where the dividing lines were. In some areas, residents would wave at us; in others, they would flip us the middle finger; and in others, if we crossed into them, we knew that we were likely to find ourselves in a gunfight. I can imagine it was somewhat like being in parts of Los Angeles in the 1970s and 1980s with the Bloods and Crips.[26] There were streets that served as dividing lines between clans and people knew not to cross the line.

Lamb: And it was hot. It was freaking hot with temperatures over one hundred degrees.[27] We took out some of the ceramic plates from our body armor because it was so hot. Often, we had to climb over walls which was also a lot easier to do without all the plates in our vests. But I put the plate back in after October 3 and I never went into combat again without all my plates ever again. I even started wearing a back plate once we had the ability to do that. Being miserably hot was not just bad for the troops on the ground, it also decreased the lift capacity of our helicopters. There were days that the Little Birds could not lift what they needed to because the air got too thin from the heat or however that works. Some of you rockets scientists will have to—

Collins: The cadets have had physics.

Lamb: One of you will have to explain it to me. I just do what I am told…rarely.

Collins: Apart from the lean-tos, what were the buildings like in Mogadishu? Were they mostly one- or two-story? How were they constructed?

Van Arsdale: I would describe it as almost adobe or mud construction. They were not particularly strong. We soon discovered that if you stood behind a wall to seek protection from incoming rounds, it would not take long for the wall to disintegrate. Nearly all the buildings were one or two-story, but there were a few four- and five-story structures, like the Olympic Hotel, that we used for landmarks.

Collins: Returning to October 3, can one of you discuss the infiltration and the initial portion of the operation?

---

[26] See, for example, Al Valdez, "Tracing the Roots of Black Gang Rivalry," *Police Magazine*, August 1, 1996, https://www.policemag.com/338634/tracing-the-roots-of-black-gang-rivalry.

[27] Roger N. Sangvic, "Battle of Mogadishu: Anatomy of a Failure" (Monograph, School of Advanced Military Studies, 1999), 38, https://apps.dtic.mil/sti/pdfs/ADA366316.pdf.

Perino: Everyone's infiltration was a little different, based on buildings and the street at their infiltration point. The isolation force infiltrated using fast ropes because there were no viable landing zones for the Black Hawk helicopters in Mogadishu. Landing a helicopter is generally preferable to hovering and sliding down a fast rope to the ground. It is quicker, meaning the enemy has less time to engage a hovering aircraft, and it carries less risk, since sliding down a rope from dozens of feet above the ground is inherently risky.

We would have preferred to go in at night, but since the intelligence came during the day, we had to conduct the operation during daylight hours. The assault force and the blocking force was to go in by air. The assault force would land and immediately move to the target building, while the blocking force would fast rope into their blocking positions. Meanwhile, a convoy of Rangers with up-armored Humvees and five sandbagged five-ton trucks would move to the target area by ground to haul off the assault force and anyone that was captured on target. At least that was the plan going in (see figure 8.1).

Lamb: As I said before, my job was to breach a big gate. But the helicopter put us in at the wrong intersection because it was tough to distinguish one intersection from another and the global positioning system did not work particularly well in Mogadishu. My team fast roped in from our Little Bird helicopter. I saw a big gate, so I ran up to it and shotgunned it since that was my job. As we entered the compound, my team leader figured out that we were at the wrong building and pointed to a different one. So, we hit the new building and cleared both floors but found nothing of significance there, so we knew that we still had not found the correct building.

When we exited the building, things started to get exciting for us. We strongpointed (established a defensive position) at an intersection behind the building, and that is when things started to heat up for my team. I saw a guy run across the street with an AK-47 assault rifle, so we engaged him. Once he was down, his buddies tried to run up to him to retrieve his rifle and use it in the fight. And that was when things like "the donkey" started happening. We shot a donkey because it had six legs: the donkey's four natural legs and two extra legs belonging to a guy with a rifle that was trying to sneak up on us from behind it.

The situation around us escalated quickly. A Somali fighter tried to place a machine gun against our position. Earl Fillmore, one of our guys

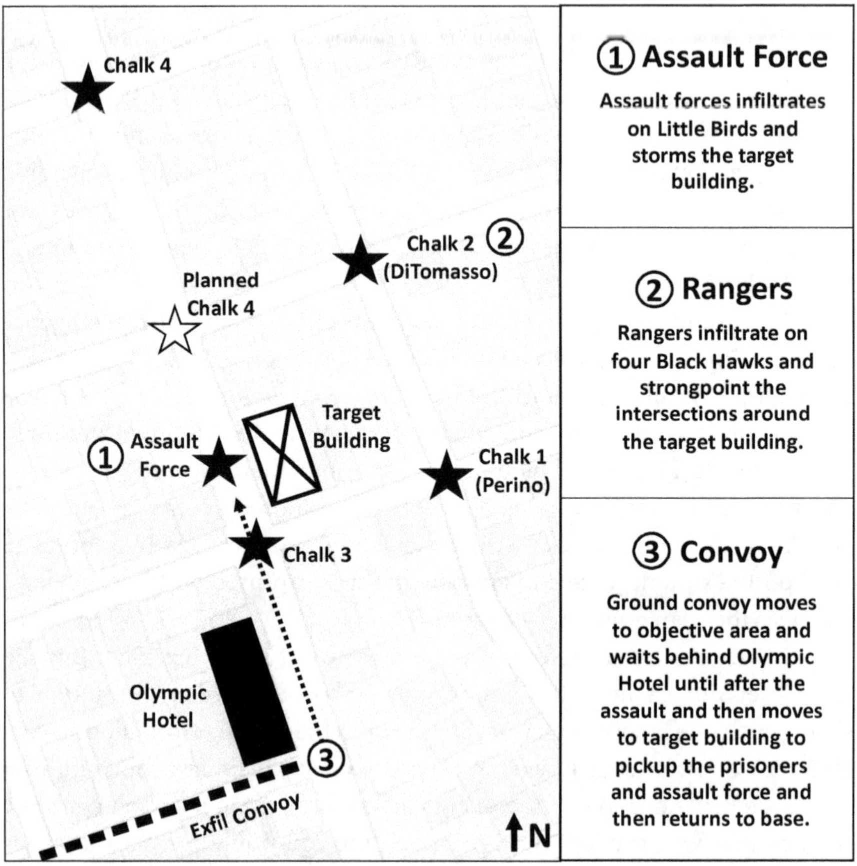

Figure 8.1. Map of the assault.
Source: Map created by Liam Collins.

who was killed later that day, stepped around the corner and killed the machine gunner using his M203 grenade launcher. At that point our team leader finally figured out where the target building was located. So, we crossed the street, climbed over four or five walls, moving along the backs of the buildings to avoid the street, before we finally reached the target building. We were not concerned at this point because we had confidence in the soldiers around us.

Perino: The four Ranger Black Hawks inserted us immediately after the Little Birds carrying the assault force departed. It was simply too dusty and the space too constrained for all of us to infiltrate at the same time. Three of the four chalks were inserted into their planned blocking positions. Immediately after hitting the ground, each chalk strongpointed an intersection to isolate the assault force from anything outside of the target area.

The fourth chalk had to be inserted a block or two away because the brownout[28] was so bad, but unlike Kyle's aircraft, the aircraft knew where it was inserting the Rangers and communicated the position to the Rangers onboard before they fast roped to the ground. It was a really high rope and that was where we suffered our first casualty.

Private Todd Blackburn fell about 70 feet when he missed the rope, so he was pretty badly wounded. They were a block or two away from where they were supposed to be, but still close enough to support the mission. We did not realize that things were starting to go sideways as a kind of Mogadishu civil defense plan was starting to be put into effect. It was not as if sirens were sounding and hundreds of Somalis were immediately running to the fight, but it was slowly building the minute we touched the ground even if we did not realize it.

Van Arsdale: I think we knew we were going into a hornets' nest, but the key was to get in and out before they woke up. And they eventually woke up. But just to take a step back and set the stage a little, the Rangers were all in Black Hawks. They are incredibly powerful birds, but they could not land on the dirt streets because all the dirt and dust would create a brownout condition, so we had to insert them using fast ropes. Additionally, as I mentioned earlier, the town was full of shacks, lean-to dwellings, and permanent structures with dubious building codes, and the roofs tended to fly off if you flew too close and you did not want anything hitting the rotors—that could take a helicopter down.

Thus, landing the Black Hawks in the city was simply not an option, and they had to hover high enough to avoid blowing the roofs off into their own rotors. The assault teams were on MH-6 Little Birds, the "sports cars" of the helicopter world. They could get in and out fairly quickly because they were smaller and produced much less rotor wash. As soon as Blackburn became injured, we had to figure out how to evacuate him. We had plans that included contingencies, branches, and sequels, but the leaders had to make decisions quickly.

[28] A helicopter brownout "occurs when downwash disturbs the dust and sand beneath the aircraft during takeoff, landing, and low altitude operations." See, "Helicopter Brownout," Naval Post Graduate School, https://nps.edu/web/rsc/brownout.

Collins: Most of you know what a fast rope is, but for those who do not, it is just a rope that hangs out the back of a helicopter that is 40 mm (1.57 inches) in diameter. Wearing gloves to prevent your hands from burning, you reach out and grab the rope with your hands, next you get your feet on the rope, and then you slide down it. The only thing keeping you on the rope are your hands and feet, and keep in mind, the Rangers had their combat load of ammunition, body armor, and other equipment, so they were carrying a lot of weight.

Van Arsdale: If you are Private Blackburn and for whatever reason you missed your handle on the rope, then you are falling 70 feet to the ground way faster than the human body is intended to.

Lamb: When my team finally arrived at the target building, the rest of the assault element had already secured the building and had captured Aidid's lieutenants. The lieutenants, along with others that had been captured, were flex cuffed and sandbagged,[29] so we were just waiting for the convoy to pull up so we could load them, get on the vehicles ourselves, and return to base. Everything was still kind of normal. The convoy pulled up and as we were loading the prisoners on the trucks, *Super Six One,* one of the Black Hawks, got shot down.

Perino: We were still engaging targets, but it was not very heavy yet. One of the Somali tactics was to push kids out into the street to point out our positions, so we had to scatter them.

Lamb: We saw one Somali, I guess he was a big John Wayne fan. He had a shirt that he had placed on a stick that he brought around a corner, and everyone's like, "Huh?" And then he stepped out. I do not know how many dudes shot him, but it was a lot.

Van Arsdale: The total time that had elapsed from the infiltration of the assault force to when the lieutenants had been captured and the assault force was ready to exfiltrate was 20 minutes. Think about that. Earlier I said that speed was of the essence. During that time, the assault force had to go into a walled compound with multiple, multi-story buildings. They had to clear every building, because each of the two lieutenants had about a dozen bodyguards whose sole focus in life was to eliminate any threat to their principal. So, there were at least two dozen well-armed guys in there who could not wait to kill the assault force, and the assault force had to enter, not knowing where anyone was, capture the two lieutenants, and get out

[29] Flex cuffs are plastic handcuffs, and sandbagging refers to placing a sandbag over someone's head as a method to better control them.

before the rest of the militia, estimated to be around 3,000, could respond. It was a textbook operation for that 20 minutes, and we were in the process of exfiltrating when *Super Six One* got shot down.

Collins: I have read about the battle and remember reading about Sergeant Jeff Struecker breaking off from the rest of the convoy and leading a three-vehicle casualty evacuation convoy with Blackburn back to base.[30] Was that before or after *Super Six One* got shot down?

Perino: I could not see because I was on the other side of the block. All I knew was that they were treating him, and I think they were moving a vehicle to pick him up. I think that is probably what happened. They knew they had to get him back.

Van Arsdale: Yes, the medic at the site said that they had to get him out immediately or he would not make it. So, they decided to break off three vehicles from the larger convoy to get him back while the rest of the convoy remained at the objective to exfiltrate the force. The three vehicles had a heck of a fight getting back to the airfield where our headquarters was located, but to answer your question, they departed the target area before *Super Six One* got shot down.

Perino: That was where Sergeant Dominic Pilla was fatally wounded.[31]

Van Arsdale: Yes, he was shot while the 3-vehicle convoy was fighting its way back to the airfield.

Collins: After *Super Six One* went down, what happened next? How far away was the crash site from your location?

Perino: Two to three blocks.

Lamb: The aircraft was close to us; we heard it get hit and watched it autorotate to the ground northeast of our position. There was no discussion because we all knew what to do. If an aircraft went down in the city, the contingency was to move by foot to the crash site and secure the site as quickly as possible. That was it. We did not need to have a discussion because that was the contingency that we had developed during planning, and everyone knew it. Why not move by vehicle to secure the site? Because vehicles would not fit down the alleyways; they were too narrow, especially for vehicles as wide as our Humvees. So, it was just a matter of

[30] Bowden, *Black Hawk Down*, 54-56 and 157-8.

[31] Twenty-five years later, Pilla was posthumously awarded the Bronze Star with Valor Device for his actions that day. See, Deborah M. Marko, "Vineland school dedicated to Army Ranger Sgt. Dominick Pilla," *Daily Journal*, April 6, 2019, https://www.thedailyjournal.com/story/news/2019/04/06/school-dedicated-honor-fallen-army-ranger/3388056002/.

determining the order of movement and Lieutenant Perino, they said, "You have the lead."

Perino: Actually, DiTomasso's chalk, which was north of mine, could see the crash. They were closer. It was almost a straight shot from where he was, and when the aircraft first went down, half of DiTomasso's element took it on themselves and immediately started running to the site. DiTomasso decided that he was not going to split his force, so he ran with the rest of his force to the crash site, and it was a great call because the Somalis were racing to get there as well.[32]

While they were running to the crash site, the combat search and rescue team fast roped into the crash site. As Lee mentioned, we had a number of contingencies built into our plan, and a downed helicopter was one of those contingencies. So, we had a combat search and rescue team that included special operations operators, Rangers, and a couple of U.S. Air Force pararescue specialists,[33] flying on a Black Hawk for this very contingency. They fast roped into the crash site about the same time that DiTomasso arrived with his element. At this time, I was still moving to the crash site with my element (see figure 8.2).

Lamb: We were not exactly sure where the downed aircraft was. We knew the general direction, so when the Rangers started heading north, my team, who was near them, started moving with them. That is when all hell broke loose. We started receiving heavy, heavy fire. One of the guys on my team, Earl Fillmore, was killed at that point. I did not see that happen—Larry, you saw it since my team had caught up to your element by that time. I was clearing a building with another guy from my assault team when we became separated from the rest of our assault team because Earl got shot and they were working on him. He was already dead, but they were trying to get him out of the street. Around this time, Corporal Jamie

[32] Thomas DiTomasso was awarded the Silver Star for his actions on October 3-4, see, "Thomas R. DiTomasso, Wall of Valor Project, *Military Times*, https://valor.militarytimes.com/hero/103486.

[33] "Pararescue specialists rescue and medically treat downed military personnel all over the world. These highly trained experts take part in every aspect of the mission and are skilled parachutists, scuba divers and rock climbers, and they are even arctic trained in order to access any environment to save a life when they're called to do so." See, "Pararescue," AirForce.com, https://www.airforce.com/careers/detail/pararescue.

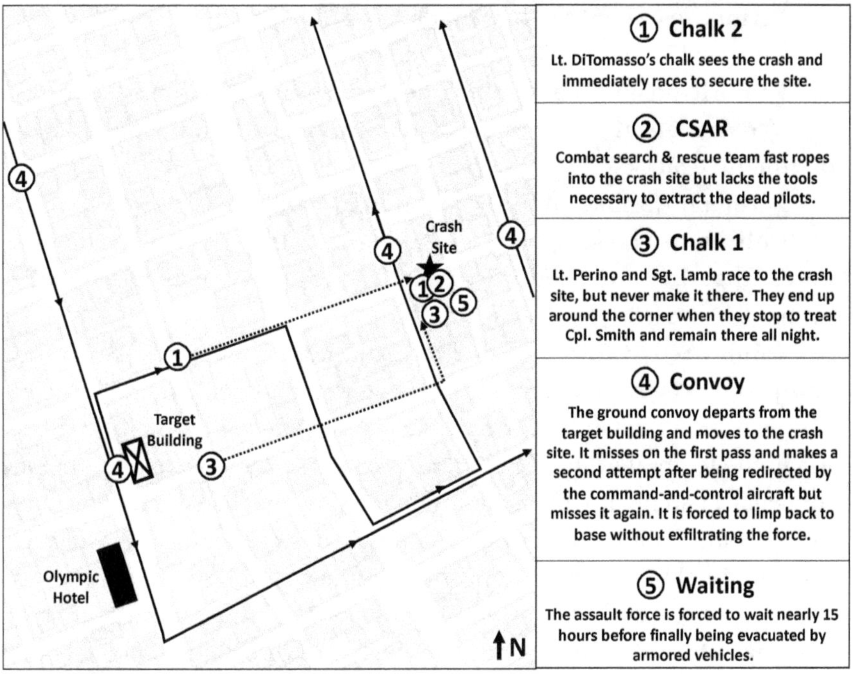

Figure 8.2. Map of the crash site.
Source: Map created by Liam Collins.

Smith was also shot.[34] I cleared a courtyard to get the element out of the street so they could treat Smith.

Perino: It was a race at that point because militia members from the Somali National Alliance were trying to beat us to the downed helicopter. Luckily, DiTomasso's element beat them. As we approached the crash site, we ran into a wall of lead from the enemy's fire. To avoid the wall of lead and to try to treat the casualties that we had already sustained, we moved into the courtyard that Kyle had cleared. We did not know it at the time, but the crash site was just around the corner from us.

Lamb: There was no smoke or fire at the crash site; nothing to indicate where the downed aircraft was. We were just around the corner from the aircraft and DiTomasso's element, maybe 30 meters, and we did not even know it.

Collins: So, you had moved about three blocks total?

[34] Jamie Smith was posthumously awarded the Bronze Star with Valor Device for his actions on October 3; see, "Cpl James E. Smith," NJ Run for the Fallen 2022, https://www.njrunforthefallen.org/cpl-james-e-smith.html.

Perino: Yes, about three or four blocks.

Collins: How long did that movement take?

Lamb: It did not take very long; despite the gunfire, we moved quickly.

Van Arsdale: It was a couple of minutes at most; I mean, how fast can you run three blocks?

Perino: It was like an "airborne shuffle"[35] with guns.

Collins: As you said, you were taking casualties as you were moving, how did that slow down the movement? Trying to move with the casualties through the street?

Lamb: Well, it forced us to stop. I did not know where the crashed helicopter was even though it was just around the corner, only 20-30 meters from where I was standing with a Ranger casualty. Smith had been shot high in the femoral artery. It also hit his pelvis so there were a lot of bone fragments. I had my fingers in him trying to slow that bleeding. Then Larry grabbed him. Smith was an M203 gunner, so he was a heavy dude. We pulled him behind cover, but incoming rounds continued to zing off the wall right beside us and we were catching some of the fragments off that. We realized this was not just random fire, they were aiming and shooting at us. So, we pulled Smith further into the courtyard, got the guys out of the street and out of the gunfire, and then tried to pull security while at the same time providing first aid since we did not have a medic with us.

Perino: We took most of our casualties, especially those that could not move on their own—what we refer to as litter patients—right in that vicinity. Although we did not know DiTomasso' s element was right around the corner, it did provide some modicum of security. One of the guys in my platoon, Sergeant Aaron Williamson, caught some shrapnel off a grenade that had flipped over a wall. At that point, we still had contact with the vehicles. Now, remember the vehicles—if you watch the movie, it was kind of confusing. All I know is they were there and then they were not. I believe they drove within a half a block of the crash site twice, but never knew it. Luckily, we were able to load Williamson into one of the convoy vehicles.

Collins: So, at this point everybody is more or less consolidated near the crash site and you are waiting to exfiltrate?

Lamb: Yes and no. We were waiting to exfiltrate but the reason we could not was because the body of Cliff Wolcott, the pilot, had been crushed

[35] An airborne shuffle is a term used to describe the slow jog that is used to move the formations of students around at the U.S. Army's airborne school.

inside the aircraft.[36] When *Super Six One* crashed, it crushed Wolcott's body and trapped him inside the wreckage, but we refused to leave without recovering his body.

Van Arsdale: Wolcott and his co-pilot, Donovan Briley.[37]

Lamb: Yes. We were not going to leave until we recovered both of their bodies. Because we had seen what the Somalis would do to our guys if we left them there.

Collins: Did the combat search and rescue crew have "jaws of life"[38] or any other recovery tools to assist in the extraction of the crew.

Lamb: No.

Collins: When I was a special forces detachment commander in 1999, I supported the combat search and rescue mission during the Kosovo War and "jaws of life" was standard equipment for us. So, we learned some lessons from your experience in Somalia.

Van Arsdale: Yes, that was one of the lessons.

Collins: So how did you pull Wolcott out? I can imagine that was not easy.

Van Arsdale: While they were attempting to recover the bodies, the vehicular convoy, which was Plan A for the exfil from the target site, was trying to move to the crash site to pick everybody up and get out. But the convoy was being vectored from a command-and-control helicopter above the city and it was nearly impossible to direct the vehicles without knowing how wide the streets were. So, they were being told, "take a right here," and they would take the right and then all of a sudden, the street necked down to the point where they could not go any further. Still taking continuous fire, the vehicles had to then back out of the unexpected dead ends.

To be fair, it was extremely challenging for the helicopter to direct the ground convoy, because the aircraft could not simply hover without becoming an easy target, so the aircraft was doing continuous loops in the

---

[36] Clifton Wolcott was posthumously awarded the Silver Star for his actions that day, "Clifton P. Wolcott," Wall of Valor Project, *Military Times*, https://valor.militarytimes.com/hero/114565.

[37] Donovan Briley was posthumously awarded the Air Medal with Valor Device for his actions in Somalia on September 18 and the Silver Star for his actions on October, "Special Operations Aviation Veterans Receive Upgraded Awards," Defense Visual Information Distribution Service, October 29, 2021, https://www.dvidshub.net/image/6931270/special-operations-aviation-veterans-receive-upgraded-awards.

[38] The "jaws of life" is a hydraulic rescue tool that is used to cut through metal to recover trapped persons.

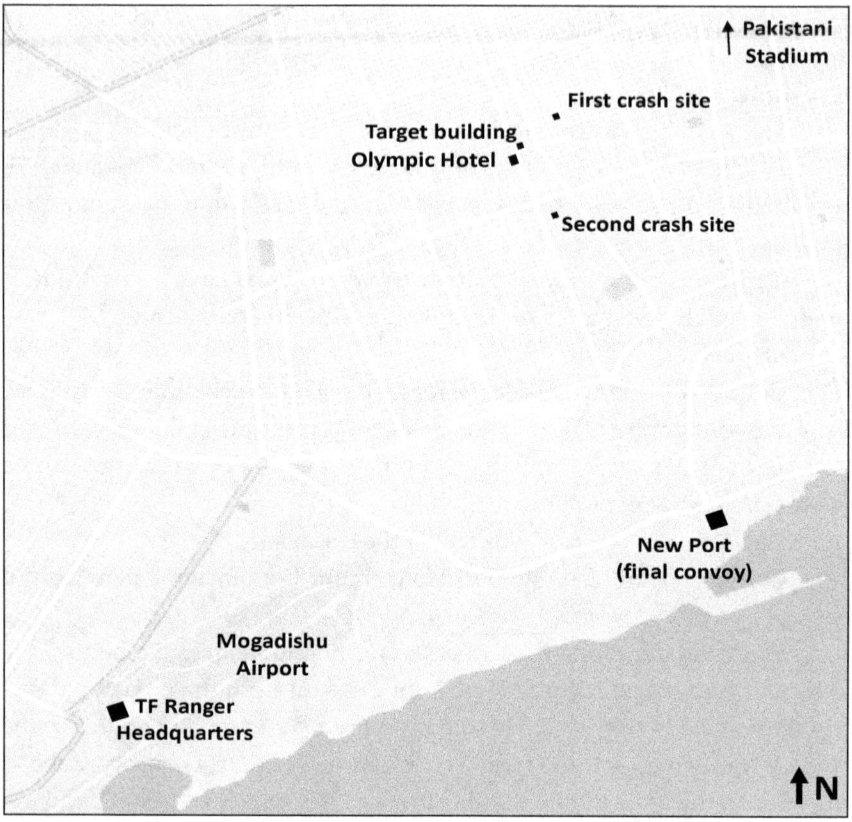

Figure 8.3. Map of Mogadishu with the crash sites.
Source: Map created by Liam Collins.

sky. But given the three-dimensional nature of a city, the buildings obscured the crew's ability to see the roads unless they were directly over them. So, what looked like a road from the sky and from the photos that had been taken only months earlier, was now closed off from all the shanties that had been built. Eventually the convoy got so shot up, that the commander decided he had to abandon the exfiltration mission and limp back to the base or the convoy would never make it. Since we still had the assault element at the crash site, we had to figure out a new recovery plan to get them out.

Perino: Minutes before the convoy abandoned its search for the crash site, *Super Six Four* got shot down, so now we had two crash sites (see figure 8.3).

Van Arsdale: We actually had five Black Hawks get shot that day. Most people are not aware of that. Three of them made it back to the airfield, but the other two went down in "less than nice" neighborhoods.

Collins: Prior to this mission, how many times had a Black Hawk been hit by a rocket-propelled grenade (RPG)?

Van Arsdale: The Somalis had hit a 10th Mountain Division aircraft a week earlier. They had broken the code on volley firing RPGs at the aircraft instead of firing single RPGs—sort of like using a shotgun, as opposed to a rifle, for bird hunting. It is tough to get a debilitating hit with a single RPG, but if there are enough fragments in the air from multiple rounds, there is a better chance of bringing an aircraft down. The Somalis figured this out when they shot down the 10th Mountain Division helicopter. When Kyle said that we knew what they did with the bodies, this was the incident that he was referring to. The Somalis pulled the bodies out and desecrated them in horrible ways.[39] So that was one of the reasons that we felt that we had to get to the crash site first. We could not leave them behind to get desecrated.

Collins: As you were learning how to fight in this complex operational environment, so was the enemy. So, now that the convoy was gone, what was the plan? What were you doing?

Perino: We were still treating Corporal Jamie Smith, strongpointing our position, and buckling down, because, with the convoy gone, we knew we were going to be there awhile.

Lamb: We had an RPG hit the side of the building near our courtyard, so I realized that we needed to get out of the courtyard and into the building for more protection. So, I grabbed Larry and I said, "I am going to kick in this door. You go left. I am going right. Do not shoot me." Because we could hear these —

Perino: It was more like, "Don't shoot me!" I got it.

Lamb: That stickpin was still up my ass, you know. [laughter] When the RPG hit the building, we heard women and children scream, but we had to go in and clear the building just to make sure there were no bad guys in there. We pushed all the women and children back into a safe room—I guess you guys would call it a "safe space"—and put a couch in

[39] See, for example, Keith Richburg, "3 GIs Killed as Somalis Down Helicopter," *Washington Post*, September 26, 1993, https://www.washingtonpost.com/archive/politics/1993/09/26/3-gis-killed-as-somalis-down-helicopter/70f8e493-6e8f-44b3-b760-7a2e00e0df17/.

front of the door to block it. They stayed in there the rest of the night; we did not hear another peep out of them. At that point, we started to get a better sense of the battlefield and realized that we were going to be there for a while. As I listened to calls on the radio, I started to get a better sense of how the battle was unfolding for everyone else as well. I heard my team leader and then the commander, Scott Miller, he came on the radio and said —

Perino: Scott Miller is now General Scott Miller, who is now the commander in Afghanistan.[40]

Lamb: Yeah, so now he is "Big Sarge," "Big Cheese," or whatever you want to call him. Big General Dude, right? Is that the proper terminology? By the way, I am probably the only non-politically correct guy up here if you have not figured that out. So, General Miller makes a call and says, "Alpha 1, what's your status?" My team leader comes on the radio and says, "I have [X] wounded and one KIA (killed in action)." And when I heard that, in my mind I justified it as a Ranger that had been killed. It was very unfortunate and very sad, but I kept telling myself that it was not one of my buddies, not someone who I had trained with every day for the past several months. Five minutes later, Miller came back on the radio and asked, "Alpha 1, say again your status." My team leader responded, "[X] wounded, one KIA." When Miller asked him who, my team leader responded with the callsign of the KIA, "Alpha 2."

When I heard "Alpha 2," it was like a punch to the gut. I had not seen Earl get shot since I had become separated from the rest of my team, but now I knew that Earl Fillmore was dead. It was extremely deflating. I started to question my motivation to fight. But seconds later, my team leader came back on the radio and said, "All elements, this is Alpha 1, this is what we're going to do." I heard that and thought, if he can get back into the fight with his best friend lying there dead, then we all have to get back into the fight. That leadership is exactly what we needed at that point of the battle. And I did some pretty serious praying at that time too, so the Good Lord was looking out for me, and we continued the fight.

[40] General Miller served as the Commander of U.S. and NATO forces in Afghanistan from September 2, 2018, until July 12, 2021. He was the longest serving commander in the war. Charlie D'Agata, "Top U.S. general in Afghanistan steps down as military mission nears end," CBS News, July 12, 2021, https://www.cbsnews.com/news/gen-austin-scott-miller-afghanistan-steps-down-us-mission-nears-end/.

Collins: At that point, it was getting dark, and the intensity of fire continued to ramp up. But we "own the night" so the darkness should favor us?

Lamb: Except we did not bring our night vision goggles (NVGs).

Perino: Well, actually some brave soul climbed into *Super Six One* and grabbed the pilot's NVGs. But, of course, we did not have a way to mount them to our helmets, so we had to hold them up with our hands and shoot at the same time.

Collins: In hindsight, you probably should have brought your NVGs, even if you expected the daylight raid to last only 20 minutes.

Lamb: Yes, obviously that was one of the many lessons that we learned, but NVGs were relatively new at the time. We could not even mount them on our helmets at the time. You could attach them to a head harness for patrolling in the woods, but we could not mount them on our helmets.

Van Arsdale: Like Kyle said, we had them, but you wore them around your neck, and they were cumbersome, awkward, and would fill up with dust from Mogadishu's dusty streets. The assault force thought it would be a 20-minute operation, and they were wearing at least 75 pounds of "lightweight" gear (including ammunition), so they looked to cut weight where they could. Unfortunately, the NVGs were cut. Ammunition, explosive charges, grenades, body armor, and communications gear were all critical, so the assault did not feel that they needed their NVGs for a daylight raid.

This was 1993, so they were not the NVGs of today. Just imagine this clunky thing bouncing off your chest and every time you do something, they bounce up and hit you in the face. And when you put them on, you have to remove your helmet, so you hope to God you do not take a shot in the head. And when you put them on, you cannot see anything because they are covered in dust, so you have to take them off to clean the dust off the lenses, before finally putting them back on. In hindsight, should we have brought them? Yes, but it is useful to understand why we did not, because many of you will have similar dilemmas in the future.

Collins: It is not like today when you have nice helmet mounts, where you can easily flip them up or down. Things have changed.

Lamb: Soon enough, we found ourselves low on water and ammunition. Hindsight is always 20:20 but it is important to remember that we passed on many of the lessons that we learned that night, they were not lost. The guys and gals that fought in Afghanistan and Iraq were direct descendants of what we experienced in Mogadishu. And the three of us coming to West Point to talk to you about our experience and the lessons that we learned will hopefully better prepare you when you find yourself in battle for the first time. Because you do not know when or where that is going to be, but if you stay in the military for ten years, I guarantee you are going to be in something.

Collins: Sometime and somewhere, but it is never when or where you thought it would be. I never thought I would be parachuting into Afghanistan in 2001, but I did.

Lamb: Yes, somewhere. You just never know what is going to happen. Six months prior, I had never heard of Mogadishu, let alone been able to find it on a map. OK, maybe I was not quite that dumb, but you get the point. You don't know what you don't know.

Collins: So, within the city, you consolidated, or attempted to consolidate, the force around the downed helicopter. Lee, what was the headquarters doing to get the assault force out, because the ground vehicle convoy had returned to base?

Van Arsdale: We were working on multiple plans. We cobbled together whatever vehicles we had at the airfield and anyone that was not on the initial assault and sent them out. They, unfortunately, were quickly shot up so badly that it was clear they just could not make it to the crash site, so they returned to base. They tried multiple routes but whenever they got close to the target site, they were shot to pieces. Also, when the first helicopter went down, General Garrison activated one of our contingencies which was the quick reaction force from the 10th Mountain Division. They were not far away but we were at different locations.

Collins: Was it pretty quick?

Van Arsdale: Yes, it was just a few minutes, so it was not bad. He showed them the target and told them they needed to pick up our people. General Garrison turned to me and said, "Go with them." It was only three words, but with those three words, I knew the mission and the commander's intent. I knew because we had worked together for years, and I echo what Kyle said: General Garrison was one of the finest leaders I have ever worked with. So, "go with them" meant that it was my job to go out there, develop the situation, and get the desired result. So that is what I did, I went out with the 10th Mountain Division's quick reaction force. Unfortunately, the 10th Mountain Division also lacked armored vehicles, so after hitting some pretty well-laid ambushes, the quick reaction force was also forced to abandon its rescue attempt.

Collins: So now the convoy, the relief convoy, and quick reaction force are all either back at base or returning to base. Larry, Kyle, what were you guys hearing on your end? What were your expectations?

Perino: The hardest part is viewing a battle from where you are located, it is kind of like watching a football game through a soda straw or a toilet paper tube. It is not like today where you can see the battle on screens from intelligence, surveillance, and reconnaissance aircraft flying overhead. Nor did we have Blue Force Tracker[41] to know exactly where the vehicles were during the battle. So, it was a constant struggle for information.

At one point I realized, "The vehicles are right behind us! What the hell!" But they were trying to consolidate and never made it to us. At another point, I realized that we were not going to get a casualty evacuation for Corporal Smith, and he ended up passing away at approximately 20:30 hours. Then I heard that they were cobbling together a relief convoy, and I started to think, "It's coming! Does anybody know when they will make it here?" And then we found out they had turned back. So, it was a constant struggle for information.

Van Arsdale: So, our Plan B, or maybe it was Plan C or D after the other failed attempts to recover the force, was to put together an international task force with Malaysian armored personnel carriers and Pakistani tanks. This was not an easy task since we had to work through at least three different languages, with units that we had never worked with—ones with different standard operating procedures—and send that ad hoc element out, for the first time, into an intense combat situation. So, it took several hours to cobble that force together.

When the convoy rolled out, I found myself riding into combat on the back of a Soviet-made armored personnel carrier—I never thought that would happen. Finally, with the tanks and armored personnel carriers, we are able to make it through all the roadblocks and ambushes to the target area. The entire convoy drove to a rally point at a bus station near

41 Blue Force Tracker is "a term coined by the U.S. military that refers to electronically monitoring the location of friendly forces' personnel and resources in a particular area of operations" using global positioning system (GPS). Icons are typically then portrayed on computer screens in command posts and later in vehicles. See *Blue Force Tracking Systems Application Note* (Washington, DC: Space and Naval Warfare Systems Center Atlantic, 2014), 1, https://www.dhs.gov/sites/default/files/publications/GPS-Blue-Force-AppN_0414-508_0.pdf.

the Olympic Hotel which was located approximately 800 meters from each of the crash sites. Part of the convoy waited there while smaller elements broke off to move to the two different crash sites.

Perino: I remember my company commander saying, "They are on their way." I just said, kind of flippantly, "Yes sir, I can hear them." Because it was just like a rolling firefight that got louder and louder and louder.

Lamb: And then all of a sudden, Matt Rierson comes walking in with a quickie saw[42] in one hand and a CAR-15 rifle in the other. And I was like, "This guy is my hero!"

Collins: So, at that point you finally loaded up the convoy?

Lamb: Not so fast, Lucky! You know his call sign is Lucky? Did you guys know that? Do they know why?

Collins: I do not think so.

Lamb: Because a cat has nine lives, and sometimes you call your cat Lucky. I was putting him through our OTC, and we were in a barn.

Van Arsdale: Tell them what OTC is.

Lamb: OTC is the Operator Training Course, the training course for our unit. I was his instructor; he looked up to me like a father. [laughter] We had a barn, and like many barns this one had the hay-out doors in the hay loft —

Collins: To be fair, how many doors have you gone out in your entire life? Have any of them —

Perino: Keep going!

Collins: — not had anything on the other side?

Lamb: I grew up in South Dakota, so I have been in a barn before. [laughter] Lucky's team cleared the first floor and then cleared the second floor. Lucky gets to the door that leads to nothing, he opens it, and then he runs out. And it is like Wile E. Coyote in the air. [laughter]

Perino: And he holds up a sign that says, "Oops."

Lamb: And he disappears.

Collins: It was like time stood still. After running through the door, I felt like I was still moving normally, turning the corner, clearing the corner, and then all I could think was, "Oh crap! There is nothing under my feet."

Lamb: So, he fell 15 feet, into a crumpled mass, and immediately popped back up and yelled, "I'm good."

[42] A quickie saw is a power tool that is used to cut concrete, masonry, etc.

Collins: Somebody pulled out the white-lens flashlight and asked, "Are you okay?" And I said, "I'm good. Turn that thing off, we have to exfil." [laughter]

Lamb: This guy, he is not real smart, but he is lucky. [laugher] So that is where he earned his nickname Lucky.[43]

Collins: So, you are loading the vehicle — [laughter]

Van Arsdale: You can never escape your call sign.

Lamb: When the vehicles arrived, I figured we would load the armored vehicles and get the heck out of there, but not so fast. Once we loaded all the dead and wounded, there was not any room for any of us.

Perino: Oh, by the way, returning to Rierson and his quickie saw, you should know that quickie saws work really well. It is like a very heavy —

Van Arsdale: It is a construction saw.

Perino: A construction saw that works on everything except —

Van Arsdale: Except for Black Hawks.

Perino: Yes, except for Black Hawks. And the hard part was we still had to get Wolcott out. So, it was a race against daylight because everyone wanted to get out under the cover of darkness because we knew that it would only get worse when the sun came up.

Van Arsdale: We did not drive the entire convoy to the crash site. As I mentioned earlier, the entire convoy moved to the bus station near the Olympic Hotel. From there, a smaller element, led by a 10th Mountain Division company, broke off to travel to each of the crash sites while the rest of the convoy remained at the rally point—the bus station—as a reserve. I went with the company that moved to the first crash site, because that was where all our troops were, and I figured one of my primary responsibilities was to prevent fratricide between the assault element and the recovery element.

As we departed the rally point, we were forced to dismount the vehicles because the Somalis had constructed several burning roadblocks and the vehicles could not drive through them. So, we dismounted, tore the first burning roadblock apart, and stayed dismounted because the next burning roadblock was only a block away. We provided mutually supporting fires between the vehicles and the dismounted troops—a technique that was used effectively in urban fighting during World War II.

Some of us had to travel the 800 meters from the rally point to the crash site on foot because of all the roadblocks. About halfway there we got

[43] While the story is true, that is only partly contributed to the origin of the call sign Lucky.

bogged down. We took some heavy fire and the convoy stopped. The 10th Mountain Division company commander told me that the lead armored personnel carrier refused to go any further. The Malaysian there said "No. That's it! We've had enough." He said he would not go forward without the tanks. So, I told Matt Rierson to get in the lead armored personnel carrier and make him go forward. I do not know what Matt said or did to get that thing moving forward but he got the vehicles moving again.

Lamb: I want to add something quickly about Matt. Matt was killed a few days later in a mortar attack, so he paid the ultimate price, too.

Van Arsdale: Matt was an assault team leader on the initial assault. He had captured the lieutenants and accompanied the lieutenants and their bodyguards back to the base in the ground convoy, and then he went in with me on the recovery mission. Not a scratch! He had been out there in the fight as much as anybody, and then two days later, on October 6, he was killed in a mortar attack.[44]

Collins: So, you finally loaded the vehicles and drove back to the base before daylight?

Perino: No. It was kind of surreal but with the vehicles filled with casualties, we were forced to run alongside the vehicles, back to the base. Or at least that is what we thought. Ultimately, we only had to run back to the rally point 800 meters away to load vehicles there, but we thought that we were going to have to run the entire five kilometers back to the base.

Collins: So, you have been fighting all night and now you think you are going to have to run all the way back!

Lamb: Well, we did not really know the plan until it unfolded in front of us. We had to get to the Pakistani stadium which was five kilometers away! We figured it was best to stay next to the armored personnel carriers, so they could provide cover as we were running down the street. But as we got ready to finally depart the crash site, the armored personnel carriers just took off. We were standing there, staring at each other as the sun was coming up, looking at the back of the vehicles that had just left us. It was kind of a surreal moment. We were told that the Little Birds were going to make one final gun-run pass and it was like something out of the movie *Apocalypse Now*. I looked over and saw a couple of Little Birds just running down the street parallel to us with their guns blazing. Then they

[44] Rierson was posthumously awarded the Silver Star Medal for his actions that day. See, "Matthew Loren Rierson," Wall of Valor Project, *Military Times*, https://valor.militarytimes.com/hero/23830.

popped right over our heads and lobbed their remaining rockets. So, we looked at each other and then took off running to the rally point.

Van Arsdale: All of you have done road marches or group runs so you understand the accordion effect. It was not that vehicles had left us, it was more of an accordion effect that resulted in the trail vehicles going from not moving at all, to a quick acceleration away from the crash site. But if you were standing there when the Little Birds emptied their pods, it sure felt like you had been left behind! So, it was an 800-meter run—trot is probably a better word to describe it because I was old and slow—back to our rally point. After consolidating at the rally point and confirming that we had everyone present and accounted for, we loaded the vehicles and departed.

Perino: Once we reached the rally point, we were able to load onto vehicles that had space.

Lamb: But the fight was not over. Just prior to arriving at my vehicle, a 10th Mountain Division soldier had been shot and they were working on him.[45] I believe he had a sucking chest wound and ended up passing away a couple of days later. He had come out to evacuate us and was, unfortunately, killed in the process.

Van Arsdale: Yes, they lost two soldiers. The story of *Black Hawk Down*, the book and subsequently the movie, was basically told through the eyes of Rangers, because the special operations soldiers and the pilots were not allowed to talk to the press. So, the story is a little biased from their perspective. Many of the Rangers were not aware that the 10th Mountain Division was even there. So, their story was largely untold, and we probably would not be having this conversation today if it was not for them. The 10th Mountain Division soldiers fought heroically that day.

Lamb: Those guys and the 160th Special Operations Aviation Regiment pilots saved our skin. They flew gun run after gun run for nearly 15 straight hours. After exhausting their ammunition, they would return to base, land, grab a bottle of water, maybe a PowerBar, refill gas, reload ammo and rockets, and then go right back to support us, despite the fact that two helicopters had been shot down.

---

[45] The soldier was Sergeant Cornell L. Houston. He was posthumously awarded the Bronze Star with Valor Device for his actions. See, Chris Flora "'Black Hawk Down' movie depicts service of local veteran," *Tucson Local Media*, July 2, 2014, https://www.tucsonlocalmedia.com/features/heroes/article_48ee96d4-01c1-11e4-af19-0019bb2963f4.html.

Collins: What is one of the key lessons that each of you learned? When we went to Afghanistan for the first time in 2001, we probably had some of the same debates that you had prior to your mission in Mogadishu. Do we carry as much ammo and everything else as possible? Or do we go in with a smaller load, the "fight light" methodology, believing that being light allows you to move faster which is key to surviving firefights? My snipers told me that they could not lay in the prone behind their sniper rifles if they were wearing a thick body armor plate. So, okay, we will just not wear the front plate, right? But then you have this history that you are up against. We all knew the history of Mogadishu, and we did not want to find ourselves stuck on the ground hours later, wishing we would have worn our front plates. There are always going to be these tough choices that you have to make out there. So, what was one of the key lessons that you learned?

Lamb: In Iraq, we did not have to carry everything because we could fight off the vehicles. We had armored vehicles that were similar to a Stryker armored vehicle.

Collins: That was one lesson from Mogadishu. Instead of having sandbagged five-ton vehicles and unarmored Humvees, you really need to have armored vehicles, because you have to plan for the worst-case scenario.

Lamb: In Iraq, we did not have to fight off of our kit and carry everything we needed. We fought off the vehicles. We had water bottles in the vehicle that we could drink while driving to the target so we could carry less water on us. We had magazines that were velcroed to the side of the vehicle so we could carry a basic load instead of a double basic load of ammunition. Beyond that, we had go-bags that included only the most essential items such as water and ammunition. We did not waste precious weight with food—the body can go a long time without food. We could carry the go-bag to the entrance of the building and then drop it there, so we had it if we needed it. It really depended on the situation. And it varied by the situation, person, and their previous experiences. You could see some guys that, if they got more than like ten feet from their go-bag, they wanted to stop and suck their thumb a little bit there. [laughter] People have experiences, and they learn from them. But you can also learn from training and education, that is why all of you are here today—you want to learn as much as you can about combat prior to experiencing it for the first time.

Van Arsdale: There were two important lessons that I learned which are important for you. The first is the concept of leading from the front. You have all heard that you need to "lead from the front," and I shared a little bit of my own story with you. I was a staff officer, so I was not in anybody's chain of command, but as a lieutenant colonel, I found myself walking point for an infantry company because somebody had to do it. Somebody had to lead from the front. So, when you become lieutenants and platoon leaders, you cannot email in your leadership. You cannot phone it in. You have to be there. Your troops have to see you. You have to be visible. You have to lead from the front.

Collins: But that does not necessarily mean taking point. I am not the first one through the breach, right?

Van Arsdale: That is exactly right.

Collins: You have to know where you have to be on the battlefield to best lead your formation.

Van Arsdale: A better term for that is "lead from the decisive point," because chances are, as the commissioned officer, you are not going to be walking point. In fact, I do not recommend it because you have soldiers who are better at that than you are. And if you are walking point, then you have no idea what is going on behind you, so you have to be at the decisive point where you can best command and control. The other lesson is to always be ready. We did not have a single contingency plan that involved me leaving the headquarters, yet I found myself in the middle of Mogadishu. So, you always have to be ready, no matter what. Your unit may be on post support detail and eighteenth in deployment priority, but the next thing you know, you are up. So always be ready.

Perino: I will talk a little about leadership. One of the things that you are going to have to accept is that you may not be the subject matter expert despite being the leader. Sure, you will have spent four years at West Point, maybe earned your Ranger tab, and completed your basic officer leader course, but you are probably still not the expert. Your soldiers understand that. When you train, it is a mutual training experience—you are learning just as much as they are, maybe more. And tough, rigorous training is how you learn to trust one another. Another thing that I want to share is that combat can be scary as well, as Kyle said. Fear gripped me so badly that I did not think I could move at times. Understand you may feel that way. But I also realized that there were a lot of people who were looking at me and if I "lost my shit" it was going to be bad for everybody else. So, the most important thing I could do was to remain as calm as possible, because if I did not, then I would become a liability as opposed to an asset. And I refused to be a liability.

Lamb: I will add to what both these guys said. As a noncommissioned officer, I was looking at these guys. Even though I may have had more time in the army than Larry, he still outranked me, even if he was not in my chain of command. Lee gained his credibility long before Mogadishu, but what do you think Mogadishu did to his credibility after that fight? It went up. It was the same with Larry. He went from being some lieutenant that I did not really know, and I was like, okay, he seems like a decent enough guy. But after Mogadishu, he proved himself to be a credible combat leader. That is what all of you guys and gals should strive for. You want to be credible. But keep in mind, it is much easier to lose credibility than it is to gain it. Establish your credibility, maintain your credibility, and remember that everybody knows what you are doing so you can lose that credibility in a heartbeat. These guys are awesome because they never lost their credibility. Your commandant never lost his credibility; he is a highly respected special operations forces combat veteran. Same thing with Lucky. So, you should strive to gain and maintain that level of credibility throughout your army career.

9

The Second Battle of Fallujah

The United States invaded Iraq in March 2003 and quickly overthrew Saddam Hussein and his Ba'ath Party government. Some mistakenly believed that the war was over when President George W. Bush declared that "Major combat operations in Iraq have ended," while standing beneath the "Mission Accomplished" banner onboard the *USS Abraham Lincoln*, but a Sunni insurgency was already simmering.[1] The insurgent stronghold was located within the "Sunni Triangle," a densely populated region in central Iraq inhabited largely by Sunni Muslims. The triangles' points included the cities of Baghdad, Tikrit, and Ramadi.

The city of Fallujah, located between the cities of Baghdad and Ramadi (see figure 9.1), became an insurgent hotbed and served as the base of operations for al Qaeda affiliate Jamat al-Tawhid wal-Jihad.[2] Fallujah rests on the banks of the Euphrates River. It is a densely populated, industrial city that grew as a way station along silk road branches that connected Baghdad to major population centers such as Aleppo, Syria.[3] In 2004 Fallujah had an estimated population of 250,000 to 300,000 residents.[4]

---

[1] Seth Cline, "The Other Symbol of George W. Bush's Legacy," *U.S. News & World Report*, May 1, 2013, https://www.usnews.com/news/blogs/press-past/2013/05/01/the-other-symbol-of-george-w-bushs-legacy.

[2] At the time, al-Qaeda in Iraq operated as Jamat al-Tawhid wal-Jihad. The group changed its name to al-Qaida in Iraq following Abu Musab al-Zarqawi's publicly pledged allegiance to Bin Ladin in October 2004. Brian Fishman, "The History of the Islamic State," in *Routledge Handbook of U.S. Counterterrorism and Irregular Warfare Operations*, eds. Michael A. Sheehan, Erich Marquardt, and Liam Collins (New York: Routledge, 2021), 57, DOI: 10.4324/9781003164500-5; and Joel D. Rayburn and Frank K. Sobchak, *The U.S. Army in The Iraq War, vol. 1* (Carlisle Barracks, PA: U.S. Army War College Press, 2019), 176, https://publications.armywarcollege.edu/pubs/3667.pdf.

[3] Coley D. Tyler, *Ghosts of Fallujah* (Athens, GA: Deeds Publishing, 2018).

[4] Timothy S. McWilliams and Nicholas J. Schlosser, *U.S. Marines in Battle: Fallujah, November-December 2004* (Quantico, VA: U.S. Marine Corps History Division, 2014), 3, https://www.usmcu.edu/Portals/218/FALLUJAH.pdf.

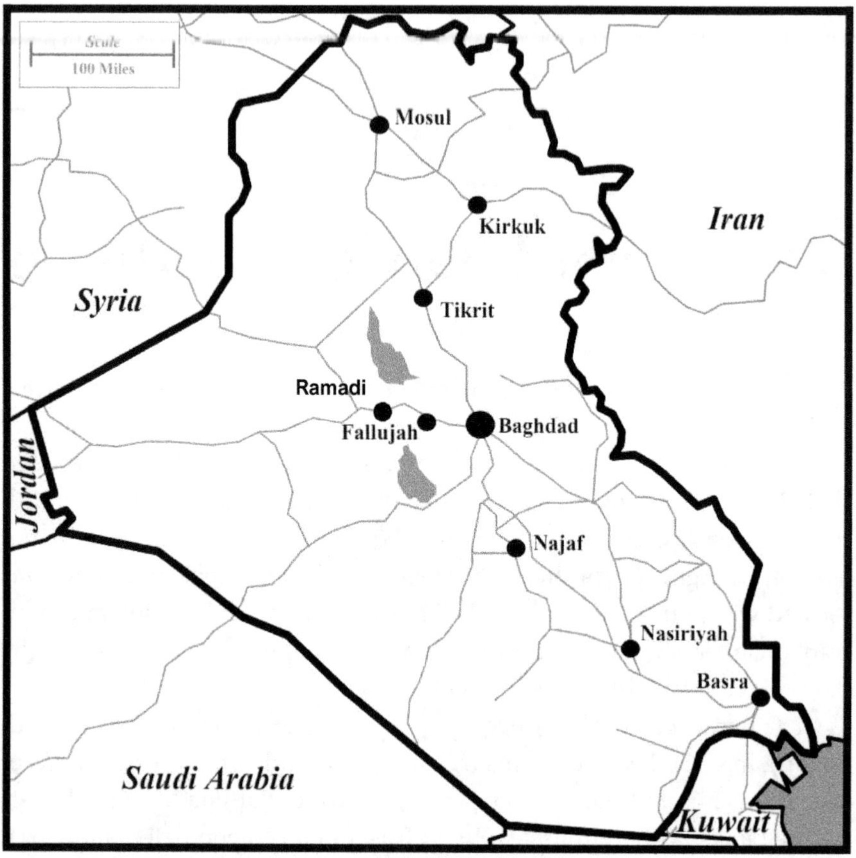

Figure 9.1. Map of Iraq.
Source: Map is modified from map 1 found in Kendall Gott, ed., *Eyewitness to War Volume I* (Fort Leavenworth, KS: Combat Studies Institute Press, 2006).

On March 31, 2004, insurgents ambushed, killed, and subsequently burned four American security contractors in Fallujah. Two of the contractors' bodies were hung from the ramparts of a bridge on the outskirts of the city.[5] Images of the hanging bodies were televised and broadcast world-wide.[6] In response, Lieutenant General Ricardo Sanchez, the commander of the coalition forces in Iraq, ordered the I Marine Expeditionary Force to launch

[5] Jeffrey Gettleman, "Enraged Mob in Falluja Kills 4 American Contractors," *New York Times*, March 31, 2004, https://www.nytimes.com/2004/03/31/international/worldspecial/enraged-mob-in-falluja-kills-4-american.html.

[6] "'Horrific' Iraq deaths shock US," BBC News, April 1, 2004, http://news.bbc.co.uk/2/hi/middle_east/3587931.stm.

Operation Vigilant Resolve into Fallujah to deny its use as an insurgent sanctuary and to arrest those responsible for the killing of the contractors.[7]

U.S. Marine Regimental Combat Team 1 (RCT-1) initiated the assault on April 5.[8] The attack was slow due to the size of the enemy, estimated to be about 2,000 insurgents; the complexity of fighting in a densely populated large city; a lack of armor capability; and a complete loss of their Iraqi paramilitary support.[9] On April 9, four days after the operation's start, General John Abizaid, the commander of U.S. Central Command, ordered Sanchez to suspend the operation.[10] Pressure from the Iraqi governing council and the Coalition Provisional Authority forced the Marines to prematurely halt the operation. International criticism of the mission sparked by footage of what was dubbed as excessive force also contributed.[11]

On April 25, a locally raised Iraqi military force called the Fallujah Brigade under the command of Major General Jassim Mohammed Saleh assumed control of Fallujah.[12] American forces withdrew to the outskirts of the city and the Marines retained a loose cordon around the city, consisting mainly of checkpoints. In total, 27 American servicemembers were killed and more than 90 wounded during the first battle of Fallujah.[13]

Elsewhere in Iraq, violence escalated as Shia leader Muqtada al-Sadr engaged in a political battle with the new Iraqi government. Al-Sadr loyalists in his armed militia known as Jaish al-Mahdi (the Mahdi Army) eventually took to the streets and seized control of the Sadr City neighborhood of Baghdad, and the cities of Al Kut, Karbala, and Najaf.[14] Coalition forces were able to regain some control of these areas by May, but in cities like Najaf, the Mahdi Army never really left. In August, coalition forces launched operations to defeat non-compliant forces and neutralize destabilizing influences in Najaf Province. The 11th Marine Expeditionary Unit, supported by army units including the 2nd Battalion, 7th Cavalry

[7] McWilliams and Schlosser, *U.S. Marines in Battle,* 2.

[8] Kenneth Estes, *U.S. Marines in Iraq, 2004-2005: Into the Fray, U.S. Marines in the Global War on Terrorism* (Washington, DC: U.S. Marine Corps History Division, 2011), 32-33. https://www.usmcu.edu/Portals/218/Estes%20Into%20the%20Fray%20Boards_Det%20One%20copy.pdf.

[9] Robert Kaplan, "Five Days in Fallujah," *The Atlantic* 294, no. 1 (2004), 116-26, https://www.theatlantic.com/magazine/archive/2004/07/five-days-in-fallujah/303450/; and McWilliams and Schlosser, *U.S. Marines in Battle,* 2.

[10] Estes, *U.S. Marines in Iraq,* 37.

[11] Ibid., 38.

[12] Rayburn and Sobchak, *The U.S. Army in The Iraq War,* 298.

[13] Estes, *U.S. Marines in Iraq,* 44.

[14] Bing West, *True Glory: A Frontline Account of the Battle for Fallujah* (New York: Bantam, 2011), 61.

(2-7 Cavalry), conducted operations in Najaf. During the three weeks of fighting, the American forces killed an estimated 1,500 al-Sadr militiamen while suffering minimal casualties.[15] As a result, Muqtada al-Sadr brokered a truce with the Iraqi government on August 27.[16]

Despite national negotiations with civilian leaders and the activities of the Fallujah Brigade, Fallujah became an insurgent safe haven from which they launched attacks around the country. Violence in and around Fallujah continued to increase throughout the summer.[17] By September, the coalition had had enough and authorized the planning for an assault into Fallujah to eradicate the extremist insurgent nest and assigned the mission to the 1st Marine Division.

In late October, the 1st Marine Division started marshaling forces and reinforcements for what would become Operation Phantom Fury: the second battle of Fallujah.[18] By the end of October, coalition forces had convinced most of Fallujah's civilian population to depart, leaving as few as 500 civilians remaining in the city when the battle began.[19] Operation Phantom Fury, later renamed Operation al-Fajr (Arabic for Dawn), took place from November 7 to December 23, 2004. Eighty-two United States servicemembers were killed and more than were 600 wounded during the operation.[20]

The 1st Marine Division's two main units that executed Operation al-Fajr included RCT-1 and RCT-7. RCT-1 was comprised of two Marine infantry battalions—3rd Battalion, 5th Marines and 3rd Battalion, 1st Marines—and Task Force 2-7 Cavalry. RCT-7 consisted of two Marine infantry battalions—1st Battalion, 8th Marines and 1st Battalion, 3rd Marines—and U.S. Army Task Force 2-2 Infantry.[21] Each RCT also included engineers, military police, logistics, and other support elements. After the

[15] Matt M. Matthews, *Operation AL FAJR: A Study in Army and Marine Corps Joint Operations* (Fort Leavenworth, KS: Combat Studies Institute Press, 2006), 11, https://www.armyupress.army.mil/Portals/7/combat-studies-institute/csi-books/Operation-AL-FAJR-A-Study-in-Army-Marine-Joint-Operations-ul.pdf.

[16] Estes, *U.S. Marines in Iraq*, 51.

[17] Ibid., 53.

[18] Ibid., 57.

[19] Some reported only 500 civilians remained; see, for example, McWilliams and Schlosser, *U.S. Marines in Battle*, 6. Yet others reported that as many as 30,000 (ten percent of the pre-battle population) remained. See, for example, Anne Barnard, "Fallujah refugees describe ordeal of life in crossfire," *The Boston Globe*, November 15, 2004, http://archive.boston.com/news/world/articles/2004/11/15/fallujah_refugees_describe_ordeal_of_life_in_crossfire/.

[20] Estes, *U.S. Marines in Iraq*, 67.

[21] Ibid., 7.

first battle of Fallujah, the coalition conducted very limited operations into the city; thus, the insurgents had nearly six months to prepare for the second battle.

In this chapter, John Spencer interviews Lieutenant General James Rainey about his role in this battle. At the time of the interview in 2021, General Rainey was serving as the commander of the Combined Arms Center at Fort Leavenworth, Kansas. He currently serves as the Headquarters, Department of the Army, Deputy Chief of Staff for Operations, Plans, and Training (G-3/5/7). In November 2004, Rainey was a lieutenant colonel, and commander of the 2nd Battalion, 7th Cavalry Regiment within the 1st Cavalry Division. For the duration of the second battle of Fallujah, Task Force 2-7 Cavalry was under the operational control of the RCT-1.

John Spencer: Sir, I know that prior to the second battle of Fallujah you were in Najaf, Iraq and that was where you and Task Force 2-7 Cavalry honed many of your urban warfare tactics, techniques, and procedures to clear terrain and defeat enemy forces. Can you talk a little about your experience in Najaf and how the first battle of Fallujah impacted operations?

James Rainey:[22] I was not involved in the first battle of Fallujah. I think it was largely political pressure from the Iraqi government that caused the attack to be called off. A lot of people have written about it, but it was my sense that it was politics that forced the battle to be ended prematurely. That is important because remembering those political concerns was always in the background as we prepared for the second battle of Fallujah. The pace, sense of urgency, and tempo that Marine Corps senior leaders set for the second battle of Fallujah was informed or influenced by their experience with that lack of political will from the Iraqi government during the first battle.

Turning to Najaf. Yes, we gained some valuable experience in urban fighting there, but the 1st Cavalry Division's pre-deployment training also prepared us as well. General Peter Chiarelli, the division commander, and our brigade commander, Colonel John Murray, who is now a general and commanding the U.S. Army Futures Command, made our pre-deployment training hard and specific. We knew that we were going to fight in the city—we were going to Baghdad—so we trained hard, and we were ready.

[22] The views expressed by James Rainey are his and do not necessarily reflect those of the United States Army, Department of Defense, or the U.S. Government.

The division's 1st Brigade got into a fight in Sadr City almost immediately upon our arrival in Iraq and acquitted themselves incredibly well. It was Colonel Robert Abrams, Lieutenant Colonel Gary Volesky, and their units.[23] So, we were all well trained, but you can, of course, always get better. In Najaf, fighting with the 11th Marine Expeditionary Unit, our staff and I learned a ton about fighting as part of a joint (multiservice) force: interoperability, how to leverage fires, how intelligence systems work, and how commanders talk to each other on the battlefield. So, the experience we gained in Najaf was extremely useful for the second battle of Fallujah.

Spencer: When did your battalion get notified that you would be supporting the Fallujah operation?

Rainey: It has been a while, but I think it was about the 3rd of November when I received the call. Task Force 2-7 Cavalry was part of the "Grey Wolf" Brigade (3rd Brigade, 1st Cavalry Division). We were attached to the 39th Infantry Brigade Combat Team and based at Taji (a rural district just north of Baghdad). We had gone down to Najaf, fought, came back, and then received the warning order to be prepared to move by ground from Taji to Fallujah to link up with the Marines.

Spencer: When did you actually get to the Fallujah area?

Rainey: The first thing I did was put our operations officer, Tim Karcher; our intelligence officer, Mike Irwin; and our fire support officer, Coley Tyler—who I believe you have talked with previously[24]—on a helicopter to the Marine headquarters to start the planning process. Cross-attaching units—especially in combat—is always sporty but doing it across services is even more complex. We had to conduct a pretty significant move from our base at Taji to Fallujah, that included going around Baghdad. There is not much stealth in moving a mechanized task force around, so the Marines masterfully acknowledged that challenge and incorporated our move into their information operations plan to get us there. It took about four days to get the entire task force into our assembly area north of Fallujah.

[23] Gary Volesky was awarded the Silver Star Medal for his actions as the battalion commander in Sadr City on August 16, 2004. See, "Gary Joseph Volesky," Wall of Valor Project, *Military Times*, https://valor.militarytimes.com/hero/29131.

[24] Coley Tyler was interviewed by the Modern War Institute's Jake Miraldi; see Jake Miraldi, "The Second Battle of Fallujah," *The Spear Podcast*, Modern War Institute, May 8, 2019, https://mwi.usma.edu/podcast-spear-second-battle-fallujah/.

Spencer: I know that you have mentioned it, but prior to being directed to Fallujah, how knowledgeable were you and your staff with the first battle of Fallujah?

Rainey: Obviously I was paying attention to everything that was going on every day within Iraq, but Fallujah was outside our area of operations, although it was inside our area of interest.[25] Fallujah mattered because it was a support zone for the enemy from which they used to launch attacks into Baghdad and Taji, where our unit was located. So, we were paying attention to Fallujah and knew that there had been a fight. With the first battle of Fallujah ending prematurely there was a sense that a reckoning was coming. So, I watched it a little bit, but I could not exactly study it, because we had a tough fight of our own going on in Taji every day. But once we received the warning order, we rapidly got up to speed on everything in Fallujah, from the first battle to the present.

Spencer: I know that the enemy situation between the first and second battles of Fallujah was vastly different. What was the enemy situation in Fallujah in November of 2004?

Rainey: Intelligence is never 100 percent accurate, but I would say we had somewhere between a 50 to 70 percent read on the enemy. We were not sure about the total number of enemy fighters within the city. However, because of an effective information operations campaign, pressure exerted by the Marine cordon, and some pretty effective humanitarian and civil affairs work, we felt that the city was largely void of civilians. By contrast, when I fought in Najaf and Baghdad, there was always a mix of friendly, enemy, neutral, and people who looked neutral or friendly but were actually enemy. Civilians on the battlefield always complicate the fighting. But we felt that the innocent civilians had moved out of the city and the only people remaining were going to fight and fight to the death. It very much had a last stand type of feel to it.

Spencer: I know it has been a while, but what was the overall mission for Operation Phantom Fury, or the second battle of Fallujah?

Rainey: The purpose of the operation was to deny or eliminate Fallujah as a safe haven for the enemy. The Sunni insurgents used Fallujah

[25] An "area of interest" is a doctrinal term defined as "that area of concern to the commander, including the area of influence, areas adjacent thereto, and extending into enemy territory. This area also includes areas occupied by enemy forces who could jeopardize the accomplishment of the mission." See, Department of the Army, *Operations*, ADP 3-0 (Washington, DC: Headquarters, Department of the Army, 2019), 4-3, https://armypubs.army.mil/epubs/DR_pubs/DR_a/ARN18010-ADP_3-0-000-WEB-2.pdf.

as a sanctuary to plan attacks, conduct meetings, and command and control their operations. They had numerous improvised explosive device (IED) factories and vehicle-borne improvised explosive device factories in the city. So, the purpose was to take away their sanctuary and their freedom of action. The specific mission, our part—well, RCT-1's part—was enemy oriented: destroy the enemy force in the initial phase; conduct a detailed clearance in the next phase; and transition to setting the conditions for good governance, restoring essential services, and establishing stability in the final phase. But the initial mission was enemy oriented.

Spencer: When I discuss major combat operations or large-scale combat operations, I often have to explain to my audience that you can have an enemy-based mission—such as defeat, destroy, or neutralize enemy forces—or you can have a terrain-based mission—such as seize an area or retain a terrain feature.[26]

Rainey: Yes, and you can also have a friendly oriented task, right? A supporting effort's purpose may be to protect the flank of the main effort formation. I always tell people that understanding your purpose is one of the most fundamental things that you have to do to fight well. You have to know if you are enemy, terrain, or friendly oriented; it makes a big difference.

Spencer: What was the main plan of attack for the entire element? I know that the attack included more than just RCT-1.

Rainey: The Marines had already established a cordon around the city, so the plan was to start with shaping operations using fires and information. The 1st Marine Division, consisting of two regimental combat teams, would then attack from north to south. Initially RCT-1 would be the main effort to seize about 25 percent of the city, and then RCT-7, as the follow-on main effort, would clear from north to south and then east to west, the remaining three quarters of the city. That, like always, is not exactly what happened, but that was the initial plan (see figure 9.2).

Spencer: And each RCT had an army mechanized battalion as the lead element?

---

[26] For a full list of tasks, see Department of the Army, *The Army Universal Task List*, ADRP 1-03 (Washington, DC: Headquarters, Department of the Army, 2015), https://armypubs.army.mil/epubs/DR_pubs/DR_a/pdf/web/adrp1_03.pdf.

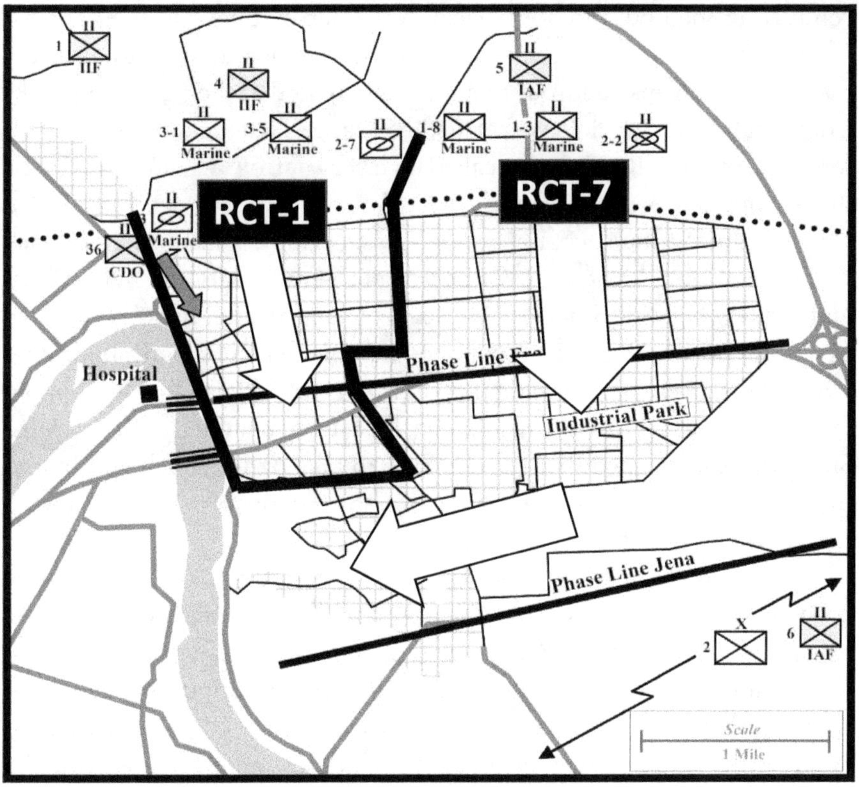

Figure 9.2. The plan for Operation al-Fajr, November 2004.
Source: Map is modified from map 3 in Gott, *Eyewitness to War*.

Rainey: They were what we would call a combined arms battalion now.[27] Our battalion, Task Force 2-7, was assigned to RCT-1, and Task Force 2-2 was assigned to RCT-7. We were basically an infantry-heavy mechanized task force with two mechanized infantry companies and one tank company.

Spencer: Were you the main- effort for the operation?

Rainey: Yes, Task Force 2-7 was the main effort for RCT-1 and RCT-1 was the main effort for the 1st Marine Division; at least for the initial phase of the attack.

Spencer: And what was your mission?

Rainey: Our task was to penetrate with its purpose to disrupt the enemy's defense so that the two Marine infantry battalions, who would be

[27] See chapter 12, "The Battle of Sadr City," in this book for a discussion on combined arms battalions.

conducting the house-to-house clearing, would not have to fight against an enemy in a strong defensive posture. So, our mission was to dislocate the enemy's defense and put them in motion so that they could not fight a deliberate delaying defense against the Marines who had to clear enemy forces in zone. But RCT-1's plan called for a very narrow avenue of approach that I thought was too narrow.

So, during the backbrief process I asked for a wider zone of attack so that we could bring more of our combat power into the fight. Tanks and Bradley Fighting Vehicles (Bradleys)[28] take up space, right? And the 25 percent of the city that RCT-1 would be fighting in had kind of a Byzantine structure and layout. It was not like some parts of Fallujah that had a more linear city block type of layout. The RCT-1 commander, Colonel Mike Shupp, was concerned that I was asking to take on too much terrain and that would slow us down, but having more experience with mechanized units, I was confident we could do it.

We had a great professional exchange throughout the planning process and because he was the commander—a great commander—it was his call, and he settled on giving me some extra space but not all that I wanted. I wanted another big road, but in hindsight, it was a great decision. Focusing our battalion on a narrower front resulted in a very deep and rapid penetration that dislocated the enemy's entire defense. It put the enemy in motion, forced them to leave many of their defenses, and gave the Marine infantry battalions the initiative.

It was a joint fight and whenever fighting as part of a joint or combined (multinational) force, it is important to remember that each unit brings its own unique capabilities, strengths, and weaknesses, and it is important to discuss them, to truly understand the capabilities of all the units in a joint or combined formation. Ultimately, with any joint or combined formation, there is a single commander who is responsible for the mission, its intent, its risks, and ensuring all the subordinate units are moving out within that intent. This mission was a textbook example of how commander-to-commander dialog, decision making, and execution within the commander's intent, should happen. It was as enjoyable as a fight could be.

[28] Bradley Fighting Vehicles are a type of armored personnel carrier used by the United States Army. For more details, see, "M2/M3 Bradley Fighting Vehicle," Military.com, https://www.military.com/equipment/m2-m3-bradley-fighting-vehicle.

Spencer: In reading about the planning process for this mission, I read that when you sent your planning team to the Marine headquarters, there were conversations about understanding the use of mechanized units in urban terrain. During the conversation your operations officer said something to the effect of, "When you unleash a heavy mech battalion like this in the city, it's going to be Armageddon," meaning there would be additional collateral damage to the terrain based on the firepower of the mechanized units.[29]

Rainey: Yes. From the onset, I felt we needed capability briefs so that we all understood each other's capabilities. The Marines had M1 Abrams tanks and knew what tanks could do but employed them differently from the army. The army fights tanks as companies of tanks as opposed to the Marines which use tanks in a direct supporting role to the infantry. So, we had to deal with some of those conceptual differences. Each service brought immense capabilities and skills and it was important to identify, discuss, and understand them.

One thing the Marines know well—and they have a bit of an advantage because they are almost joint by nature—is ground-air integration. They had significantly more experience employing joint fires, with the exception of army attack aviation, than us. There was a lot of mutual learning of how to get our targets prosecuted when we had a joint stack (many layers of aircraft above an objective area) of seemingly unlimited assets. But it all came down to whether our joint fires guys had the right radios and the right procedures to leverage those fires. So, there was a lot of back and forth.

When I think about the second battle of Fallujah, to me it was a textbook example of how to bring together a joint force. What right looks like. We all go to school, we see the line of block charts of joint task organizations, and we all inherently know that we fight as part of a joint force. But in my career, that fight was the most effective that I have ever seen come together. It was a great example of when the United States military comes together and puts all the pieces together as a joint force, the result can be incredibly powerful.

Spencer: You said that you had one tank company and two Bradley companies. What did they consist of?

Rainey: The tank company included fourteen M1A2 system enhancement package tanks, the most modern tanks we had at the time.

[29] Kendall Gott, ed., *Eyewitness to War Volume I: The US Army in Operation AL FAJR: An Oral History* (Fort Leavenworth, KS: Combat Studies Institute Press, 2006), 131-2, https://apps.dtic.mil/sti/pdfs/ADA460458.pdf.

I had about thirty M2 Bradley Fighting Vehicles. In total, our task force had three tank platoons and six Bradley platoons. Each Bradley platoon had two rifle squads. The 1st Cavalry Division was fully modernized so we also had 120 mm mortars. We had taken some casualties and had soldiers constantly rotating out on mandatory leave, but we were in pretty good shape, probably around 90 percent strength in terms of personnel.

Spencer: After you had received your mission and started conducting your mission analysis, you requested some additional assets. What were those?

Rainey: The 1st Cavalry Division leadership, Colonel Murray and General Chiarelli, told me that they would give me whatever I needed. I knew the rules of engagement and the law of armed conflict, but I also knew that in an urban environment, questions—such as what is acceptable in terms of collateral damage—are more common and more profound, so I asked for an operational law lawyer since army battalions do not have a lawyer assigned, and I received one. I also asked for another company. The division did not have one that they could give me, but I did get permission to bring a sizable portion of a company from the Oregon National Guard's 162nd Infantry Regiment, that was located in Fallujah, with me. I also asked for Joint Terminal Attack Controllers form the air force. They were a precious asset—experts at calling in bombing missions from air force, navy, and Marine aircraft—and there were never enough for everybody. True to his word, General Chiarelli provided me with some Joint Terminal Attack Controllers that were assigned to the division. So those were three big ones.

Spencer: That makes a lot of sense to me. Now, I would like to get into the nitty-gritty of fighting a tank and mechanized element during a penetration operation like this. When reading an oral history of the battle, I was surprised to see how the operation was conducted.[30] In many other battles—such as the battle of Mosul in 2016-2017, the battle of Marawi in 2017, or even the first battle of Fallujah—the military conducted some form of an isolation of the city, breached the exterior, secured a foothold, and then started clearing buildings from the outside to the center.[31] But instead, you conducted this deep penetration. Can you describe your scheme of

[30] Gott, *Eyewitness to War*.

[31] See chapter 13, "The Battle of Mosul," and chapter 14, "The Battle of Marawi," in this book; and John Spencer and Jayson Geroux, "Urban Warfare Project Case Study Series: Case Study #2 – Mosul," Modern War Institute, September 15, 2021, https://mwi.usma.edu/urban-warfare-project-case-study-2-battle-of-mosul/.

maneuver and how you positioned your forces along those main avenues of approach to conduct the kind of penetration that you did?

Rainey: Let me start a little bigger. Urban operations are a bit of a paradox. There is a consistent trend toward urbanization. Cities grow in important places—at major-road intersections—or lines of communications in military vernacular—along rivers, at ports, and places like that. And just about anybody we may fight is likely going to try and hide and mix into populations, which tend to be in urban areas. After seeing the U.S. military decimate the Iraqis in Operation Desert Storm, nobody wants to face the U.S. military in the open plains or the open desert. So that all argues for an increased probability that we will fight in urban areas.

Fighting in urban areas is something that we should do only as an absolute last result, because we lose many of our advantages. We give up our ability to sense the enemy in the deep area,[32] it limits our ability to mass fires at speed and scale because of collateral damage and the risks to civilians, and things like that. So, there are many reasons why you do not want to fight in urban areas. Another reason is the correlation of forces and means, meaning we require a large number of forces to conduct military operations in extremely large urban areas. That is something I spend a lot of time thinking about. I think the doctrinal five-to-one advantage to fight in an urban area is conservative at best.[33] I mean it really is even higher than that.

By our own doctrine, anything that you clear, you must also secure, right? If you do not secure it, you will find yourself clearing it again. You do not want the enemy popping up behind you—not to get too far down into the tactical level. But we saw that in Fallujah. My point is, if you look at the force ratios and the requirement to secure everything you clear, I do not care how big your military is, there are many urban areas in the world where clearing and securing the entire city simply is not feasible. You would run out of combat power before you could clear and secure everything.

[32] U.S. Army doctrine describes the commander's area of operations including the close, deep, and support areas. The deep area is "Where the commander sets conditions for future success in close combat." The close area is "the portion of the commander's area of operations where the majority of subordinate maneuver forces conduct close combat." But in an urban area, it is more difficult to engage the enemy in the deep area, meaning more engagement occurs in the close area. See, Department of the Army, *Operations*.

[33] According to Army doctrine, "History suggests urban offensive missions require three to five times greater troop density than for similar mission in open terrain." See, Department of the Army, *Urban Operations*, ATP 3-06 (Washington, DC: Headquarters, Department of the Army, 2017), 2-3, https://armypubs.army.mil/epubs/DR_pubs/DR_a/pdf/web/ARN6452_ATP%203-06%20FINAL%20WEB.pdf.

In the future we will probably be asked to do something that is kind of in between that. I think we have to be able to temporarily fight in an urban area for a specific purpose. For example, if a line of communication (supply route) passes through a large urban area, it is reasonable to expect that a United States, coalition, or partner unit could be told to clear and secure the line of communication so that logistics can move through it. That is a much different requirement from clearing and holding a gigantic urban area.

Another example would be conducting a raid. You might need to dominate or control a small sector of an urban area so that you can conduct a raid, but it is only a small area and for a short period of time. You go in, conduct the raid, and get out. So, saying we will not fight in an urban area is probably not realistic, but on the other hand, the idea that anybody would build, man, train, and equip a ground force solely for the purpose of getting consumed in a giant urban area, I do not think that is a part of the future either. Does that make sense?

Spencer: One hundred percent. I often ask people, especially when they try to make a correlation between the size of the force required for a mission versus the size of a city and its population, to tell me what the mission is. If the mission is to go in and clear every structure in a megacity or large urban area, that is probably not feasible, but there are plenty of scenarios where the military will find itself conducting operations in major cities. Everything from, like you said, limited objectives such as stopping rocket attacks to securing supply routes. There are plenty of scenarios that would require a military force to deploy into a very large urban area. In fact, fighting in cities has been a consistent part of war throughout history and there is no reason to believe that we can somehow avoid it in the future. It is often said that "war is human endeavor" and since humans live in cities, cities cannot be avoided. Our doctrine, which is great, does cover this.[34] But we have to be ready for this very unique and difficult operating environment.

Rainey: Yes, and I think our modernization efforts in the army are well informed by what we think the future fight is going to be. I think

[34] For example, U.S. Army doctrine states, "Military forces operate in urban areas for several reasons: The urban environment offers defensive advantages. The urban area harbors threats which can attack friendly forces elsewhere. The urban environment's people (their allegiance and support), infrastructure, capabilities, or other resources have operational or strategic value. The urban area has significant symbolic importance. The urban area's geographical location dominates a region or avenue of approach." Department of the Army, *Urban Operations*, 1-2.

we are making significant progress. For example, the idea of developing unmanned, sensor robotics that would allow us to make first contact with the enemy with something other than our soldiers. I think technology is going to increase our capability and those technologies will help if we do find ourselves fighting, for limited purposes, in an urban area. But none of that existed 16 years ago. In the second battle of Fallujah, we faced a dug-in, booby-trapped, mobile enemy who was largely committed. The enemy was there to fight to the death. Some of them eventually ended up surrendering or running away, but initially everybody that was there, was there to fight.

The enemy also had a significant amount of time to prepare its defenses. Every house was booby-trapped and laced with mouseholes (holes in the walls that allow fighters to pass between rooms and between houses). We would go through a breach and there would be a machine gun bunker built inside the room waiting for us. We would enter the first floor and there would be loopholes in the ceiling two floors up that they would use to drop grenades on us. They also built booby-traps on an industrial scale. We would kick in a door and the whole building would blow up. So, it was a tough, nasty fight.

Spencer: One thing that I found interesting about this battle was how the mechanized infantry and tanks were used. It may not be completely reminiscent of the drive to Baghdad, but as I go through the steps of a doctrinal deliberate attack—isolate, breach, seize a foothold, clear the objective—that is not what I saw happening here.[35] Your battalion penetrated deep into the city without doing a house-by-house search to get there. And, I believe, you did not dismount your infantry unless necessary; the tanks led the penetration. In other battles that I have studied, like the Battle of Aachen in World War II, dismounted forces were employed with or even in advance to protect the armored vehicles.[36]

---

[35] In the battle for Baghdad in April 2003, a single Armored Brigade penetrated deep into the city, secured key road intersections, and established a defense—ultimately leading to the capture of Baghdad overnight. See, Army University Press, "OIF: The Fight for Baghdad," YouTube video, August 14, 2020, https://www.youtube.com/watch?v=d8uaFZAxzpw&t=6s. For the doctrinal phases of a deliberate attack, see Department of the Army, *Combined Arms Operations in Urban Terrain*, ATTP 3-06.11 (Washington, DC: Headquarters, Department of the Army, 2011), 2-12, https://armypubs.army.mil/epubs/DR_pubs/DR_a/pdf/web/attp3_06x11.pdf.

[36] William Robertson and Lawrence Yates, *Block by Block: The Challenges of Urban Operations* (Fort Leavenworth, KS: U.S. Army Command and General Staff College Press, 2003), 75, https://cgsc.contentdm.oclc.org/digital/collection/p16040coll3/id/88/.

Rainey: Yes. The Marines had a great plan. I cannot overstate the fact that the operation was a well-planned, well-rehearsed, integrated, joint operation and the two large infantry battalions basically cleared in zone. They cleared the buildings and destroyed the enemy. So, if you take a mechanized force, whose strength are tanks and Bradleys—mobility, protection, and firepower—why would we dismount those couple squads and clear two floors of a building? It would exhaust our combat power and slow the penetration. That was a job for the Marines. They had large infantry squads, were very good with mortars, and they were optimized for close-in fighting.

One reason the plan was so successful is because of exactly what you said. Our job was to penetrate. We penetrated rapidly to the middle of the city during the first period of darkness, leading with tanks because that was where the firepower, survivability, and speed of our formation was located. If the enemy wanted to hide, we bypassed them; if they wanted to fight, we killed them.

Our goal was to reach Jolan Park, a big giant park in the middle of the city (see figure 9.3). The park even had a giant Ferris wheel which was kind of creepy in a combat zone. The intelligence team determined that the enemy's most likely course of action was to defend against us from all around the city. Once they figured out where our main attack was occurring, they would delay us there while the remainder of their force would fall back to that park and defend there. So Alpha Company, 2-7 Cavalry attacked along four or five roads right towards that park, destroyed the enemy force that was there, was there before the main enemy groups could fall back, and we were waiting to destroy them when they did. At the same time Charlie Company, 3-8 Cavalry, our tank company, attacked right down the main road, destroying the enemy as they came out to fight.

By morning, we had secured the park. The enemy was then facing a large, mechanized formation that commanded the center of the city and controlled all the roads leading into it with fires. One of my mechanized infantry companies was located at what I believe was their decisive point. Instead of falling back to defend, they now had to deal with multiple threats.

Simultaneously, the same thing happened in RCT-7's sector. Multiple small units of very lethal and well-trained Marines were pressing the enemy in multiple directions. It disjointed the enemy. What I learned from the attack and what we also discovered in Najaf, was that if you could penetrate an urban area and get somewhere you could defend, the enemy must come

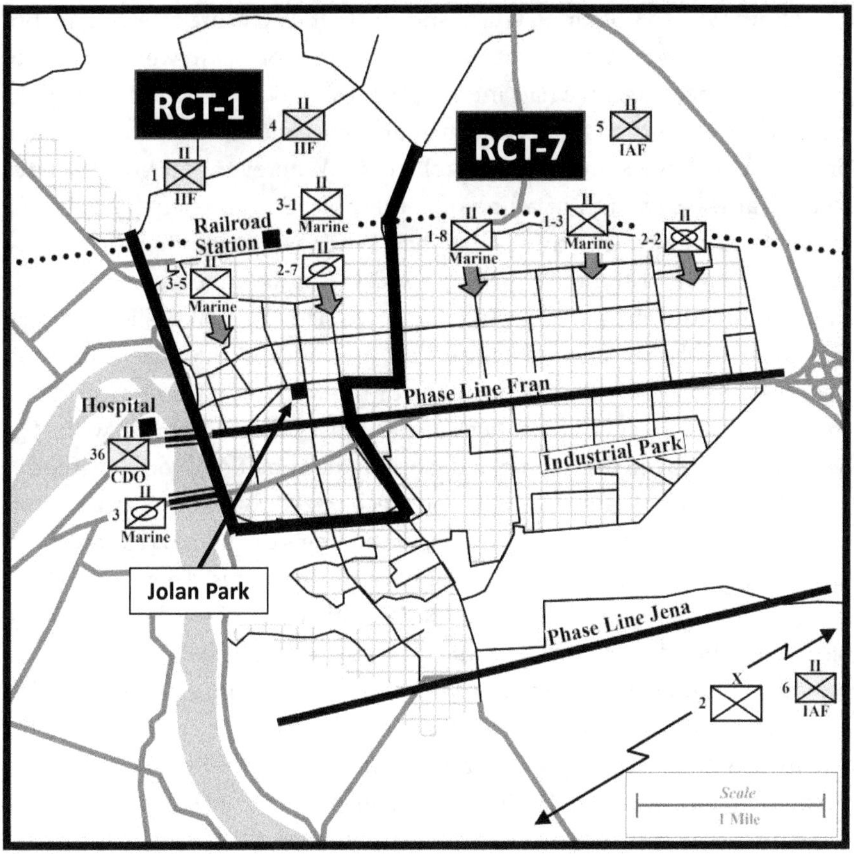

Figure 9.3. Operation al-Fajr initial assault, November 9, 2004.
Source: Map is modified from map 4 in Gott, *Eyewitness to War*.

to you. And that is what you want. You want the enemy moving to you versus you having to root them out of well-prepared defensive positions.

Spencer: I think that is something that I am learning the more I study these battles. Your battle in Najaf, this one, the fight for Baghdad with General Perkins were all similar: conduct a deep penetration to get in there and hold terrain within the enemy zone. It does start to shape my own thinking about the successful tactics of an operation like this. But to execute a deep penetration requires vehicles with the speed, protection, and firepower. In other battles that I have studied, where armies lacked that mobile protective firepower, they were forced to start at the exterior, conduct a shallow penetration, and then start clearing rather than getting inside and seizing this key terrain.

Rainey: I do not know which specific battles you are referring to, but it is important to remember that we were a joint force. We were successful because we had fires, we had intelligence, we had all of it. There are risks with penetration-type tactics in urban areas because if your force gets cut off then you have some serious problems.[37] Wherever you go, you have to hold at least enough of a line of communication to resupply yourself, evacuate casualties, and things like that. When we penetrated the city, we continued to hold the road that we used to conduct the penetration. So, in a combined arms battalion like ours, we had the tanks' shock, speed, firepower, and protection. We also had mechanized infantry which had vehicles that basically exist to position our rifle squads. It was very different from a light infantry unit, or the Marines that we were fighting with. The Marine squads were breaching and clearing floor by floor, room by room.

The infantry that comes with the mechanized formation largely exists to protect the tanks and Bradleys from the enemy's dismounted infantry. So, we would penetrate and then dismount our rifle squads. They had to fight in the buildings too, but only for a specific purpose and we did not have to clear every building along our route. The infantry breached into buildings, cleared what they had to get observation of the road, and emplaced snipers to support the advance. All of this was to protect the vehicles from the enemy's fighters who might be maneuvering against the armor formations, because tanks are vulnerable in an urban area.

If the enemy can get close, they can hurt the tanks. The urban space gave the defender all kinds of "keyhole" shots. We lost six tanks in Fallujah. A couple ran over mines, but most of them were disabled by rocket-propelled grenade (RPG) volley fire. They did not, however, get a catastrophic kill on any of the tanks or hurt any of the soldiers inside them, but they did penetrate gun tubes, blew off tracks, blew vision blocks[38] into the turret, and things like that. Urban warfare requires combined arms capabilities.

Spencer: The battle I was thinking about, the one where the military lacked armor and was instead forced to clear block-by-block, was 2017's Battle of Marawi in the Philippines. The Armed Forces of the Philippines lacked mobile protected firepower, so they experienced a much different fight.[39]

[37] See, for example, chapter 8, "The Battle of Mogadishu," in this book.
[38] Vision blocks are periscopes that the driver uses to drive a tank when the driver is forced to close his hatch. See, "M1 Abrams Main Battle Tank," Global Security, https://www.globalsecurity.org/military/systems/ground/m1-intro.htm.
[39] See chapter 14, "The Battle of Marawi," in this book.

Rainey: Yes. You have to have the ability to rapidly penetrate, which takes speed, protection, and firepower. There were IEDs in Fallujah but not to the extent that we would see if someone was defending against an attack by the United States in an urban area today. For example, there were no drone swarms back then. There are going to be a lot of new challenges in the future. If you have to fight in an urban environment, you want everything. You want tanks that can penetrate. If you want tanks, then you have to have infantry to protect them. If you want infantry to protect them, then you have to have infantry fighting vehicles to position your rifle squads. You need a mix between a lot of firepower, a lot of mobility, and a lot of protection. It requires both tanks and rifle squads on the ground. Rifle squads have to be lethal; they have to be well trained; and they have to know how to fight and execute battle drills.

It also requires great intelligence. It requires fires, which can be challenging to employ in urban environments, so you may have to be artful in their employment to minimize collateral damage and the risk to civilians. There are also medical and logistic challenges, which are a challenge in any fight, but they are like ten times more complex in an urban area. There are many challenges associated with urban warfare.

Spencer: I think the use of fires in support of maneuvering forces is really important. As you mentioned, we have a Modern War Institute podcast with your fire support officer, Captain Coley Tyler, about the use of fires in the second battle of Fallujah.[40] He talked about the shaping operations—the operations "that create[d] and preserve[d] conditions for success of the decisive operation through effects on the enemy, other actors, and the terrain create and preserve[d] conditions for the success of the decisive operation"—that were conducted prior to your advance.[41] He also discussed the fires that supported your maneuver, as well as some of the very different capabilities that he was exposed to during the battle. Could you talk about some of the joint fires that you had available to you, everything ranging from AC-130 gunships to unmanned aerial systems. Some people might point to this battle as one of the first times that the United States used unmanned aerial systems to such an extent to identify enemy positions and provide the intelligence to conduct effective fires during an operation.

[40] Miraldi, "The Second Battle of Fallujah."

[41] Department of the Army, *Operations*, 4-5.

Rainey: Like I said, we were but one battalion in the division. Fires were very centralized for the shaping operations and the division and the RCTs planned and executed the fires. Because the enemy was doing things like building fighting positions, digging in, and emplacing IEDs, that created the conditions within the law of armed conflict and the rules of engagement to use fires during the shaping operations. You cannot just blow up a building, but if the enemy builds a fighting position in the building and is occupying it, then it becomes a legitimate target.

The battle of Fallujah was the biggest priority in Iraq for the United States military, so we were given an array of assets. Our battalion had priority of fires from an AC-130 gunship, for example, during the penetration that first night. As opposed to not having enough fires capability, we had so much that the ability to deconflict it was our challenge—an interesting and good problem to have. We have made significant progress on joint interoperability over the past decade and a half, but at the time, Captain Tyler and his crew had a lot on their hands. The Joint Terminal Attack Controllers were great at talking to the air force aircraft and they were pretty good at talking to navy aircraft, but the army forward observers had to rapidly learn how to effectively get targets prosecuted through the Marines' systems.

It also worked the other way around with army systems supporting the Marines. The 120 mm mortars that we had during the fight were major killers on the battlefield. The Marines have a lot of mortars, but at the time, they did not have the 120s. The challenge with shooting artillery into a city is the low trajectory of the round. You may not be able to fire artillery rounds at your desired target because it may be blocked by another building. Mortar rounds travel at a higher trajectory and can go over the buildings that are in the way of the artillery. Mortar rounds can come down right on the top of a building or in between two buildings. Over the course of the battle the Marines learned how to get fire missions to the army 120 mm mortars and became very proficient.

Spencer: I will go ahead and read you my favorite quote from you about fires, because it really motivates me. In one of your post-battle interviews you said, "You can't win anything without infantrymen on the ground physically taking it, but the easiest, most effective and least casualty producing on friendly forces way to fight an enemy, even in urban environments, is with fires."[42]

[42] Gott, *Eyewitness to War*, 131.

Rainey: Yes, I grew up as an infantryman, but I like to think of myself as a combined arms maneuver officer now as opposed to just a single branch officer. Infantry and fires are not mutually exclusive. Everybody in the army ultimately exists to put a rifle squad in a position of advantage to close with and destroy the enemy. Nobody wants to do that, but at the end of the day—and I think the history of war supports the idea—somebody always has to do the close-in, tough fight. Someone has to close that last 100 meters. In a contest of wills between humans, sometimes people must get killed. But, at the same time, any good maneuver officer is going to exhaust every resource at his or her disposal before they commit men and women to close-in fighting. So yes, my first solution to every problem is to exhaust other means before I decide to commit infantry or even armor forces. If I can get accurate intelligence and employ fires to destroy the enemy from a distance, I will.

Spencer: I really appreciate your comments about the 120 mm mortars because I was an infantry mortarman in my enlisted days. But I also know that within dense urban terrain, like Fallujah, or in even denser urban areas, you are going to have tall buildings that obstruct fires and the responsiveness you get from having organic mortars is huge in an urban fight.

Rainey: Yes, so the key for younger folks coming up is to understand our profession and understand how to fight. They have a responsibility. If they are going to command a formation in combat, they have a responsibility to be knowledgeable, proficient, and an expert in many things. Maneuver commanders have to understand the intelligence warfighting function, they have to know logistics, they have to be extremely competent in the employment of fires. And they have a moral responsibility to do that in an ethical way. We often talk about leadership and command in combat, the weight and responsibility that comes with that, how dedicated and committed you need to be, how hard you have to aspire to be competent, and why being a high character individual as a leader matter. You can find all those lessons, not just in 2-7 Cavalry's fight, but in other units throughout the second battle of Fallujah.

Spencer: I hope people can see the complexity—all warfighting is complex but urban is another level—of this battle. I also found it interesting that the 1st Cavalry Division's 4th Aviation Brigade commander who moved troops around—and conducted the aerial outer cordon, attack, and lift missions—was then Colonel James McConville who is now the Chief of Staff of the Army.

Rainey: Yes, we had a great team. We trained together and we were close. I mentioned earlier that every fire system we could ever want was there for us. But there was one particular night when we were in a really, really bad situation. Coley Tyler and our fires team were working through deconflicting fires, trying to get them cleared. We were taking fire from a protected target. We had an infantry formation pinned down that was taking fire from a barricaded position that was on the no strike list (a list that classifies protected sites such as hospitals, churches, mosques, etc.). So, we were having trouble trying to get the fire mission approved. Then we got a "check in" over the radio from an air weapons team of Apache attack helicopters from the 4th Aviation Brigade. The pilots dropped to our radio channel and told us that they could solve our problem. It took a couple of shots, but they absolutely saved the lives of a whole bunch of soldiers, and they let us get back on the attack and finish the mission that we were on that night. One of the Apache pilots was Colonel McConville. His combat credibility is unquestionable, I am an eyewitness to that.

Spencer: My last question relates to leader development. As one of your former subordinate commanders, I know that you firmly believe in self-development to better understand the warfighting functions.[43] If a brigade combat team was training for an attack on a city like Fallujah, what recommendations would you provide to the commander and staff that is planning to execute a mission like that?

Rainey: This may be unsatisfying to you, but I believe that urban fighting is a condition of a battlefield. It does not mean that standards change. What I tell people is that the best thing you can do is have fit, disciplined, and cohesive teams. Good small unit formations that are well led, that can execute battle drills, and that are absolutely lethal in terms of employing their weapon systems. That is your best bet. There are some unique things that you have to adjust based on the environment, be it the desert, jungle, or city. But the most important advice I would give anybody is that fit, disciplined, cohesive soldiers; and small unit teams that are very, very good at battle drills, converging direct and indirect fires, and the ability to shoot, move, communicate; survive on the battlefield. Make sure they are incredibly well led, by officers and noncommissioned officers at the platoon, company, troop, and battery level. I think that is going to

[43] The five warfighting functions are: command and control, movement and maneuver, intelligence, fires, sustainment, and protection. To learn more about the Army's warfighting functions, see Department of the Army, *Operations*.

carry the day in any fight. And if you cannot do that, then you are in a bad place on any modern battlefield. The people—soldiers, leaders, and commanders—and the way we train with passion and seriousness are the real asymmetric advantages that we have as an army today.

Spencer: Well, sir, I really appreciate your time. There were some very important lessons that can be learned from the second battle of Fallujah that would support anyone's leader development program or self-study. I found our discussion helpful as I continue my own development. Thanks a lot, sir.

Rainey: Yes, and thank you, John. One last thing; I would ask everybody to remember that we, the United States Army and Marines, paid a high price for a significant victory. The victory came at the cost of the soldiers and Marines who lost their lives in the battle, and I would just ask everybody to keep them all in their thoughts and prayers. I will and I hope others always remember that.[44]

[44] Eighty-two servicemembers were killed and more than 600 were wounded during the Second Battle of Fallujah. See McWilliams and Schlosser, *U.S. Marines in Battle*, 67.

10

Rebuilding Fallujah

Studies of most urban battles tend to focus on the tactical fighting, where the bullets are flying and the explosions are occurring. However, an important aspect of urban operations is the fact that civilians inhabit the battlefield, which means militaries must be prepared to support civilians caught in the middle of urban conflict and support them after the battle has concluded. This chapter also discusses Operation al-Fajr, discussed in the previous chapter, but does so from the perspective of a civil affairs officer. As such, this chapter is unique among those in part II of this volume because it focuses on the important role that civil affairs operations play in urban warfare.

Civil affairs tend to be associated with what the military typically refers to as phase IV, or stability, operations.[1] People tend to think that civil affairs are waiting for phase III (dominate) operations to end, and then civil affairs comes in to help rebuild a city and reestablish governance. This chapter demonstrates that civil affairs can and should play an important role throughout all phases of an operation. If employed effectively, civil affairs operations directly contribute to the success of decisive operations.

In this chapter John Spencer interviews Colonel Leonard J. DeFrancisci, a retired officer who served for 32 years in the United States Marine Corps. He spent 14 years in civil affairs, conducting operational deployments to Iraq, Afghanistan, Nicaragua, Korea, Liberia, Albania, Jordan, and the Horn of Africa. He served in many different positions within civil affairs to include serving as commander of the 2nd Civil Affairs Group and the 4th Civil Affairs Group. In 2004 he served as a civil affairs detachment

[1] In United States military doctrine, there are six operational phases: phase 0 (shape), phase I (deter), phase II (seize initiative), phase III (dominate), phase IV (stabilize), phase V (enable civil authority). See Joint Chiefs of Staff, *Joint Operations*, JP 3-0 (Washington, DC: Joint Chiefs of Staff, 2018), https://www.jcs.mil/Portals/36/Documents/Doctrine/pubs/jp3_0ch1.pdf.

commander for Regimental Combat Team 1 (RCT-1) during the second battle of Fallujah.

In that role, DeFrancisci commanded the civil affairs detachment that provided civil affairs support throughout Operation al-Fajr. During the shaping operations prior to the assault, his detachment developed local contacts, built lists of key individuals, mapped critical infrastructure, supported displaced residents, and supported the division's deception operations. During the assault, his detachment conducted humanitarian assistance, population control, infrastructure assessments, critical repairs, and on one occasion, figured out how to stop the city's flooding so the assault could continue. Finally, his detachment played a critical role in the stability, or post-assault phase, of the operation. Beyond the typical things of getting the city functioning again, his detachment, along with mortuary affairs, helped facilitate the removal of more than 2,000 dead enemy fighters.

John Spencer: I would like to start by asking if you could provide us with a brief summary of your military career?

Leonard DeFrancisci: I started my career as an infantry officer, spending eight years in the active and reserve component. I then served in a motor transport company because that was the closest reserve unit to where I lived. When I went to a staff training course as a captain, I was exposed to civil affairs for the first time. The commanding officer of one of the Marine Corps' civil affairs groups discussed the unit and some of the missions that he had conducted in the Balkans. It sounded interesting, but I did not join until later. One thing that really stood out to me was the reconstruction aspect of the civil affairs, because I built pasta machines in my civilian job at DEMACO.

After 9/11, I called the commander of the 4th Civil Affairs Group and told him that I was interested in becoming a civil affairs officer. The commander, Colonel Steve McKinley, laughed and said, "I interview everybody, and I will size you up in five minutes." He said that he only conducted in-person interviews, so I had to fly to Richmond International Airport from my home in Florida. I never left the airport; we conducted the interview at a coffee shop there and I caught a flight home that same day. A couple days later, he offered me a job. I suspect it had less to do with anything that I had said and more to do with the fact that he was short officers. Regardless of the reason, I started my training in civil affairs. At the time, the Marine Corps outsourced its civil affairs education to the army, so I had to complete a six-module correspondence course with the army to become a civil affairs officer.

Spencer: I think most people are familiar with infantry, but many are less familiar with civil affairs. What is civil affairs and what are civil affairs operations?

DeFrancisci: According to United States military doctrine, civil affairs operations are "actions planned, coordinated, executed, and assessed to enhance awareness of, and manage the interaction with, the civil component of the operational environment; identify and reduce underlying causes of instability within civil society; and/or involve the application of functional specialty skills normally the responsibility of civil government."[2] Civil affairs forces are "trained, organized, and equipped to plan, execute, and assess civil affairs operations."[3] Whether a commander is conducting combat or non-combat operations, civilians are often present so civil affairs serve as experts in that realm.

Spencer: Great. Let us jump right into the second battle of Fallujah. Can you discuss your actions leading up to Operation al-Fajr?

DeFrancisci: I was mobilized with the 4th Civil Affairs Group in February 2003 and deployed to Iraq in March to support the ground war. During this deployment I served as a team leader for 26th Marine Expeditionary Unit in Mosul until we were demobilized around September 2003. When the group was remobilized in July 2004, I was assigned to RCT-1. My detachment, I had been promoted to detachment commander, replaced a detachment from the 3rd Civil Affairs Group in August 2004. The first battle of Fallujah occurred during the 3rd Civil Affairs Group's deployment, but we started planning for Operation al-Fajr shortly after I arrived.

Fallujah was a city of 300,000 smack in the middle of the "Sunni Triangle" a densely populated region inhabited predominantly by Sunni Muslims (see figure 9.1 in the previous chapter).[4] It was known as the "city of mosques" because the city and its surrounding communities contained

[2] Department of the Army, *Civil Affairs Operations*, FM 3-57 (Washington, DC: Headquarters, Department of the Army), Glossary-3, https://armypubs.army.mil/epubs/DR_pubs/DR_a/ARN33094-FM_3-57-000-WEB-1.pdf.

[3] Ibid., iv.

[4] In 2004, Fallujah was estimated to have a population of 250,000 - 300,000. See, Timothy S. McWilliams and Nicholas J. Schlosser, *U.S. Marines in Battle: Fallujah, November-December 2004* (Quantico, VA: U.S. Marine Corps History Division, 2014), 3, https://www.usmcu.edu/Portals/218/FALLUJAH.pdf.

200 mosques.[5] The residents were very proud of their staunch honor codes and conservative social conformity. Historically they were a restive people.[6]

Coalition forces had bypassed Fallujah during the initial march to Baghdad in 2003, so it had been largely unaffected by the ground war.[7] But the city soon became a hotbed of insurgent activity and served as an insurgent base of operations. Fallujah had a troubled history with the coalition dating back to April 2003, when soldiers from the 82nd Airborne Division returned fire after being engaged by hostile militants hidden among an angry crowd of locals which resulted in several casualties.[8]

After coalition forces prematurely ended the first battle of Fallujah in April 2004, the Fallujah Brigade was given control of the city, but it proved to be completely ineffective.[9] By the time I returned in August 2004, the insurgency had coalesced into a more radical movement led by local militants and al Qaeda in Iraq leader Abu Musab al-Zarqawi. They controlled the city.[10] The insurgents believed that they had won the first battle of Fallujah and were confident, I would say overconfident, that they could do it again if attacked. Since the first battle of Fallujah, few American forces had entered the city, and the few that had, conducted short duration raids.

When I arrived, Fallujah fell within RCT-1's area of operations, or sector. For Operation al-Fajr RCT-1 was comprised of 3rd Battalion, 5th Marines (3/5 Marines)), 3rd Battalion, 1st Marines (3/1 Marines), and Task Force 2-7 Cavalry. I had a civil affairs team assigned to each infantry battalion, two in general support of the regiment, and teams at the Abu Ghraib Prison and the Mahmudiyah area, outside of Baghdad.

---

5 "Fallujah: Embattled City of Mosques," BBC News, May 30, 2016, https://www.bbc.com/news/world-middle-east-25658586#:~:text=Thousands%20of%20buildings%2C%20including%20almost,its%20300%2C000%20residents%20were%20displaced.

6 Coley D. Tyler, *Ghosts of Fallujah* (Athens, GA: Deeds Publishing, 2018), 12-13.

7 Joel D. Rayburn and Frank Sobchak, eds., *The U.S. Army in the Iraq War, vol. 1* (Carlisle Barracks, PA: U.S. Army War College Press, 2019), https://digitalcommons.usmalibrary.org/cgi/viewcontent.cgi?article=1018&context=books.

8 Ibid., 123-4.

9 Ibid., 297-8, 344.

10 At the time, al-Qaeda in Iraq operated as Jamat al-Tawhid wal-Jihad. The group changed its name to al Qaeda in Iraq following Abu Musab al-Zarqawi's publicly pledged allegiance to Bin Laden in October 2004. Brian Fishman, "The History of the Islamic State," in *Routledge Handbook of U.S. Counterterrorism and Irregular Warfare Operations*, eds. Michael A. Sheehan, Erich Marquardt, and Liam Collins (New York: Routledge, 2021), 57, DOI: 10.4324/9781003164500-5; and Schlosser, *U.S. Marines in Battle*, 4.

Prior to the second battle of Fallujah, we conducted counterinsurgency operations. The Marines isolated the city, rooted out militants, and built relationships with locals. The insurgency used hit and run tactics, conducted improvised explosive device attacks against coalition forces throughout Iraq, and conducted fairly effective mortar attacks against RCT-1's headquarters at Camp Fallujah, a base located 2–3 kilometers southeast of Fallujah. The insurgents also intimidated the local population to further their cause and strip away support for the coalition.

Isolating the city was difficult because it required the support of the smaller towns and suburbs that surrounded Fallujah. The regiment attempted to isolate the city by conducting security patrols to keep the insurgents off balance and meeting with town officials to gain their support. The main towns surrounding Fallujah were Saqlawiyah, Karmah, and Nassir wa Salam near the Abu Ghraib Prison.

Civil affairs operations preceded combat operations into Fallujah. Over time, residents started leaving the city. Some went to live with extended families elsewhere in Iraq, but many had no place to go, so they looked for places in neighboring towns. Many ended up living in schools, so the schools stopped functioning as places of education and instead served as temporary living spaces.

We met with local civilians in a compound we called the "Fallujah liaison team," which was in essence a Civil Military Operations Center in military vernacular, located between the city of Fallujah and Camp Fallujah, just outside the camp's gate. The regular meetings with city's residents and the proximity to the city offered us a good window into Fallujah before the battle. After the battle started, people continued to visit the Fallujah liaison team compound, and they proved to be an excellent source of information as to what was going on in the city. So as a civil affairs unit, we were able to provide intelligence about the city.

Another important civil affairs mission was humanitarian related. There was a shortage of critical goods and services, so we helped facilitate the distribution of food, water, blankets, and other essential supplies to the dislocated people who needed them, as well as medical treatment. Once again, this offered us access to people and people meant information.

We also built a map of the city that proved to be invaluable on numerous occasions throughout the battle. We could not physically enter the city prior to the battle to conduct infrastructure assessments, but the many civilians that we dealt with at the Fallujah liaison team compound, temporary living spaces, or elsewhere, could. But few knew how to read a military or civilian map. Thus, we had to be inventive and learn the streets by the names that they used, so it took us a while to build a map of the city using their street names. Even if they could not read a map, they could talk us onto specific locations on the map, based on the streets or well-known areas, such as known buildings or parks. We used this information to map out critical infrastructure and critical systems before the battle started. We crossed-referenced the street names that the locals used with the phase lines and street names we used.

I will provide an example, which turned out to be important later during the battle itself. Water run-off needed to be pumped out of the city through a series of sewer lift stations because Fallujah's elevation was not much higher than the neighboring Euphrates River. When the battle started, the city flooded because the sewer lift stations did not work. If one lift station in the series failed, then parts of the city would flood. Flooding negatively impacted maneuver within the city, and I will discuss that a little later.

Over time, we built a pretty good understanding of the city and developed relationships with many people but focused on the people who actually ran the city: technocrats, municipal workers, business leaders, electrical and water plant technicians, doctors and hospital staff, garbage and maintenance workers, teachers, firefighters, and police. We needed to identify these people prior to the battle because they would be absolutely critical to getting the city quickly functioning after the battle.

As we developed a list of key infrastructure, we shared it with the regimental staff so that they could get the structures added to the restricted fire list, to minimize the chance that these structures would take damage unnecessarily during the battle.

By late October, the security situation in Fallujah had devolved to such a level that it was clear that the Marines would have to enter and clear the city of its estimated 4,500 hard-core fighters.[11] A city of this size required a division, so other units were moved to support the operation. But it was not like units were sitting around Iraq doing nothing. Any units moved to support Fallujah created gaps or thinned the lines in other parts

[11] Donald P. Wright and Timothy R. Reese, *On Point II: Transition to the New Campaign* (Fort Leavenworth, KS: Combat Studies Institute Press, 2008), 44, https://www.armyupress.army.mil/Portals/7/Primer-on-Urban-Operation/Documents/OnPointII.pdf.

of the country, so the Marines needed to clear the city quickly to free up the concentrated forces for other missions. Likewise, everyone remembered how the first battle of Fallujah ended, so no one wanted to be forced to end a second battle prematurely. This was yet another incentive to finish the battle quickly. Prime Minister Ayad Allawi ultimately provided the coalition 60 days to complete the battle, but the coalition could never be sure that he would not get cold feet and prematurely end it.[12]

The army's task force 2-7 Cavalry was attached to RCT-1 for the battle to provide it additional armor capability. RCT-7 was also moved to support the operation and it was likewise augmented with an army task force, 2-2 Infantry. A total of 18,000 Soldiers, Marines, and special operators were committed to the battle.[13] The Marines attacked from east to west during the first battle of Fallujah, so we decided to attack from north to south this time around.[14] Civil affairs helped shape the deception plan to confuse the enemy about the direction of our attack. We hired contractors to build an assembly area south of the city. We did not occupy it, but the insurgents, on hearing about this project, oriented many of their defenses toward the south. This is an example of how civil affairs helped to shape the battlefield.

We had about a month to plan for the operation, and we planned out just about every contingency imaginable. We split the operation into four phases: shaping operations, isolation, assault, and reconstruction.[15] During the first two phases, we stockpiled civilian supplies into what we called the "green mountain" and identified technocrats that we would need for the reconstruction phase.

We did a lot of planning, but no matter how much you plan, you always forget something. One of the things that we forget to consider was dislocated animals. We had a pretty good plan for dislocated civilians. During the shaping operations, we warned civilians to leave the city. We did not tell them when the battle would start, but they took the warnings seriously and many left. As few as 500 civilians were estimated to have

[12] Rayburn and Sobchak, *The U.S. Army in the Iraq War*, 345.

[13] Wright and Reese, *On Point II*, 44.

[14] Stanley McChrystal, *My Share of the Task* (New York: Penguin, 2013), 127, https://doi.org/10.1080/23296151.2015.1017990.

[15] Unlike the previous mention of large-scale joint operation phases, services execute operational and tactical missions and use phases as planning and execution tools to divide an operation in duration or activity. See Department of the Army, *The Operations Process*, ADP 5-0 (Washington, DC: Headquarters, Department of the Army, 2019), 2-4, https://armypubs.army.mil/epubs/DR_pubs/DR_a/ARN18126-ADP_5-0-000-WEB-3.pdf; and Rayburn and Sobchak, *The U.S. Army in the Iraq War*, 348-9.

remained in the city when the battle started.[16] But many animals were left there and actually became problematic. That was something that I had not considered, and it surprised us.

Spencer: I would like to dig into that last part a little more. How were you able to convince the population to leave without totally giving away the fact that the operation was about to kick off? I think that is a very important part of the operation because I think allowing the city to be largely devoid of civilians facilitated the Marines being able to clear the city so quickly. In other operations that I have studied, I have seen anything from air dropping leaflets, to broadcasts, to key leader engagements asking civilians to leave. How did you do it for this battle and where did you tell them to go?

DeFrancisci: We used multiple methods and a multi-pronged approach to encourage civilians to leave and the belligerents to cease hostile actions. We dropped leaflets and used army psychological operations forces to go to the edge of the city with loudspeakers and broadcast messages, such as "get out of the city" or "we're coming to get you."[17] We conducted other direct and indirect intimidation tactics. We also conducted what we called a "whisper campaign" where we said certain things during meetings with civilians at the Fallujah liaison team building and in some of the outlying towns.

Lieutenant General John Sattler, the commanding general of the I Marine Expeditionary Force, along with State Department officials, met with insurgent leaders at the Fallujah liaison team building where they told the insurgents to basically knock off this nonsense or we would go into the city and take them down. I believe that the construction of the assembly area also convinced people to leave because it was concrete proof that an attack was imminent. It demonstrated that Sattler's warnings were not simply cheap talk. Collectively, these actions convinced most of the population to leave and many of the less hardcore fighters to hang it up.

Since I discussed Sattler, I would like to take a step back to discuss the task organization. The I Marine Expeditionary Force was responsible for the Al Anbar province, the largest governorate in Iraq which included Fallujah

---

[16] Some reported only 500 civilians remained; see, for example, McWilliams and Schlosser, *U.S. Marines in Battle*, 6. Yet others reported that as many as 30,000 (ten percent of the pre-battle population) remained. See, for example, Anne Barnard, "Fallujah refugees describe ordeal of life in crossfire," *The Boston Globe*, November 15, 2004, http://archive.boston.com/news/world/articles/2004/11/15/fallujah_refugees_describe_ordeal_of_life_in_crossfire/.

[17] Rayburn and Sobchak, *The U.S. Army in the Iraq War*, 349.

and Ramadi. The 1st Marine Division, commanded by Major General Richard Natonski, led the ground combat element. The 4th Civil Affairs Group headquarters was located with the Marine Expeditionary Force headquarters. My civil affairs detachment was in direct support of RCT-1, a second detachment was in Ramadi, a third detachment supported RCT-7 in the west, and the fourth supported the division. A Marine civil affairs group is similar to an army civil affairs battalion, so my detachment was similar to an army civil affairs company.

The Marines typically deploy as a Marine Air-Ground Task Force (MAGTF), an extremely versatile and self-sufficient expeditionary unit that includes ground, air, logistics, and command elements. A MAGTF ranges in size from a Marine Expeditionary Unit of about 2,200 Marines to a Marine Expeditionary Force of 46,000 to 90,000 Marines.[18] A civil affairs group is not normally part of a MAGTF, but we had great leaders, so the command-and-control relationship was excellent because of the people involved. I was shocked when General Natonski stated that the civil affairs group would be the main effort for phase four, the reconstruction phase, of the operation. I would not have believed it if I did not hear it with my own ears during the delivery of the operations order. He did not make his ground combat element his main effort, he made a civil affairs group his main effort.

Spencer: Can you now describe how civil affairs supported the assault?

DeFrancisci: The assault started on November 7. From a civil affairs perspective it was like General Charles Krulak's "three-block war," meaning within three city blocks, we would be required to conduct different actions, including full-scale combat operations, stability operations, and humanitarian aid.[19] As planned, we cleared from north to south, but even with 18,000 forces, that took time. We had 35,000 buildings to clear.[20] In the Jolan area, where the heart of the enemy's defenses was located and

[18] "Marine Air-Ground Task Force," Marine Corps, https://www.marines.com/about-the-marine-corps/marine-corps-structure/air-ground-task-force.html.

[19] General Charles Krulak, the Commandant of the United States Marine Corps, published the idea of the "three-block-war" in a 1999 article. See, Charles C. Krulak, "The Strategic Corporal: Leadership in the Three Block War," *Marines Corps Gazette* 82, no. 1 (1999), 14-17, https://mca-marines.org/wp-content/uploads/1999-Jan-The-strategic-corporal-Leadership-in-the-three-block-war.pdf.

[20] DeFrancisci remembers it being 35,000 buildings, but some reported as many as 50,000 buildings. See, "U.S. Troops Occupy Fallujah," CBS News, November 14, 2004, https://www.cbsnews.com/news/us-troops-occupy-fallujah/.

where 3/5 Marines and 2-7 Cavalry conducted their penetration, enemy personnel occupied roughly one out of every 15 houses. In other parts of the city, it was more like one out of every 70 homes.

If one element moved faster than another, then the enemy could move in behind the faster moving element. Thus, the division conducted a well-orchestrated and synchronized movement and controlled its approach using phase lines. As the infantry cleared an area, we evacuated and provided humanitarian assistance to the handful of residents who had remained while at the same time we assessed the damage. The battalion commander of 3/1 Marines emphasized to us that we needed to entirely clear any building we entered, even if we had just left it.

Spencer: That was probably because tunnels provided the enemy the ability to maneuver into spaces that had already been cleared?

DeFrancisci: You got it! They knew the city better than we did and it was a three-dimensional battlespace. They knew the seams—tunnels, alleys, whatever—to get behind us, but eventually we cleared the city. We literally cleared the entire city of every single person. By the end of the operation, we had detained 2,052 insurgents and killed an estimated 2,175.[21]

During the battle, we evacuated any of the remaining civilians who had not fled the city. We also started the infrastructure assessments and damage recovery while the battle was still raging. As soon as an area was clear, we conducted assessments and, in some cases, began repairs. By some estimates, as much as 60 percent of the city's buildings had been damaged or destroyed, so the level of destruction was significant.[22] It was important for us to determine the level of damage to critical infrastructure so that we could prioritize the rebuilding effort. Civil affairs teams with support from army engineers and navy Seabees[23] conducted the assessments. The engineers conducted bridge assessments to determine their structural integrity, especially bridges that would be required to hold tanks.

We confirmed the locations and conducted assessments of all critical infrastructure within the city. Our detachment supported RCT-1 in

---

[21] Rayburn and Sobchak, *The U.S. Army in the Iraq War*, 354.

[22] Kendall Gott, ed., *Eyewitness to War Volume I: The US Army in Operation AL FAJR: An Oral History* (Fort Leavenworth, KS: Combat Studies Institute Press, 2006), 14, https://apps.dtic.mil/sti/pdfs/ADA460458.pdf.

[23] The Seabees are the Navy's construction battalion that "is responsible for building temporary and permanent infrastructure at U.S. military locations around the world." See, Sandi Gohn, "What is a Seabee? 9 Things to Know About the Navy's Construction Battalion," USO, March 4, 2021, https://www.uso.org/stories/3009-what-is-a-seabee-9-things-to-know-about-the-navy-s-construction-battalion.

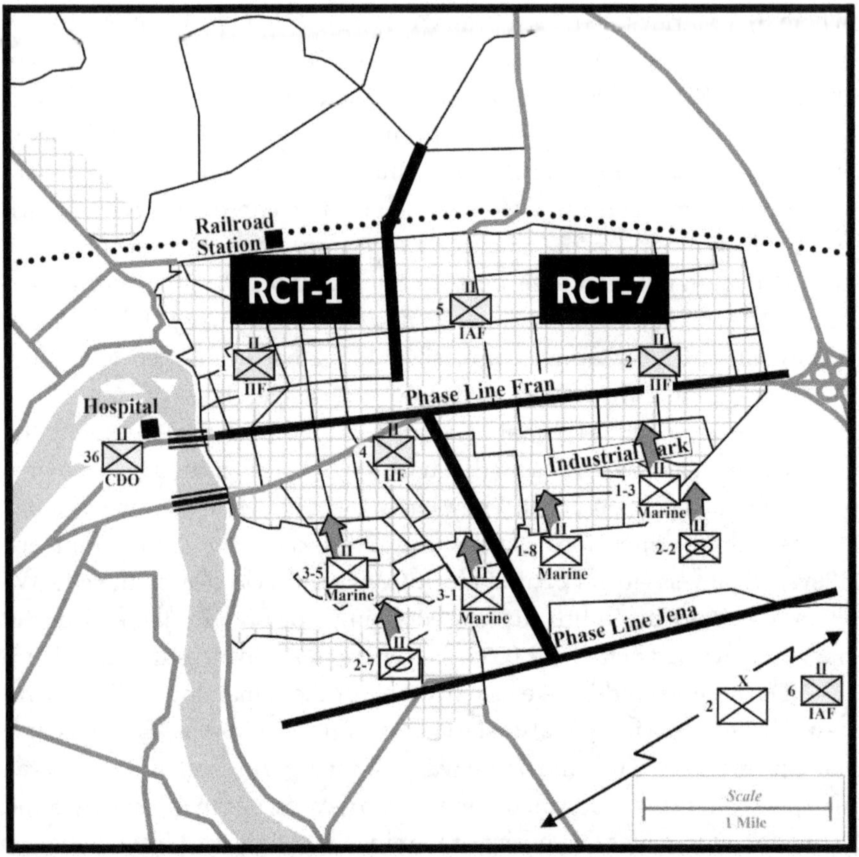

Figure 10.1. Map of the second battle of Fallujah, November 15, 2004.
Source: Map is modified from map 6 found in Gott, *Eyewitness to War*.

Fallujah's west, while another detachment supported RCT-7 in Fallujah's east (see figure 10.1). The detachment attached to RCT-7 conducted some of their assessments using unmanned aerial systems. I thought this was pretty clever. Remember, this was 2004 so unmanned aerial systems were still relatively new. While aerial imagery was helpful, there was no substitute for on-the-ground assessments.

We also organized the removal of dead bodies. There were more than 2,000 of them and if we did not, they would have become a major public health problem. As I mentioned earlier, there were dislocated animals running around the city, and as they were mixing in with the dead bodies it became kind of grisly. So, we worked with the neighboring cities, leveraging the relationships that we had established weeks earlier, to bring in local clerics with mortuary teams, so to speak. They provided the proper honors to the dead and then removed them. It was very moving to watch them conduct their prayers and ceremonies. Captain Alex Henegar, one of our civil affairs team leaders, was instrumental in getting this going, and later, the Marine Corps mortuary affairs took over that mission.

After the battle started, parts of the city flooded. As I mentioned before, some of the sewer lift stations failed so water and raw sewage flooded the streets and created a toxic mess. Marines were wading in waist-deep water in certain areas, and some vehicles were having trouble getting through. The flooding soon became one of General Natonski's biggest concerns and he turned to civil affairs to solve the problem. At first, we assumed a water line was broken, but we really had no idea of what was causing the problem. In fact, we did not even know about the sewer lift stations. Keep in mind, we were Marines with some civil affairs training, not structural engineers, city planners, or flooding experts. But we figured it out with the help of the engineers, the Seabees, and by leveraging some of the important civilian connections that we had made.

The freshwater production and pumping station was located on the western side of the Euphrates River, so the first thing we did was go to the station and have the municipal workers turn the water off. They told us that they turned it off as soon as the battle started, so that was not the problem. The municipal workers, who we had met with several times in the past, told us the problem was the sewer lift stations. Prior to the battle we had mapped out the infrastructure prior to the battle, we mapped out much of the critical infrastructure, but we did not know about the sewer lift stations nor their importance.

I showed them my map and asked them to show us where the sewer lift stations were located. They looked at the map and said, "I don't know how that thing works. I have never seen a map before in my life, but I know where they are." After much discussion about the importance of the issue, we convinced some of them to go into the city, in the middle of the battle, to show us the sewer lift stations so that we could get them running again. As a side note, I was impressed with the bravery of the few that accompanied us, and later I heard one of them bragging to their friends about what they had done during the battle to help save the city. For them, it was a badge of honor. After we identified the damaged stations, the Seabees set to work on getting them operational. Some of the pumps were damaged during the battle and broken, so we had to install new pumps.

That helped, but that was not the sole answer to the water problem because after getting them all running again, there was still significant flooding in some areas.

We were out of answers for General Natonski when pure luck struck. While driving over a bridge to get to the water station on the other side of the Euphrates, I happened to look out the window and see the Plimsoll line[24] of the dam and realized that the water was extremely high, like nine meters or something on the mark. We stopped and I asked the dam operator if that was normal. He said no and that he did not have any gas to run the generator to open the dam, so I immediately called the regimental commander and told him that we needed gas at the dam so we could open the gate. I told him that I thought this would help the flooding problem.

Sure enough, within 12 hours of us opening that dam, the city was bone dry. I think that really demonstrated the value that civil affairs could provide during an assault, and it really cemented our value to General Natonski. I think people understand the role that civil affairs play in shaping or stability operations, but I think they often forget, or do not even realize, they can play an important role during offensive combat operations.[25] During the assault we conducted a wide variety of tasks including dead body removal; infrastructure assessment, protection, and repair; civilian control and evacuation; and leveraging civilian assets, resources, and our civilian experience as reservists to solve problems.

Spencer: That is fascinating. I had not heard that story. I understand the valuable role that civil affairs can provide to the maneuver force, but even I was surprised by how much you contributed during the assault!

DeFrancisci: Civil affairs can provide value throughout all phases of an operation if they are included in the planning process and understand how they can contribute to achieving the commander's intent.

Spencer: I would now like to turn to the reconstruction phase of the operation, when the division transitioned from offensive combat operations

---

[24] A Plimsoll line is the mark on a ship that shows when it is fully loaded. In this case, DeFrancisci is using the term to describe the high-water line for the dam. Near the dam, there was an actual huge vertical "ruler" with markings by meter.

[25] The U.S. Army and Marines have four basic operations: offense, defense, stability, and defense support of civilian authorities. Department of the Army, *Operations*, ADP 3-0 (Washington, DC: Headquarters, Department of the Army, 2019), 3-1, https://armypubs.army.mil/epubs/DR_pubs/DR_a/ARN18010-ADP_3-0-000-WEB-2.pdf.

to stability operations. Can you describe your role in what was effectively a rebuilding of the city?

DeFrancisci: Our objective during the reconstruction phase was to stabilize the area, reestablish the economy, restore the infrastructure, and transition control back to the Iraqi government. And we wanted to do this as quickly as possible because running the city was manpower intensive and we needed to get some of the units back to their normal sectors. As I said, this was a classic "three-block-war." We had multiple types of missions in the same areas, often with different missions on adjacent streets. Marines were clearing buildings and eliminating the enemy in some zones, conducting assessments and providing humanitarian assistance in others, and repairing infrastructure and buildings in other areas.

I have a short anecdote that describes the day when I realized that we, as a regiment, had transitioned to the reconstruction phase of the battle. Colonel Michael Shupp, the commander of RCT-1, was a great commander. He was extremely demanding and really pushed us. He was a real energizer. He seemed to be everywhere at once and he never seemed to tire. Each day, the regiment conducted a daily battle brief where each subordinate commander and staff member stood in front of him and the rest of the staff and let everyone know what his or her unit or section had accomplished and were going to accomplish to support the mission.

During the assault phase, my brief typically lasted 2-3 minutes. Then one day, he kept me in front of the staff for 45 minutes. I was not expecting it, but he just kept asking me all kinds of questions. After he finally finished, I asked, "Sir, are we in the reconstruction phase now?" The entire staff laughed, but that was a clear signal that we had transitioned to the reconstruction phase and the commander wanted to succeed in that phase just as much as he had wanted to succeed in the assault phase.

I want to reinforce once again how important it was to have a good command relationship. As I mentioned before, we were not organic to the task force, but they fully integrated us into their team. The key to our success was that everybody was working toward the same objective. Civil affairs were not simply an afterthought—they did not build an operational plan and then figure out where civil affairs fit in. They viewed us as an enabler of the combined arms element and provided us clear direction, just as they did their subordinate battalions. This allowed us to stay synchronized with the commander and support his priorities throughout the operation. What really helped was that we had an excellent operations section that provided specific guidance to synchronize all of us. The operations section provided excellent taskings through fragmentary orders, which were written with a clearly stated task and purpose. It was an excellent example of what the U.S. Army calls "mission command." This really helped us know what to do and why.

The other obvious sign that we had entered the reconstruction phase was that our task organization changed. The assault had been a division operation, but when we transitioned to reconstruction, RCT-7 had to return to its normal sector in the west. Thus, the reconstruction of Fallujah fell to RCT-1 with some support from Iraqi military forces and the civil affairs group command element.

As we transitioned to reconstruction, we continued with our assessments. Some sections of the city looked like a war zone because that is what it was. In some areas, there was nothing left standing. We used the assessments to develop a plan which prioritized what needed to be fixed so we could repopulate the city. For critical infrastructure, we identified and assessed individual nodes and then figured out how all the nodes worked together to create a system.

For example, electrical power plants were nodes and the distribution of power to the end users was the system. Some of the damage was so severe that we had to develop temporary solutions while waiting for longer repairs to be completed. For example, we brought in close to 25 gigantic containers and filled them with water to provide the civilian population a water source while repairs on the water pipes continued.

We had to do something similar for the electrical system. We brought in a bunch of generators and distributed them throughout the city to provide a temporary power grid as a stopgap measure until repairs to the power station and power lines could be completed. Nearly all of the electrical lines had been destroyed during the two battles of Fallujah, so they had to be replaced.

There was rubble everywhere, so we hired contractors from Baghdad to come in with heavy equipment to clear the rubble. We also hired unemployed military-age males to help clear the rubble with wheelbarrows and shovels. We joked that previously they had probably been fighters who laid down their arms and picked up shovels, but they were there, and they wanted to work. They became part of the solution, and they were out sweeping the streets. There was even debate as to why we were giving them shovels. In 2004, improvised explosive devices remained a major problem for the coalition and the insurgents buried many of them. So, we provided them flat shovels, not spades. It is very hard to dig holes with flat shovels, so we felt it would be difficult for them to use the shovels against us if any chose to do so. We also collected the shovels at the end of the workday to keep track of them.

We provided humanitarian support to every single head of household in Fallujah. Each home was given a public distribution system ration card. Each month, for the next several months, they received an allotment of rations that was funded by the Oil-for-Food Program.[26] We also provided every head of household a $300 payment, which was quite a bit of money by Iraqi standards. In a one-week period, I believe we injected nearly $6.6 million into the local economy. Almost overnight, the city came back to life: businesses reopened, and people were cleaning the city and rebuilding their homes. The cash payments were a real success story for getting the city going again.

This brings up the idea of employing money as a weapon system. General David Petraeus has described how "money can be 'ammunition' as the security situation improves" and using "a targeting board process to ensure the greatest effect for each 'round' expended to ensure that each engagement using money contributes to the achievement of the unit's overall objective."[27] We managed many different projects, but we were very deliberate in our approach. We did not want to get involved with building new things, or what we considered "development." Instead, we focused our limited assets on reconstruction, to help rebuild or repair things that were already there but had been broken during the battle. It did not matter who caused the destruction, what mattered was getting things rebuilt so they could be used by the population.

Spencer: That is pretty impressive. How did you prioritize what needed to be done? Did you have assessment tools like SWEAT MSO (sewage, water, electricity, academics, trash, medical, safety, other) to help with that prioritization?[28] I know from my previous research that

[26] For information on the Oil-for-Food Program, see "Office for the Iraq Programme Oil-for-Food" webpage at https://www.un.org/depts/oip/.

[27] David H. Petraeus, "Multi-National Force-Iraq Commander's Counterinsurgency Guidance," *Military Review* 88, no. 5 (2008), 3, https://www.armyupress.army.mil/Portals/7/military-review/Archives/English/MilitaryReview_20081031_art004.pdf.

[28] Department of the Army, *Stability Techniques*, ATP 3-07.5 (Washington, DC: Headquarters, Department of the Army, 2012), 1-13, https://armypubs.army.mil/epubs/DR_pubs/DR_a/pdf/web/atp3_07x5.pdf.

the reoccupation of Fallujah was very controlled. That requires significant manpower to control checkpoints while at the same time trying to conduct assessments. How did you prioritize all this work?

DeFrancisci: We did not have acronyms like that at the time, or if they existed, we did not know about them. We simply used Maslow's hierarchy of needs.[29] There were essentials for human life that we had to do first, such as the provision of water that I discussed. Things like hospitals were higher on the list while things like reopening schools were lower on the list. As you mentioned, we had to man many checkpoints, but they were not all manned by men. We needed to search women, so the Marine Expeditionary Force created the Lioness Program, who were female Marine volunteers attached to all-male combat units to search Iraqi women and children.[30]

One funny story about prioritization involved the fire department. The fire department was fairly high on the priority list, so every day I briefed its status to the regimental commander. Early on, the commander asked, "Why do we need a fire department?" Many of their buildings were constructed using cinderblock or similar types of materials. I think it was a common belief that they did not have a lot of fires that required a fire department. But I figured that every good city needed a fire department, so it became kind of a running joke within the regiment.

Then one day we got a call at the headquarters about a vehicle fire. So, we called the fire department and they rolled up with a regular fire truck just like you would see in any large city. The fire department had six firehouses. Only one was destroyed during the battle and luckily someone was smart enough to drive all the fire trucks out of the city before the battle started. Thus, we did not have to worry about getting new trucks after the battle. At the meeting with the regimental commander that day, I told him that we were able to scramble the fire department to put out the fire and everyone, including the commander, got a good laugh out of that. That is one funny little anecdote that happened in the middle of this whole operation.

---

[29] For more on Maslow's hierarchy of needs, see, Neel Burton, "Our Hierarchy of Needs," *Psychology Today*, May 23, 2012, https://www.psychologytoday.com/us/blog/hide-and-seek/201205/our-hierarchy-needs.

[30] Regina T. Akers, "Marine Corps Lioness Program: 'In and out of harm's way'," *Flagship*, March 25, 2009, https://www.militarynews.com/norfolk-navy-flagship/news/from_the_fleet/marine-corps-lioness-program-in-and-out-of-harm-s-way/article_f60a2c18-9ff8-539c-8df6-f9787338c45b.html.

Reviewing my notes, I see that we accomplished much during this time. We—meaning the regiment with a lot of help from division, the Marine Expeditionary Force, and other agencies such as the Department of State and the Iraqi civilians—restored five water treatment plants, one hospital, three health clinics, three electric distribution centers and electrical wires throughout the city, nine sewer lift stations, one liquid propane gas distribution center, five gas stations, one fuel storage depot, and two banks; we reestablished cellular telephone service; and we reopened 25 schools.

As I mentioned earlier, schools were one of the lower priorities and I remember that our detachment assessed schools on Christmas Day 2004. In terms of emergency services, we restored one fire station, one highway patrol station, and one traffic police station. In terms of public services, in addition to the working parties that I previously mentioned, we restarted garbage collection and public transportation. We cleared a lot of rubble and also repaired many damaged roads.

Prior to the battle we helped create Iraqi public order battalions and they were very helpful during the reconstruction effort. They helped with law enforcement and security in general and became a base of the Iraqi government on the security side.

Spencer: I study urban warfare, but so much of the focus is on the offensive operations, there is so much less out there relating to the rebuilding effort, so I find this fascinating. According to Maslow's hierarchy, safety needs are near the top of the list, falling just behind physiological needs like food and water. I was not aware that you had created an Iraqi security element before the battle began.

DeFrancisci: As I said, they were called public order battalions.[31] They were a special police unit that included a number of law enforcement functions such as city policing, building security, and traffic management. After the first battle of Fallujah, the insurgents controlled the city, so there had not been a legitimate city police in months. To help reestablish the police department we also brought in contractors to help provide the necessary organizational and training expertise. The people from the public order battalions were generally not from Fallujah, so that helped curb intimidation.

I was responsible for developing the priority list, but the regimental commander was responsible for approving it. He had a really good sense

[31] Rayburn and Sobchak, *The U.S. Army in the Iraq War*, 325.

of what was going on in the city and he knew that he needed to get it transitioned over to the Iraqis as soon as possible and the entire regiment contributed to this goal. For example, the Marine infantry continued to provide security until the Iraqis could get the police up and running. It was a team effort between the civil affairs, the public order battalions, the infantry, Department of State, and you name it; everyone contributed.

Spencer: Earlier you discussed a technocrat list that you created. How did this list help?

DeFrancisci: We created a list of key individuals that we would need to go back to work immediately after we had cleared the enemy and established basic security. We focused on the municipal workers, electrical department, water department, fire dependent, sanitation, and medical services. We were not focused on the government because that was being handled at a higher level. We built this list before the battle started, and that was a good thing because when the battle started, everyone disappeared. They hunkered down in the neighboring towns to avoid the raging battle. I mentioned the water treatment plant earlier; we had established some really good connections with workers and administrators there.

There were thousands of dislocated people. The city of 250,000-300,000 had been almost completely cleared, so all these people had gone somewhere. Many went to the surrounding towns and some of these places probably were not much better than Fallujah prior to the battle. In fact, some places looked like a scene out of the wild west with a lot of rough looking characters. I remember visiting Saqlawiyah, which was a large town eight kilometers northwest of Fallujah. I wanted to talk to them about law enforcement, so we went to the police station and found many sketchy looking, military-age males that did not look like they were from the area. Nonetheless, we had a job to do, so I walked up and started talking to one. Apparently, he did not want to talk, so he pulled out a knife, looked me in the eye, and slit the throat of a goat that he had with him, and it bled out on the ground next to my boots. You talk about a psychological operation! I got the message.

We also established two humanitarian assistance distribution centers, one with 3/1 Marines and one with 3/5 Marines. It was an organized way to distribute supplies to the residents. Some of the supplies we contracted for delivery to the distribution center while others came from the "green mountain" that we had built prior to the battle. It was December and winter was approaching, so one of the things that we distributed was blankets.

On January 30, 2005, Iraq held parliamentary elections to elect a new national assembly. We were aware of the election during planning, but never considered that Fallujah might participate. But in January, this became a real possibility, so we used the humanitarian assistance distribution centers as voting sites for the national election. Roughly 8,000 of Fallujah's residents cast votes during Iraq's first election. That was unthinkable before the battle.

Spencer: I think the elections likely contributed to the decision to conduct the second battle of Fallujah. A newly elected Iraqi government did not want to be in the position of having a rogue city only 60 kilometers outside its capital.

DeFrancisci: The election was one of the driving reasons for the battle to take place.[32] The other, of course, was the fact that Fallujah was an insurgent stronghold that was being used as a giant base from which to plan, prepare, and conduct attacks. They built improvised explosive device factories and used the city as a safe haven to treat wounded fighters.[33] Zarqawi also used the city as a place to kill and torture hostages.[34] Thus, the city had to be cleared at some point, but I do think the upcoming election helped drive the timing. The validity of the election could have been questioned if a large city not far outside of Baghdad had not been allowed to participate.

Spencer: You discussed how you prioritized providing for the citizens' physiological needs, safety, and security, but what about governance and administration? I realize that is secondary to the other two, but when did you start to reestablish governance?

DeFrancisci: My detachment was focused on the technocrats, but that did not mean there was not any action being done to reestablish governance. That was the responsibility of our higher headquarters, and they were a great help in that regard since we simply lacked the capacity to focus on everything. There were individuals who identified themselves as the mayor and city leaders, so they had to be vetted, but that was happening at the civil affairs group and division level. They were working with the interim government, Department of State officials, and other officials who were

[32] Gott, *Eyewitness to War*, 7.
[33] See chapter 9, "The Battle of Fallujah," in this book.
[34] See, for example, Jackie Spinner, "More Fallujah Hostage Sites Found," *Washington Post*, November 22, 2004, https://www.washingtonpost.com/archive/politics/2004/11/22/more-fallujah-hostage-sites-found/fe6c2727-9665-4947-8c8b-3811d0360ae1/.

involved in reestablishing governance and getting a mayor and city council in place. I believe there were at least two individuals who claimed to be the rightful mayor. So, to answer your question, reestablishing governance was being done simultaneously as the provision of basic needs and security, it was just being done at a higher level.

Spencer: I was a part of the 2003 invasion and after the invasion my unit moved into Kirkuk, a city in Iraq's north. So, I know how complex this is. One of the things that we always struggled with was how to measure progress and what baseline we should be comparing that progress to. For example, prior to the battle, was the power on in Fallujah only eight hours each day, intermittently, or was it on 24 hours? How did you determine your goals as you were trying to reestablish essential services and how did you measure your progress?

DeFrancisci: I think that is always challenging. To use your example of power, before the battle, not everyone had power and they often rationed it. So, during the reconstruction, how did we determine that our goal had been met, even though not all residents had power 24 hours a day? I think we tend to try to do too much at times. But that was really outside our scope. Our job was reconstruction: to restart essential services, reestablish a suitable level of security, and reestablish some governance, even if limited. Larger development projects belonged to the United States Agency for International Development, the Department of State, or higher coalition headquarters.

Nonetheless, there were ways to measure progress using measures of performance and measures of effectiveness. Measures of performance are easier to measure, for example, you can count the number of police in Fallujah. Measures of effectiveness, however, are more important, but they are harder to measure. For example, is crime actually decreasing? Do people feel safe? That is harder to measure but it is more important than the number of police.

Returning to the police example, let us say the rule of thumb is that you need approximately 20 police for every one thousand residents. Thus, if Fallujah had 300,000 residents, then they needed approximately 6,000 trained police. Thus, it was possible to set a goal number and measure progress toward that goal. You can do the same thing with hospital beds or ambulances, for example. Also, as a rule of thumb, you can say you need about one megawatt of electricity for every 2,000 homes.

These goals are not arbitrary; they are based on data and an assessment of the environment. So, these are examples of goals that you could set, but again, that was beyond the scope of our mission which was simply to get essential services back up and running and then to turn it over to the Iraq people. And I think this is the way it should be. I believe that the military should stay out of development and stick with the reestablishment of essential services and the provision of security.

Spencer: I know that is not an easy question, but your answer makes perfect sense. Based on some of my own research, I understand where some of these metrics or goals come from. I interviewed the Colorado Springs police chief and some of those same statistics came out: you are supposed to have this many policemen for this size population or this is how you determine patrols.[35] I have one final question for you. Based on your experience in one of the biggest urban battles of the modern generation and subsequent experience rapidly rebuilding a city, what advice would you give to future soldiers and Marines that are thrust into a similar situation, that may not have the training and education to fully prepare them for the challenge?

DeFrancisci: When it comes to civil affairs, I cannot emphasize enough that you really have to focus on the commander's mission and what you can do to help the commander accomplish that mission. It is not simply building a well to build a well. How can you enable the mission? What is your role in that? You may not be an expert, but if you analyze the problem, conduct a good mission analysis, and leverage outside experts when required, you can succeed even if you are less knowledgeable on the particular problem.

Ultimately, you have to have a seat at the planning table so that you can understand how civil affairs fits into the mission. But at the end of the day, the commander is balancing many competing requirements, such as risk to the force or any other number of factors—so he or she may not always follow your recommendations. When this happens, you cannot get your feelings hurt; it is your job to provide options and recommendations so the commander can make informed decisions. Sure, you will make mistakes, but the key is to learn from your mistakes and focus on the mission.

I do want to say one more thing. I was lucky enough to go to the U.S. Army Ranger School as a Marine. It really helped me with small unit leadership and the importance of mission planning and preparation. It

[35] John Spencer, "Urban Policing 101," *Urban Warfare Project Podcast*, Modern War Institute, August 21, 2020, https://mwi.usma.edu/urban-policing-101/.

helped me understand the importance of providing clear mission orders so that everyone understands the mission, the commander's intent, and their role in achieving it. You cannot take shortcuts. You have to conduct inspections and do the rehearsals so that everyone is on the same page when you are on the move or at the objective. Our civil affairs teams led patrols that included many other people, and the inspections and rehearsals we did prior to each patrol built confidence within our teams and our attachments. The skills that I learned in Ranger School helped me throughout my entire career to include my time as a civil affairs detachment commander in Fallujah.

Spencer: As a former Ranger instructor I cannot tell you how proud I am to hear a career Marine emphasize the value of Ranger School. Throughout our conversation you mentioned the importance of understanding the mission, the commander's intent, and your task and purpose. That is music to my ears, and like you, I believe that it applies no matter what your job is. It was our profession, it was who we were, and it was what makes us so powerful.

DeFrancisci: I learned a lot from the Ranger instructors. They really pushed me to the limit at Ranger School, and I always drew upon that. I am thankful to people like you who provided that training for people like me. It really helped. Thank you for that.

11

The Battle of Ramadi

The 2006 Battle of Ramadi was a turning point in the Iraq War. Three years into the conflict, the coalition had experienced few successes. Coalition casualties continued to increase, al Qaeda in Iraq (AQI) grew increasingly deadly, the insurgency had expanded, and the country was on the brink of a civil war. Political leaders were calling for the United States to withdraw and the military needed to demonstrate that it possessed the ability to stem the growing insurgency.[1] While the battle of Ramadi was still unfolding when President George W. Bush made the decision to implement a major increase in the number of soldiers deployed to Iraq, dubbed the surge, the success of operations in Ramadi offered hope that an increase in the number of troops employing a counterinsurgency strategy could work. Ramadi demonstrated that when coalition troops employed counterinsurgency doctrine coupled with Iraqi civilian resistance to al Qaeda, they could defeat the insurgency and provide the stability required for Iraq to govern itself.

The United States initiated the Iraq War on March 20, 2003. The initial invasion was exceptionally fast and effective and less than six weeks later, Bush declared that "Major combat operations in Iraq have ended," while standing beneath the now infamous "Mission Accomplished" banner onboard the *USS Abraham Lincoln*.[2] Few realized at the time that the war was just beginning. It took months for most American political and military leaders to recognize that an insurgency was brewing. A few military leaders—including General Jack Keane, Vice Chief of Staff of the Army, and General John Abizaid, Commander of U.S. Central Command—recognized the insurgency for what it was as early as July 2003, but they

---

1 Linda Robinson, *Tell Me How This Ends: General David Petraeus and the Search for a Way Out of Iraq* (New York: Public Affairs, 2008), 18-20.

2 Seth Cline, "The Other Symbol of George W. Bush's Legacy," *U.S. News & World Report*, May 1, 2013, https://www.usnews.com/news/blogs/press-past/2013/05/01/the-other-symbol-of-george-w-bushs-legacy.

were the exception.[3] As late as November 2003, Secretary of Defense Donald Rumsfeld directed those within the Bush administration not to refer to it as an "insurgency."[4]

By the spring of 2004, the evidence was irrefutable and by the end of the year, the number of American servicemembers killed in Iraq had nearly doubled from the previous year, from 486 to 846; and the number of wounded had more than tripled, from 2,422 to 8,002.[5]

By 2006 at least four major problems existed. The first was an insurgency led by foreign fighters known as AQI. The second was an insurgency led by former Baathist and Sunni Iraqi nationalists centered in Al Anbar province. The third was an insurgency built around Shiite militias taking place in Baghdad and other major cities such as Karbala, Najaf, and Basra. The fourth was a criminal element that aligned itself with the various insurgencies at different times.[6] From a strategic standpoint, this situation was one of the most complex insurgency challenges faced by an army in modern times. In the summer of 2006, the United States had not yet published its counterinsurgency doctrine, nor had it developed an effective strategy to counter the insurgencies in Iraq. So, despite being three years into the war, the United States lacked the doctrine and strategy that its forces required to be effective.

At that time, the United States was rotating brigade-sized army and Marine units into Iraq on 12-month cycles and assigning them to predominantly urban areas with the mission of providing security. In most cases, the brigades' ability to establish security was marginal at best.

One notable exception was Colonel H.R. McMaster's 3rd Armored Cavalry Regiment which operated in the city of Tal Afar in 2005. At the time, most units were consolidated on large forward operating bases located on the periphery of cities and provided only part-time security, primarily through limited mounted patrols, through the cities. By contrast, McMaster implemented a unique counterinsurgency strategy. He cleared Tal Afar of

[3] Thomas E. Ricks, *Fiasco: The American Military Adventure in Iraq* (New York: Penguin Books, 2007), 170-2; and "Abizaid Briefing 7/16/2003," HistoryCentral.com, July 16, 2003, http://www.historycentral.com/freeIraq/Iraqinfo/Abiziad.html.

[4] Ricks, *Fiasco*, 170-2.

[5] Defense Casualty Analysis System, "U.S. Military Casualties – Operation Iraqi Freedom (OIF) Casualty Summary by Month," last updated March 18, 2014; and Defense Casualty Analysis System, "U.S. Military Casualties – Operation New Dawn (OND) Casualty Summary by Month and Service," last updated March 18, 2014.

[6] James A. Russel, *Innovation, Transformation, and War: Counterinsurgency Operations in Anbar and Ninewa Provinces, Iraq, 2005-2007* (Stanford, CA: Stanford University Press, 2011), 96.

the insurgents, held the city, and then passed the city off to indigenous Iraqi forces. He is widely credited with being the first commander to successfully implement the "clear, hold, build" strategy in Iraq, and the army's 2006 counterinsurgency manual even included the excerpt, "Clear-Hold-Build in Tal Afar" as an example of how to implement the strategy effectively.[7]

In this chapter, Lieutenant Colonel (retired) Louis DiMarco provides a short summary of the 2006 Battle of Ramadi. DiMarco is a professor of military history at the U.S. Army Command and General Staff College. In 2020, Dr. DiMarco gave an online lecture on the Battle of Ramadi at the Dole Institute for Politics. In the lecture, DiMarco described the context of the battle, the key players involved, and the techniques used to turn the city of Ramadi, widely known as the "the deadliest city in Iraq," into a comparative model of stability that guided the way counterinsurgency operations were conducted across Iraq and Afghanistan.[8] Questions from John Spencer were added later.

Louis DiMarco:[9] In 2006, Ramadi was considered the deadliest city in Iraq.[10] With the exception of forward operating bases located in and around Ramadi, the city and its population of 300,000 were controlled by AQI.[11] Foreign fighters roamed the city at will, openly displaying their weapons and intimidating the local population. The area averaged three times more attacks per capita than any other region in Iraq.[12]

Ramadi is located 70 miles west of Baghdad (see figure 11.1) and covers nearly 100 square miles, spanning eleven miles from east to west and nine miles from north to south. The city straddles the Euphrates River. The densely populated central city is bounded by the Euphrates River to

[7] Department of the Army, Counterinsurgency, FM 3-24 (Washington, DC: Headquarters, Department of the Army, 2006), 5-22 to 5-23.

[8] John Spencer, "The Battle of Ramadi, 2006-2007," Modern War Institute, February 5, 2021, https://mwi.usma.edu/the-battle-of-ramadi-2006-2007/.

[9] The views expressed by Louis DiMarco are his and do not necessarily reflect those of the United States Army, Department of Defense, or the U.S. Government.

[10] Mark Kukis, "The Most Dangerous Place in Iraq," *Time*, December 11, 2006, http://content.time.com/time/world/article/0,8599,1568724,00.html.

[11] Ann Scott Tyson, "Troops Fight to Expand Foothold in Ramadi, U.S., Iraqi Forces Move Block-by-Block To Retake Western City From Insurgents," *Washington Post*, August 2, 2006, https://www.washingtonpost.com/archive/politics/2006/08/02/troops-fight-to-expand-foothold-in-ramadi-span-classbankheadus-iraqi-forces-move-block-by-block-to-retake-western-city-from-insurgentsspan/908aff56-3baa-4f29-99a8-97b6cd8f900c/.

[12] Sean MacFarland and Neil Smith, "Anbar Awakens: The Tipping Point," *Military Review* 85, no. 2 (2008), 65-66, https://apps.dtic.mil/sti/pdfs/ADA486828.pdf.

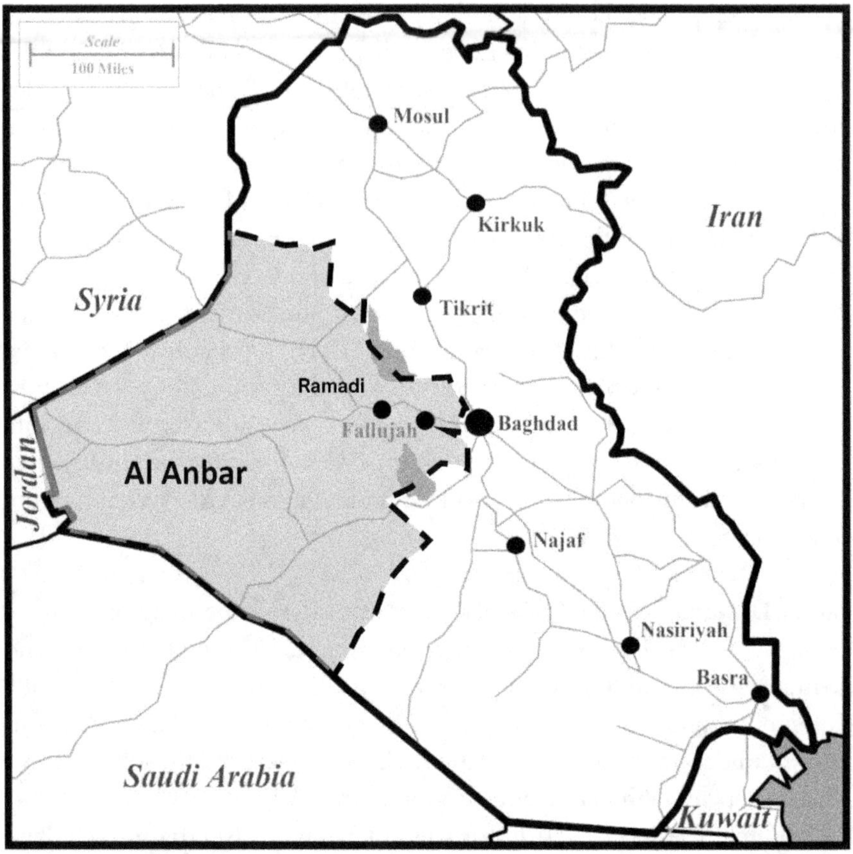

Figure 11.1. Map of Iraq with Al Anbar province.
Source: Map is modified from map 1 found in Kendall Gott, ed., *Eyewitness to War Volume I* (Fort Leavenworth, KS: Combat Studies Institute Press, 2006).

the north, the Habbaniyah Canal to the west, a railway line to the south, and residential areas to the east (see figure 11.2).[13]

Ramadi is a fairly new city, established by the Ottoman Turks in 1869, so it was not built in a traditional Arab manner. There is no kasbah (citadel), old city, or downtown area typical of older Middle Eastern cities. The city itself was designed more along European standards, with wide paved streets and modern concrete buildings. The city hospital, seven stories tall, was the largest building, though there were other five- and

[13] Louis DiMarco, *Concrete Hell: Urban Warfare from Stalingrad to Iraq* (Oxford: Osprey Publishing, 2012), 190.

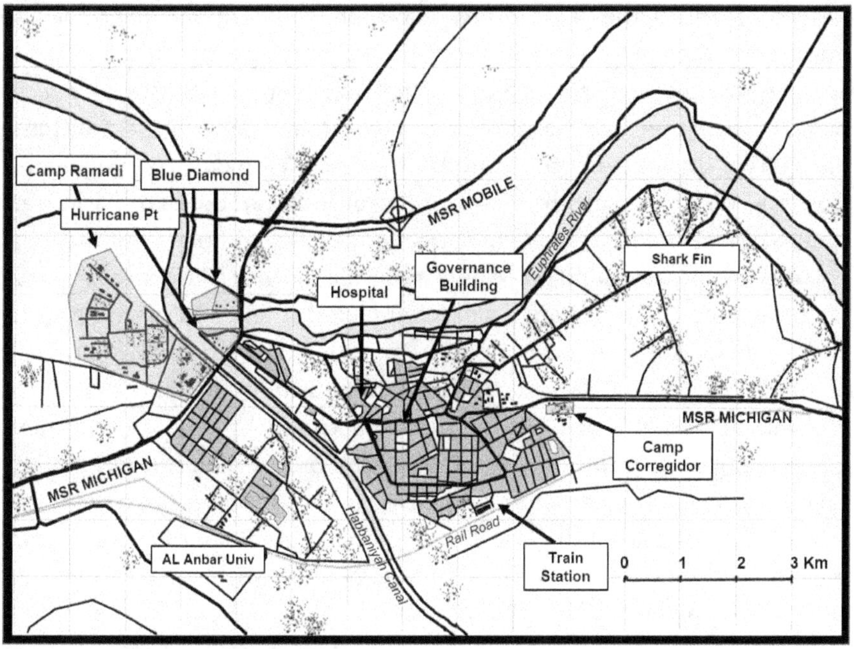

Figure 11.2. Map of Ramadi.
Source: Map created by Louis DiMarco.

six-story buildings in the downtown area.[14] The dominant architecture was the typical low, two- to three-story tall, concrete, flat-topped roof, Middle Eastern building surrounded by a concrete privacy wall. These buildings dominated the architecture in the suburbs and residential areas.

The downtown area included a modern hospital and a central government building complex. Both were important to the Americans. The area to the west of the canal included the suburb of Al Tamim and Ramadi University. A railroad line running from Syria to Baghdad was south of the city. Two main supply routes (MSRs) ran through the city. Highway 1, codenamed MSR Mobile by the Americans, was a major four-lane highway that ran from east to west on the north side of the Euphrates River. Highway 11, codenamed MSR Michigan, also an east-west highway, ran just south of Camp Ramadi, across the Habbaniyah Canal, through the downtown area, and past Camp Corregidor (see figure 11.2).

When McMaster's 3rd Armored Cavalry Regiment rotated out of Iraq in February 2006, it was replaced by Colonel Sean MacFarland's 1st Brigade

[14] Ibid., 191.

Combat Team of the 1st Armored Division.[15] MacFarland continued to implement McMaster's "clear, hold, build" strategy in Tal Afar in the ensuing months. But with the city of Tal Afar relatively pacified—attacks had dropped from 170 per month to fewer than 20 (with most of those ineffective) during the 3rd Armored Cavalry Regiment's deployment—General William Casey, the Commander of Multinational Forces in Iraq, ordered MacFarland to move his brigade south to reinforce Multinational Force-West, the command that was responsible for all of Al Anbar province.[16]

On June 6, 2006, MacFarland completed a relief in place with the 2nd Brigade, 28th Division of the Pennsylvania National Guard, and assumed control of Ramadi and the responsibility to pacify the city and its surrounding areas.[17] The National Guard brigade did not have the experience or training of a regular army or Marine unit. In addition, it suffered from manpower and logistics shortages; thus, its capacity was fairly low.[18] Its mission had been to secure the governance center, secure the main supply routes (MSR Mobile and MSR Michigan), and protect coalition supply and other types of units that moved along those routes.[19]

In early 2006, most American military and political leaders were focused on leaving Iraq as opposed to expanding operations. The Multinational Force in Iraq commander's strategy at the time was to turnover responsibility for security operations and pacification to the Iraqi army and police and begin the drawdown of American forces. Under these circumstances, the National Guard unit's priority was force protection.[20] Thus, the 1st Brigade Combat Team had to expand the scope of operations in the city.

When the brigade arrived, Ramadi had no power, no running water, no garbage disposal, no functioning city government, and no landline or cell phone service. The lack of garbage disposal created both hygienic and transportation issues. After several years of little to no city services, the city streets had become covered with dirt and debris. Also, due to the number of terrorists operating from within the city, Ramadi had been subjected to

---

[15] Jay B. Baker, "Tal Afar 2005: Laying the Counterinsurgency Groundwork," *Army* 59, no. 6 (2009), 67, https://www.ausa.org/sites/default/files/Baker_0609.pdf.

[16] Ibid.

[17] Steven Clay, *On Point III: Interview with Colonel Sean MacFarland* (Fort Leavenworth, KS: Combat Studies Institute, 2008), 22, https://cgsc.contentdm.oclc.org/digital/collection/p4013coll13/id/2364.

[18] Russel, *Innovation, Transformation, and War*, 97-100.

[19] DiMarco, *Concrete Hell*, 197-198.

[20] Ibid.

numerous artillery and air attacks, leaving the streets rubbled. The dirt, debris, and rubble made it very easy for insurgents and terrorists to hide improvised explosive devices (IEDs) to target coalition forces as they traveled through the city. These attacks added to the destruction of the city's roads. Although Ramadi was one of Iraq's most modern cities before the war, by the time the 1st Brigade Combat Team arrived, it was a war-torn city with no modern services.[21]

John Spencer: What units comprised the 1st Brigade Combat Team?

DiMarco: When the 1st Brigade Combat Team arrived in Ramadi in June 2006, it had five maneuver battalions: two light infantry battalions, one of which was a U.S. Marine Corps battalion; a mechanized infantry battalion; and two M1 Abrams-equipped tank battalions. The brigade also had an engineer battalion, a self-propelled artillery battalion, a general support logistics battalion, two Navy SEAL platoons, and Navy riverine boats. The boats, however, did not come with Navy crews, so the boats were assigned to the Marine battalion. Only one of the five maneuver battalions was part of the brigade when it deployed, the others were assigned to the brigade in Iraq.[22]

Two combat battalions, the engineers battalion, the artillery battalion, the support logistics battalion, and the brigade headquarters were based out of Camp Ramadi, a large forward operating base located in the northwest part of the city. The 1st Battalion, 35th Armor Regiment (1-35 Armor), was responsible for the area west of the canal and south of Camp Ramadi, which included the Tamim suburbs and Al Anbar University. The 1st Battalion, 37th Armor (1-37 Armor) was responsible for the downtown area south of the governance center (see figure 11.3).[23]

The 3rd Battalion, 8th Marines (3/8 Marines), was based out of Camp Hurricane Point, a base located southeast of the canal and river junction. The battalion was responsible for providing security to the governance center and the northern portion of the downtown area to the Euphrates River.[24]

The mechanized infantry battalion, the 1st Battalion, 6th Infantry (1-6 Infantry), was based out of Camp Blue Diamond, on the north side of the river. It was responsible for the suburban area north of the river. The 1st Battalion, 506th Infantry (1-506 Infantry), based at Camp Corregidor,

---

21 William Doyle, *A Soldier's Dream: Captain Travis Patriquin and the Awakening of Iraq* (New York, NY: Penguin Group, 2011), 65.

22 DiMarco, *Concrete Hell*, 197-198.

23 Ibid.

24 Ibid.

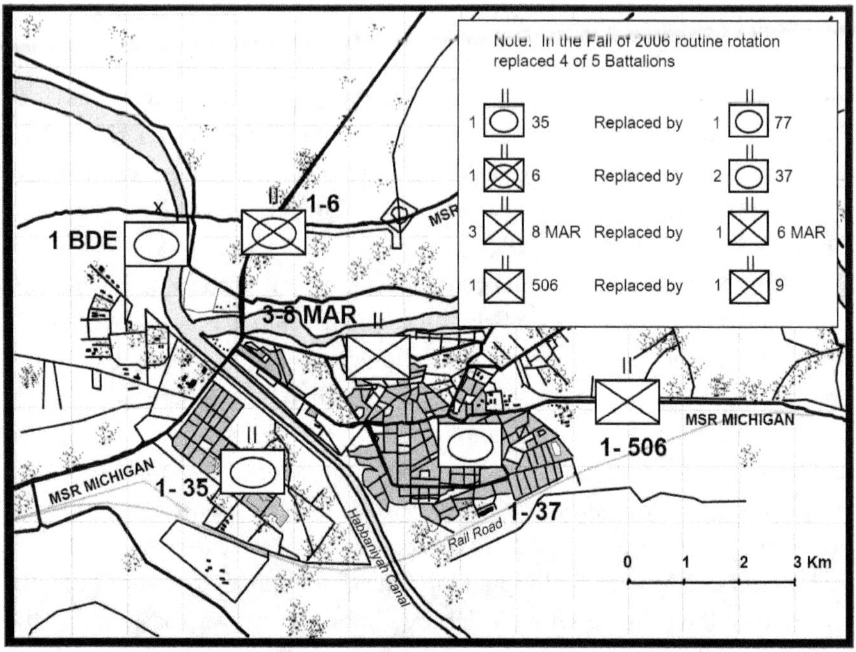

Figure 11.3. Disposition of the 1st Brigade Combat Team, 2006.
Source: Map created by Louis DiMarco.

located to the east of the city, was responsible for the east side of the city. The bases were located in good positions to interdict traffic entering the city from the east, west, and north. South of Ramadi is an open desert and generally uninhabited, so this was where most insurgents infiltrated into the city.[25]

John Spencer: Who were some of the important personalities for the battle?

DiMarco: Colonel Sean MacFarland commanded the 1st Brigade Combat Team. By the time he arrived in Ramadi, he had been in command for about a year. MacFarland was an experienced officer with over 20 years of service, including operational deployments to Saudi Arabia for Operation Desert Shield / Desert Storm and Kosovo for Operation Joint Guardian. He had previously served in Iraq earlier in the war as a member of the V Corps staff.[26]

---

[25] Ibid.

[26] "Command and General Staff College: Who is Sean MacFarland?" AllGov, November 28, 2010, http://www.allgov.com/news/appointments-and-resignations/command-and-general-staff-college-who-is-sean-macfarland?news=841817.

Sheikh Abdul Sattar Abu Risha (Sattar) was the key Iraqi leader in Ramadi. Sattar was the informal leader of the Sunni tribal forces that allied with the Americans during the battle. He started as a minor sheikh within the tribal hierarchy of the Sunni tribes that made up 95 percent of the population of Al Anbar province but had grown in prominence. When AQI's power grew, most of the senior sheikhs departed and exercised their leadership by telephone and messenger from outside of Iraq where they were safe from threats and reprisals.[27]

Sattar, although young, possessed a very charismatic personality. By remaining in Iraq when most sheikhs fled, he filled the power vacuum among the city's population. There was no formal city council or mayor, so the limited governance that did exist rested with the traditional families and tribal leaders of which, at the time of the arrival of the 1st Brigade, Sheikh Sattar was one of the most prominent.[28]

A third figure worth noting is Captain Travis Patriquin, who some referred to as the "Lawrence of Arabia" in Ramadi.[29] He had a unique background. Patriquin had served as special forces support officer, commanding a signals intelligence team in combat during Operation Anaconda in Afghanistan, but then transferred back to the regular army as an infantry officer. Patriquin was a fluent Arabic speaker and had immersed himself, on his own initiative, in Arab culture. MacFarland recognized Patriquin's unique skillset and assigned him to be the brigade's civil affairs officer and later dubbed him the "brigade liaison to the Iraqi Civil Forces." The Iraqi Civil Forces were organized informally under Sheikh Sattar and more formally as the Iraqi police. MacFarland, Satter, and Patriquin were the key leaders responsible for developing the military and the political strategy which ultimately achieved success in the Battle of Ramadi. [30]

On the enemy side, there were three different insurgent forces operating in Ramadi: AQI, Sunni insurgents who often allied with AQI but were not under their control, and criminal elements who, for hire, would work with either group. The American forces, with the help of the recently allied local tribes, gradually pacified most of the Sunni insurgents. Key to this effort was the growth of the local Iraqi police forces and its improved

[27] Jim Michaels, *A Chance in Hell: The Men Who Triumphed Over Iraq's Deadliest City and Turned the Tide of War* (New York: St Martin's Press, 2010), 91-92.

[28] Doyle, *A Soldier's Dream*, 23, 48, 53-56, 73-76.

[29] Chad Pillai, "A Tribute to Captain Travis Patriquin," *Small Wars Journal*, December 8, 2010, https://smallwarsjournal.com/blog/a-tribute-to-captain-travis-patriquin.

[30] Doyle, *A Soldier's Dream*, 23, 51, 75.

morale, capability, and legitimacy. Creating a large and capable indigenous police force also helped mitigate the threat of local criminal activity. Thus, AQI was the most significant threat that the Brigade had to clear from the city.

Abu Musab al-Zarqawi was the leader of AQI, and his network had grown more powerful since the invasion. Zarqawi's strategy was to isolate the United States from its allies, deter Iraqi cooperation with the United States, target reconstruction efforts, and start a civil war to force the United States to leave Iraq.[31] To accomplish this strategy, Zarqawi embraced a bombing and terror campaign targeting the United Nations mission, police stations, recruitment centers, and Iraqi politicians to discourage them from cooperating with American forces. Although much smaller in number than some of the other insurgent groups in Iraq, Zarqawi's network was having a strategic impact since his foreign fighters accounted for 75 percent of all suicide bombings in Iraq.[32] A U.S. airstrike killed Zarqawi the day after MacFarland assumed command of Ramadi; nonetheless, the group remained a powerful adversary and his lieutenants in Ramadi planned to resist the new American brigade vigorously.[33]

Spencer: Can you describe the battle? How was MacFarland able to clear al Qaeda from the city?

DiMarco: Major General Richard Zilmer, the commander of the Multinational Force-West, assigned Colonel MacFarland the mission to "Fix Ramadi but don't do a Fallujah."[34] During the second battle of Fallujah in 2004, in what would be the bloodiest battle of the war for American troops, army and Marine forces secured the city but in the process, essentially destroyed the city to save it.[35] Zilmer wanted the brigade to pacify Ramadi, but to do it with a minimal loss of life and minimal collateral damage. That was an extremely challenging mission for a city like Ramadi.

---

[31] M.J. Kirdar, Al Qaeda in Iraq (Washington, DC: CSIS, 2011), 3-4, https://csis-website-prod.s3.amazonaws.com/s3fs-public/legacy_files/files/publication/110614_Kirdar_AlQaedaIraq_Web.pdf.

[32] Joseph Felter and Brian Fishman, *Al-Qa'ida's Foreign Fighters in Iraq: A First Look at the Sinjar Records* (West Point, NY: Combating Terrorism Center, 2007), 18, https://ctc.usma.edu/wp-content/uploads/2010/06/aqs-foreign-fighters-in-iraq.pdf.

[33] Sean Naylor, *Relentless Strike: The Secret History of Joint Special Operations Command* (New York: St. Martin's Press, 2016), 279-89.

[34] DiMarco, *Concrete Hell*, 196.

[35] Dan Lamothe, "Remembering the Iraq War's bloodiest battle, 10 years later," *Washington Post*, November 4, 2014, https://www.washingtonpost.com/news/checkpoint/wp/2014/11/04/remembering-the-iraq-wars-bloodiest-battle-10-years-later/.

MacFarland quickly determined that he could not pacify a city the size of Ramadi with a single army brigade, he also required Iraqi forces and the support of the local population. Thus, he needed to ally with the sheikhs so that he could grow the local police force. The police and Iraqi Army could provide the troop strength that the brigade needed to hold areas that the American forces would clear. Serendipitously, many of the Sunni insurgents that had previously allied with AQI were breaking from the terrorist group just as the brigade moved to Ramadi.[36]

The Sunni insurgents had grown tired of AQI's brutal methods and its attempt to replace the traditional tribal political structure with their own. The two groups had different goals: AQI's goal was to establish a religious state under *Sharia* law. By contrast, the Sunni insurgents, while not completely secular, were motivated more by Iraqi nationalism and economic concerns. The Sunnis of Al Anbar province had been the force behind Saddam Hussein's Iraqi government. After the invasion, they had lost most of their political power and economic opportunities, and they were more focused on increasing their political and economic opportunities as opposed to establishing a religious state.[37]

AQI was not tolerant of some of the secular economic activity of the Sunni insurgents and determined that they could consolidate their grip on the population if they eliminated some of the Sunni insurgent leaders. So, as the 1st Brigade arrived, a mini civil war between the indigenous Sunni insurgents and the foreign fighter dominated AQI had started.[38]

The 1st Brigade took advantage of this situation to convince the Sunni sheikhs that their best chance of achieving some political autonomy and regaining economic power was to break from AQI and join with the Americans. The Americans offered protection to the sheikhs who were being targeted by AQI while the sheikhs facilitated the enlistment of young Sunni men, some of them former insurgents, into the Iraqi police. The Americans trained and equipped the Iraqi forces which then provided the necessary force to support the combat operations of 1st Brigade.[39]

---

36 Clay, *On Point III*, 27-28.
37 Russell, *Innovation, Transformation, and War*, 114-116.
38 Ibid., 115.
39 Clay, *On Point III*, 36.

Previously, recruiting efforts in Ramadi were largely unproductive due to Zarqawi's effectiveness at bombing police stations and recruitment drives.[40]

Integrating the sheikhs, their tribes, and particularly their military-age men to support the American effort was a new approach in Iraq. Captain Patriquin created a PowerPoint presentation that explained the Sunni Awakening, the motivation of the Iraqis, and how those motivations could be aligned with American goals. He distributed the presentation throughout the brigade, and it eventually spread to other U.S. forces, their headquarters, and the Pentagon. In very simple terms the presentation made the point that if they understood the Iraqi culture—particularly the tribal structure and motivations of the local Sunnis—they could win them over.[41]

The first component to achieving victory in Ramadi was winning over the local population. Gaining the support of the Sunni population was critical because it solved two of the coalition's critical needs: intelligence to identify the terrorists and police to provide the necessary security after American forces had cleared an area. The locals knew who among the population were AQI—something extremely difficult for an outsider to determine. So, in effect, the intelligence allowed the coalition to "clear," and the police provided the necessary force to "hold."

The second component was employing effective counterinsurgency tactics. Learning from its experience in Tal Afar, the brigade adopted an operational concept called "COP in a Box" that used a counterinsurgency "inkblot" method to secure the city. Similar to how inkblots spread to cover most or all of a piece of paper, combat outposts serve as "inkblots" that slowly expand security outward until the entire area is secure. Combat outpost (COP) in a box involved six steps to seize control of small parts of the city and then slowly expand and link those spots together until the entire city was secure.

Spencer: Can you describe how the COP in a Box worked?

DiMarco: A COP in a Box had six steps. The first step was identifying the best locations to establish a combat outpost. The brigade staff analyzed the enemy and determined areas that were most important to the insurgents. These areas were prioritized for control. In some cases, gaining access to these important areas required establishing other combat outposts first to

---

[40] See, for example, "Car bomb kills at least 17 at police station in Iraq," *Baltimore Sun*, December 15, 2003, https://www.baltimoresun.com/news/bal-te.attack15dec15-story.html; or "At least 6 die in car bombing near Ramadi," CNN, November 29, 2004, http://www.cnn.com/2004/WORLD/meast/11/29/iraq.main/.

[41] Doyle, *A Soldier's Dream*, 184-187.

secure routes to these areas and control dominating terrain. The brigade used its Navy SEALs to conduct detailed reconnaissance throughout the city to identify specific locations and buildings that offered the best security to establish a new combat outpost.[42]

The second step was seizing the physical structure. Once the location was selected, SEAL teams infiltrated at night to secure the buildings and evacuate anyone that was living there. They brought interpreters, and as they seized the buildings, they compensated any displaced Iraqi families $2,500 a month for the use of their building.[43]

The third step was clearing a route from the base camp to the combat outpost. After securing the building, the brigade cleared the route between the base camp and the combat outpost of IEDs using armored vehicles equipped with mine and road clearing equipment.

The fourth step was securing the combat outpost. As the company team occupied the combat outpost, the SEALs moved out and established sniper positions to cover approaches to the combat outpost while the conventional infantry and armor units reinforced it. Simultaneously, combat engineers moved to the outpost with sandbags, concrete T-walls, and HESCO barriers,[44] to reinforce and fortify the position.[45]

The fifth step was expanding the secure area. Once the combat outpost was firmly established, the U.S. Army or Marine company was reinforced with an Iraqi Army company and began patrolling the area near the outpost while the SEALs retained observation posts and sniper positions on the perimeter. When insurgents counterattacked the coalition and Iraq patrols, they exposed themselves to the hidden SEAL snipers and American and Iraqi forces fighting from their fortified combat outpost. With the Sunni insurgents switching sides, the tactical advantage transitioned from AQI to the coalition forces. The coalition eliminated most of the insurgents during this step.

The sixth step was turning the outpost over to the Iraqis. Once the insurgents had exhausted themselves trying to defeat the combat outpost, coalition patrols extended the perimeter and Iraqi police moved into the combat outpost to establish civilian control over the area freeing the Iraqi Army and American units to start the process anew.

[42] Clay, *On Point III*, 31.
[43] Michaels, *A Chance in Hell*, 67.
[44] HESCO barriers are prefabricated collapsible wire mesh containers with a heavy-duty fabric liner when can be easily transported, opened, and then filled with sand for protection.
[45] Clay, *On Point III*, 31.

The brigade used this COP in a Box technique to establish a small footprint on key terrain, and then slowly expand until they secured the entire city. This was a vastly different approach from what the coalition had been doing which was basing out of large forward operating bases on the perimeter of the city and running relatively infrequent patrols into the city. With the help of the Iraqi police and military, the brigade slowly expanded the number of combat outposts. Through the summer and fall, the brigade established new combat outposts in a systematic, yet seemingly random interval, sometimes establishing combat outposts in quick succession—2-3 in a week—and other times pausing for several weeks before establishing a new position. The goal was to keep the insurgents from guessing when and where the American forces would establish and fortify a new position.

Spencer: Can you describe how the battle unfolded?

DiMarco: The National Guard unit's mission was fairly limited: securing the government center and the main supply routes. First Brigade's mission was far more expansive—clearing the entire city—so MacFarland's first step was to isolate the city to control access in and out. The brigade believed that AQI was infiltrating the city from the desert to the south, so they realized they had to seal off the southern portion of the city.

MacFarland wanted to take a more deliberate approach, but when Zarqawi was killed the day after the brigade assumed control of the area, MacFarland accelerated his timetable and the aggressiveness of the plan to take advantage of the disruption caused by the death of AQI's leader. Instead of securing the perimeter of the city first and then building up Iraqi police and army forces to support the brigade, he decided to quickly strike the center of the city.[46] To accomplish this, he modified the brigade's task organization, weighting his main effort, 1-37 Armor, to have the forces necessary to conduct the strike. Many of the additional assets came from 1-35 Armor which would wage an economy of force operation on the western approaches into the city in the suburb of Tameem and around Al Anbar University.[47]

Prior to striking the center of the city, MacFarland needed to secure the southern side. Thus, Bravo Company, 1-36 Infantry established the

[46] Timothy S. McWilliams and Kurtis P. Wheeler, eds., *Al-Anbar Awakening: U.S. Marines and Counterinsurgency in Iraq, 2004-2009* (Quantico, VA: U.S. Marine Corps University, 2009), 178.
[47] Anthony E. Deane, *Ramadi Declassified: A Roadmap to Peace in the Most Dangerous City in Iraq* (Havertown, PA: Pretorian Books, 2016), 126.

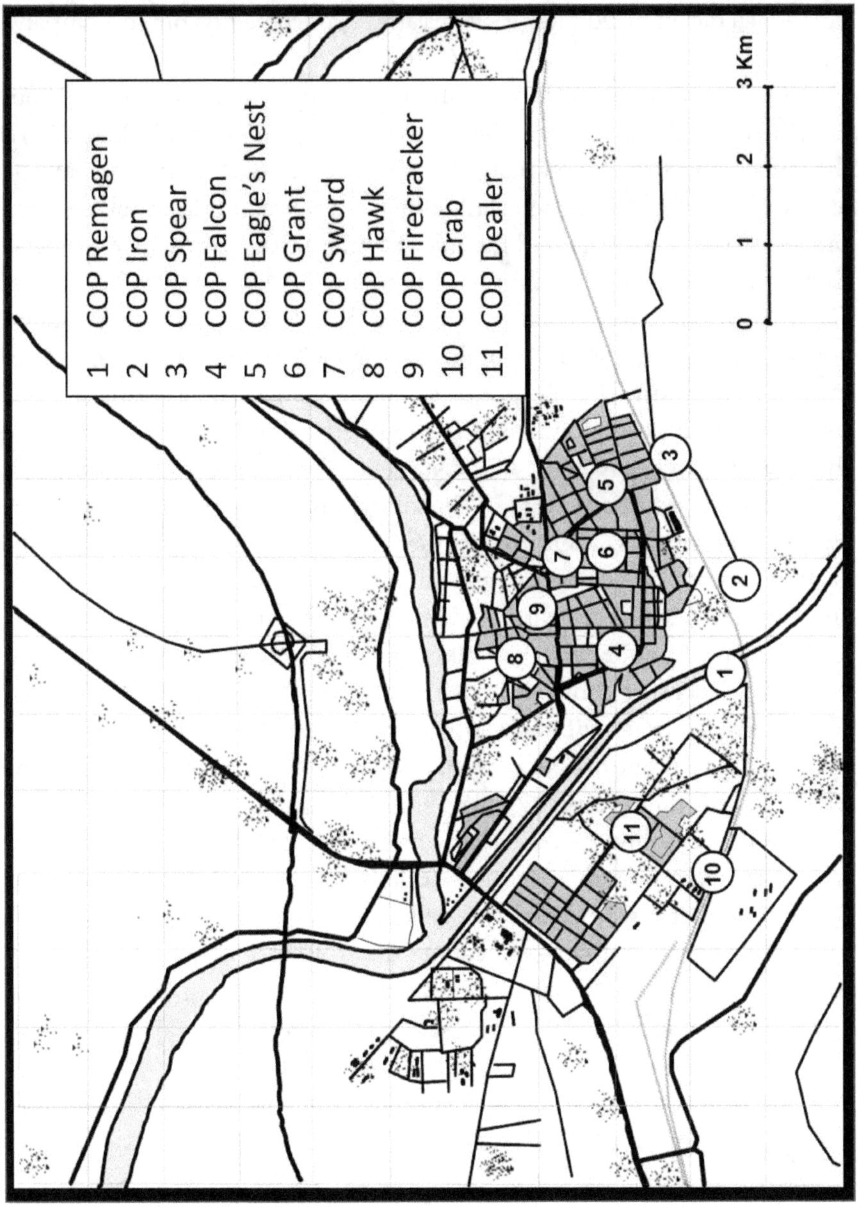

Figure 11.4. Major combat outposts in Ramadi, 2006.
Source: Map created by Louis DiMarco.

first combat outpost, COP Remagen, on June 16, at the railroad bridge that crossed the Habbaniyah Canal (see figure 11.4). Previously the bridge had been lightly guarded by the Iraqi military, but now it had the full combat power of a mechanized infantry company.[48] Concurrently, Bravo Company, 1-37 Armor repositioned to Camp Corregidor.[49]

On the morning of June 17, two company teams from 1-37 Armor launched attacks to secure the southern edge of central Ramadi. At the same time Bravo Company, 1-35 Armor, supported by Iraqi armor, attacked from COP Remagen across the Ham railroad bridge and seized a predesignated group of buildings about a mile east of the canal to establish COP Iron.[50] Simultaneously, Bravo Company, 1-37 Armor attacked out of Camp Corregidor and seized a compound of buildings on the southeastern side of Ramadi and began to build COP Spear.[51] Twenty-four hours later they had transformed both groups of buildings into small fortresses. The two combat outposts dominated the southern edge of the city to include the train station, a haven for AQI insurgents, that sat between them.[52]

With the southern edge of the city secure, the brigade prepared to strike north, pushing AQI toward the Euphrates River and east into the eastern suburbs of Ramadi, an area known as the Shark Fin. On June 26, a company from 1-37 Armor attacked into the heart of the city and established COP Falcon.[53] Simultaneously, 3-506 Infantry seized an area just southwest of the large soccer stadium and established COP Eagle's Nest, which linked the eastern perimeter with the southern perimeter. By mid-July, the brigade had established control over most of Ramadi's south, an area that had been practically off-limits to previous American units.

AQI leadership realized they were losing control over the city. They had been expecting an American offensive against Ramadi, but they anticipated that it would look more like 2004's second battle of Fallujah.[54] They were somewhat slow to react to the COP in a box strategy, but on July 24 they launched a major counterattack. More than 200 insurgents participated in more than 20 near-simultaneous attacks against all five of

[48] V.J. Tedesco, "Task Force 1st Battalion 37th Armor 'The Bandits' in OIF 05-0," PowerPoint presentation, no date, https://www.tedescos.com/resources/Documents/Bandits-in-OIF-05-07-Road-Show-Brief.pdf.

[49] Clay, *On Point III*, 29.

[50] 1-36 Infantry Battalion was part of the 1st Brigade organization. Soon after arrival in Iraq it was detached from the brigade and assigned to support the Marines in the city of Hit. B Company had previously been cross-attached with TF 1-37 Armor and remained with that task force as part of the 1st Brigade throughout the deployment.

[51] Tedesco, "Task Force 1st Battalion 37th Armor."

[52] Clay, *On Point III*, 29.

[53] Tedesco, "Task Force 1st Battalion 37th Armor."

[54] Joel D. Rayburn and Frank K. Sobchak, *The U.S. Army in the Iraq War*, vol. 1 (Carlisle Barracks, PA: U.S. Army War College Press, 2019), 605.

the brigade's maneuver battalions. The brigade repulsed all of them and killed at least 30 insurgents and wounded many more.[55]

Over the ensuing months, the brigade slowly expanded its control of the city by establishing more combat outposts. In August, 1-35 Armor built its first combat outpost, COP Crab, in western Ramadi to control the area around Al Anbar University. In the northern part of the city, 3/8 Marines established COP Hawk near the hospital. In September they were relieved by 1/6 Marines who built COP Firecracker near the governance center.[56] In central Ramadi 1-37 Armor continued to expand its stable of combat outposts. In September the task force established COP Grant and COP Sword in the very heart of the city while turning over earlier positions to the Iraqi Army and police.[57] By the end of October, the central and northern parts of Ramadi had been largely pacified. The last bastion of AQI power remained in Ramadi's eastern suburbs known as the Shark Fin.[58]

The brigade realized that AQI had suffered a significant blow after its failed counterattack because the frequency and sophistication of its attacks had declined. In June, al Qaeda typically attacked using remote control detonated IEDs, a fairly sophisticated system. After the failed attack, subsurface wire command-detonated systems became more and more common. They were less sophisticated and more difficult to emplace because the bomber had to expose himself while burying the wire. Eventually, the most common IED was a simple pressure-plate system: non-command-detonated systems that were much easier to detect. The size of the IEDs also gradually decreased as AQI's ability to build and transport large bombs declined.[59] By the fall, the attacks were both fewer and less dangerous.

Another sign that the brigade was winning was the steady increase in tribal sheikhs joining the alliance against AQI. By the fall of 2006, over 50 percent of the sheikhs had joined the coalition led by Sheikh Satter with the notable exception of those located in eastern Ramadi. Those sheikhs were willing to join but feared reprisals since AQI was concentrated in their part of the city. This situation came to a head in November.[60]

[55] Clay, *On Point III*, 26
[56] Deane, *Ramadi Declassified*, 231.
[57] Tedesco, "Task Force 1st Battalion 37th Armor."
[58] Rayburn and Sobchak, *The U.S. Army in the Iraq War*, 615.
[59] Clay, *On Point III*, 27.
[60] William Knarr et al., *Al Sahawa—The Awakening, Vol. IV-A: Area of Operations Topeka, East Ramadi and the Shark Fins* (Alexandria, VA: Institute for Defense Analysis, 2016), 2-1–2-7, https://apps.dtic.mil/sti/pdfs/AD1020274.pdf.

One of the important tribes in the eastern part of the city, the Albu Soda tribe, was led by Sheikh Jassim Mohammed. On November 27, AQI attacked Sheikh Jassim's outnumbered militia. Fortunately, Captain Patriquin had previously given Jassim a cell phone to contact the brigade in the case of an emergency. Jassim called the Americans and 1-9 Infantry, who had replaced 1-506 Infantry when they rotated back to the United States, responded from Camp Corregidor. The ensuing daylong firefight resulted in the deaths of more than 60 insurgents and 50 militiamen. In the days that followed, 1-9 Infantry established an additional combat outpost in the Shark Fin, destroying AQI's last bastion of power in the city.[61]

Securing a city is never a fast process. Yet, by January 2007, seven months into operations in Ramadi, most of the city had been secured. In March 2007 the 1st Brigade was relieved by a division from the 3rd Infantry Division having achieved its mission: AQI was all but eliminated in Ramadi, the Sunni insurgency had been co-opted through the Sunni Awakening to ally with the Americans, and the criminal element had been suppressed or integrated back into the legitimate forces allied with the Americans and the Iraqi government. Ramadi was so secure, that in September of 2007, far from being the most dangerous city in Iraq, coalition forces could walk the streets without their combat equipment and the civilian population was able to stage a 5-kilometer road race in Ramadi's downtown without significant security concerns.[62]

The 1st Brigade estimated that it killed 1,500 insurgents and captured another 1,500 during the nine-month operation before turning the city over to the Iraqi government.[63] But success came with a heavy price. The 1st Brigade lost 86 soldiers killed in action, including, unfortunately, Captain Patriquin, who was killed by a roadside IED.[64] In addition, several hundred American soldiers and Marines were wounded, and a significant number of vehicles were destroyed.[65]

Spencer: What are some of the lessons that can be learned from the battle?

---

[61] Ibid.

[62] Richard H. Shultz, *The Marines Take Anbar: The Four-Year Fight Against Al Qaeda* (Annapolis, MD: Naval Institute Press, 2013), 230.

[63] Clay, *On Point III*, 25.

[64] Monte Morin, "'Raider Brigade' takes over Ramadi," *Stars and Stripes*, February 19, 2007, https://www.stripes.com/news/raider-brigade-takes-over-ramadi-1.60517; and Pillai, "A Tribute to Captain Travis Partriquin."

[65] DiMarco, *Concrete Hell*, 208.

DiMarco: The legacy of the Battle of Ramadi is important. Using less than half of the combat power that was used in the second battle of Fallujah, the brigade was able to secure a city that was almost twice Fallujah's size.[66] The combat outpost counterinsurgency strategy combined with the Sunni Awakening became part of the political and tactical model that General Petraeus implemented during the surge in 2007, which was successful in establishing relative security throughout the country.

In comparison to the second battle of Fallujah, the battle of Ramadi offered an alternate approach to pacifying a large urban area. In Fallujah, successful pacification occurred through the overwhelming use of combat power and rapid military action, but at the cost of significant destruction to the city. Additionally, it failed to win over the population to the political cause of the coalition or the Iraqi government, so the tactical success did not lead to strategic success. In contrast, the battle of Ramadi showed that a limited, but still sizable amount of combat power employing an effective counterinsurgency strategy may take longer, but it can also defeat an enemy while suffering similar casualties, inflicting less collateral damage, and simultaneously winning over the population.

It is important to note that Colonel MacFarland had the benefit of learning from Colonel McMaster in Tal Afar prior to moving to Ramadi. As mentioned previously, McMaster was one of the few implementing effective counterinsurgency methods prior to the development of the army's counterinsurgency doctrine in December 2006. Colonel MacFarland, thus, deserves credit for the brigade's success, but it may not have been possible had he not been able to learn the combat outpost methodology during his transition with Colonel McMaster and the 3rd Armored Cavalry Regiment.

While the strategy used in Ramadi is widely hailed as a success, there are some that criticize it because the gains were relatively short-lived.[67] Some point to the Islamic State of Iraq and al Sham's (ISIS) success

---

[66] More than 13,000 Allied troops participated in the second battle of Fallujah in November 2004. This estimate is based on the known personnel strengths of the units involved in the battle described in Louis DiMarco, "The Battle of Fallujah," *Desperta Ferro Contemporánea*, June 20, 2015, https://www.despertaferro-ediciones.com/2015/faluya-en-el-centro-de-la-lucha-en-el-n-o10-de-desperta-ferro-contemporanea-insurgencia-en-irak-2003-2011/. The population of Fallujah was approximately 200,000 in 2004, the pre-war population of Ramadi was 400,000 but dropped to roughly 300,000 by 2006. "Fallujah Urban Area Population Graph," Populationstat.com, https://populationstat.com/iraq/fallujah.

[67] See, for example, Michael R. Gordon and Bernard E. Trainor, *The Endgame: The Inside Story of the Struggle for Iraq, from George W. Bush to Barack Obama* (New York: Pantheon Books, 2012), 240-252; and Carter Malkasian, Illusions of Victory: The Anbar Awakening and the Rise of the Islamic State (New York, Oxford University Press, 2017).

at rapidly capturing large swaths of land in Iraq and Syria and declaring a caliphate in 2014.[68] They argue that in the process of liberating Ramadi in 2006 with the help of the Sunni Arabs, the Americans empowered the Sunni Arab leadership and fostered a separatist view among the Sunnis vis-à-vis the Shiite dominated Iraqi government. Then, when the American forces departed Iraq and the national government came to be even more dominated by Shiites, the Sunni population and its leaders felt disenfranchised and when looking for an ally, the only readily available ally was ISIS. This allowed ISIS to quickly gain control of Ramadi and other cities, and seriously threaten the Iraqi government.[69]

Spencer: Do you agree with those authors? Do you think empowering Ramadi tribal leaders furthered the divide between the Sunni and Shiite powers in Iraq?

DiMarco: The authors who maintain that the Sunni population turned to ISIS after the United States left are essentially correct. However, I would argue this was probably more due to the Shiite dominated Iraqi government pursuing sectarian policies which essentially disenfranchised the Sunni population after the United States left. A more inclusive policy aimed at Sunni-Shiite reconciliation may have won the cooperation if not the allegiance of the Sunni population. However, Sunni-Shiite relations are strategic political issues that extend beyond the scope of the battle for Ramadi and beyond the scope of military leaders, particularly those at the tactical level.

Spencer: What are some other lessons that remain relevant for future urban operations?

DiMarco: Two simple but critical lessons are obvious from the battle of Ramadi. First, urban operations are possible with a relatively small force. However, the smaller the force, the longer the operations will take. Ramadi had to be conducted in a systematic manner, establishing combat outposts, expanding beyond the outpost, and then establishing subsequent outposts because the brigade lacked the combat power to engage across the city simultaneously. Another limiting factor was engineer support. The engineers who fortified the combat outposts could only support a limited number of operations at the same time. The combat outposts ink-spot method was a much slower process but, this approach, though still creating

[68] "Timeline: the Rise, Spread, and Fall of the Islamic State," Wilson Center, October 28, 2019, https://www.wilsoncenter.org/article/timeline-the-rise-spread-and-fall-the-islamic-state.
[69] Malkasian, Illusions of Victory.

collateral damage, resulted in less damage to the city's infrastructure than a more traditional block by block assault across the entire city.

The second lesson learned is the criticality of indigenous partners, particularly in an urban insurgency. Without the support of the local sheikhs and through them the civilian population, the American forces would not have been able to execute the "hold" and "build" phases of the "clear, hold, build" tactical approach. The support of the local population gave the American forces the manpower, intelligence, and legitimacy it required to be successful. At the same time, it denied those same assets to the AQI insurgents.

Spencer: What happened to the key figures from the battle?

DiMarco: Colonel MacFarland went on to a distinguished army career and retired as a three-star general. Sheikh Sattar, who facilitated the Sunni uprising in Ramadi, was assassinated by AQI on September 13, 2007, only one week after meeting President George W. Bush.[70] As was mentioned earlier, Captain Patriquin was also killed. So, as is far too common in war, and especially in urban battles, not all of the heroes from the battle of Ramadi survived.

[70] Alissa J. Rubin, "Sunni Sheik Who Backed U.S. in Iraq Is Killed," *New York Times*, September 14, 2007, https://www.nytimes.com/2007/09/14/world/middleeast/14iraq.html.

12

The Battle of Sadr City

The battle of Sadr City took place from March 26 to May 11, 2008, nearly 15 months after President George W. Bush announced the surge in Iraq in January 2007.[1] The Baghdad Security Plan was a key element of the surge, which called for an addition of five brigades to Baghdad and extended the deployment time of the deployed units. The plan also included an estimated 25,000 Sons of Iraq—a local militia program—to guard neighborhoods throughout the city.[2] The entire plan was contingent on employing new counterinsurgency doctrine to *clear* the insurgents from an area, *hold* the area with their Iraqi counterparts, and finally *build* enduring stability to ultimately hand off control of the area to the Iraqis.[3]

The Baghdad Security Plan called for coalition troops to leave their large forward operating bases and instead occupy smaller outposts throughout the city. The one exception, however, was Sadr City. Sadr City was an extremely densely populated and low-income neighborhood within Baghdad that contained an estimated two million inhabitants in its

---

[1] David Johnson et al., *The 2008 Battle of Sadr City: Reimagining Urban Combat* (Santa Monica, CA: RAND, 2013), xvi, https://www.rand.org/pubs/research_reports/RR160.html; and David Sanger, "Bush Adds Troops in Bid to Secure Iraq," *New York Times*, January 11, 2007, https://www.nytimes.com/2007/01/11/world/middleeast/11prexy.html.

[2] Farook Ahmed, *Backgrounder #23: Sons of Iraq and Awakening Forces* (Washington, DC: Institute for the Study of War, 2008), https://www.understandingwar.org/sites/default/files/reports/Backgrounder%2023%20Sons%20of%20Iraq%20and%20Awakening%20Forces.pdf.

[3] The U.S. Army and U.S. Marine Corps published its counterinsurgency field manual in December 2006. Department of the Army, *Counterinsurgency*, FM 3-24 (Washington, DC: Headquarters, Department of the Army, 2006); Kevin T. Ryan, "What the surge means," *Boston Globe*, January 12, 2007, http://archive.boston.com/news/nation/washington/articles/2007/01/12/what_the_surge_means_boston_globe/; Kimberly Kagan, "Enforcing the Law: The Baghdad Security Plan Begins," *Iraq Report*, February 10, 2007–March 5, 2007, 3, https://doczz.net/doc/1397458/enforcing-the-law--the-baghdad-security-plan-begins; and George W. Bush, "Press Conference by the President," February 14, 2007, https://georgewbush-whitehouse.archives.gov/news/releases/2007/02/text/20070214-2.html.

roughly 22 square kilometer area.[4] Due to agreements between Muqtada al-Sadr, who exerted significant control over Sadr City, and Prime Minister Nouri al-Malaki, coalition troops were prohibited from establishing outposts in the district, although they did conduct periodic raids.[5] But even these raids ceased following an airstrike in October of 2007 which was reported to have killed a number of Iraqi civilians.[6] As a result, Malaki placed Sadr City off limits to U.S. ground operations, leaving the United States effectively blind in the insurgent stronghold.[7]

By March 2008, the Baghdad Security Plan had been fairly successful, which set the conditions for the battle of Sadr City. First, the coalition had severely degraded al Qaeda in Iraq, which allowed the coalition to focus on other destabilizing elements, such as those led or associated with Muqtada al-Sadr. Al-Sadr was a young Shia cleric who had established a substantial following after the invasion. His followers, Sadrists, and its armed militia, Jaish al-Mahdi (JAM), often opposed the Iraqi government and attacked coalition forces. Second, the increased security strengthened the position of Prime Minister Nouri al-Maliki, which gave him the strength to confront the Sadrists in many Iraq cities. Third, the coalition had largely contained JAM to Sadr City.[8]

On March 23, 2008, JAM fired a barrage of rockets from Sadr City into the international zone (the "Green Zone") of Baghdad in response to Maliki's impending offensive into Basra, a Shia dominated city 500 kilometers southeast of Baghdad. This ended al-Sadr's self-imposed cease-fire that had been in place since August. Over the next couple of days, JAM launched additional rocket and mortar attacks into the Green Zone and seized checkpoints in and around Sadr City. In response, Malaki directed the coalition to stop the rocket attacks and defeat the criminal militias within Sadr City, and so began the 2008 battle of Sadr City.[9] The battle, named Operation Gold Wall, centered on building a five-kilometer-long concrete wall to seal off Sadr City. The ensuing 7-week battle, would cost the United States six soldiers killed in action, while JAM lost an estimated 700 fighters.[10]

---

[4] David Johnson et al., *The 2008 Battle of Sadr City*, 17.

[5] Ibid.

[6] Steven R. Hurst, "US: Raid of Baghdad's Sadr City Kills 49," *Associated Press*, October 21, 2007, https://www.washingtonpost.com/wp-dyn/content/article/2007/10/21/AR2007102100159_pf.html.

[7] Johnson et al., *The 2008 Battle of Sadr City*, xiv.

[8] Ibid., 17.

[9] Ibid.

[10] Ibid., xvii.

In this chapter, John Spencer interviews Lieutenant Colonel (retired) Rob MacMillan. MacMillan was the operations officer for the 1st Combined Arms Battalion, 68th Armored Regiment (1-68 Armor), during the battle of Sadr City. John Spencer was also assigned to the 1-68 Armor at this time. When he deployed in March 2008, he was serving as an assistant operations officer and then took command of Alpha Company in July.

John Spencer: I would like to start by asking you to discuss your deployment to Iraq. What was your job and when did you deploy?

Rob MacMillan: I was the operations officer for the 1st Combined Arms Battalion, 68th Armored Regiment (1-68 Armor). I assumed the position only two weeks prior to the deployment. We were originally slated to deploy in February 2008, but our deployment was moved up, so we deployed to Iraq in December 2007. We owned a sector just north of Baghdad. On March 1, 2008, our battalion was reassigned to a new sector northwest of Sadr City. The sector included the Sha'ab, Ur, and Beida neighborhoods within the Adhamiyah district in northern Baghdad and an area north of Baghdad that included the town of Hussaniyah. Baghdad is divided into nine security districts containing over 70 neighborhoods.[11] Adhamiyah and Sadr City are two of the nine security districts (see figure 12.1).[12]

Spencer: I was a company commander in your battalion, so it may seem odd that I ask, but were you part of the surge? Was our unit part of the Baghdad Security Plan which was sent to help secure the country and to quell the sectarian violence in Baghdad?

MacMillan: Yes, that is correct. But the surge involved more than simply sending five additional brigades into Baghdad. Deployments were also extended from 12-months to 15-months. This not only allowed a greater number of brigades in Iraq, but it also allowed each brigade more time to work within its area before redeploying.

Spencer: What comprises a combined arms battalion?

MacMillan: Combined arms battalions were a relatively new construct in 2007. Previously, battalions were organized as pure: for example, three or four armor companies comprised an armor battalion,

[11] "Baghdad City: Baghdad Security Districts," Institute for the Study of War, https://www.understandingwar.org/region/baghdad-city-%D9%85%D8%AF%D9%8A%D9%86%D8%A9-%D8%A8%D8%BA%D8%AF%D8%A7%D8%AF.

[12] When people and this chapter refer to Sadr City, they usually refer to the square portion indicated on the map, not the entirety of the Sadr City security district; thus, "Sadr City" does not include New Sadr City, Jamila, or Thawra shown on figure 12.1.

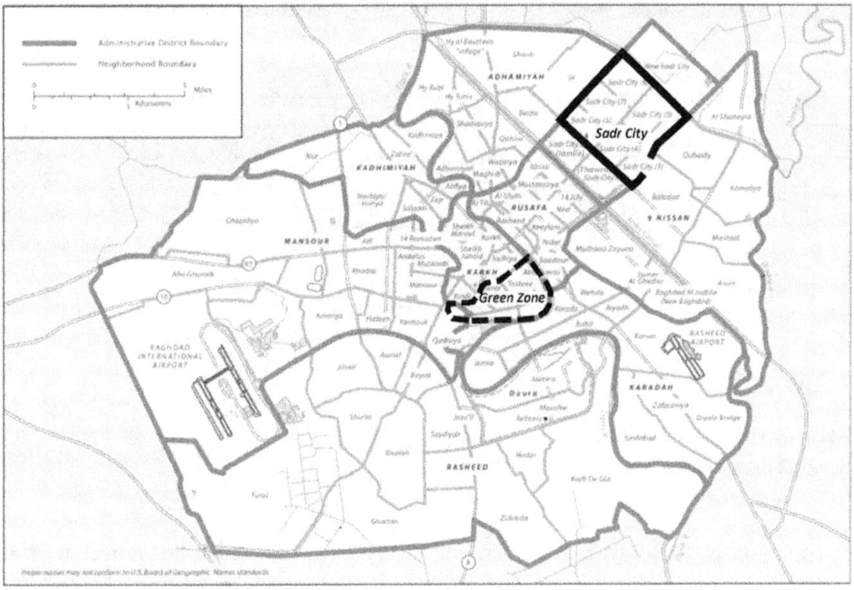

Figure 12.1. Map of Baghdad's security districts and neighborhoods.
Source: Joel D. Rayburn and Frank K. Sobchak, *The U.S. Army in the Iraq War, vol 2.* (Carlisle Barracks, PA: U.S. Army War College Press, 2019), ii.

and each mechanized brigade consisted of an armor-pure battalion and an infantry-pure battalion. While organized this way, they did not fight this way. When a unit deployed to the National Training Center, for example, they would have to change their organic task organization and swap infantry and armored companies to provide a combined arms team within each battalion.

In the mid-2000s, when it became clear that the army would maintain a large presence in Iraq for the foreseeable future, the army transitioned to brigade combat teams with new battalion task organizations—combined arms battalions—consisting of two armored companies and two infantry companies (see figure 12.2).[13] To make them more self-sufficient, a forward

[13] In 2016 the composition of the Combined Arms Battalion was changed. One of the maneuver battalions was dropped, so Combined Arms Battalions consisted of either 2 mechanized rifle companies and 1 armor company or 2 armor companies and one mechanized rifle company. Brandon Matsuyama, "U.S. Army Combined Arms Battalion (2016)," Battle Order, https://www.battleorder.org/usa-cab-2016; and Pete Norris, "Maximizing the Lethality of Armored Forces," *Strategy Bridge*, June 28, 2016, https://thestrategybridge.org/the-bridge/2016/6/28/maximizing-the-lethality-of-armored-forces.

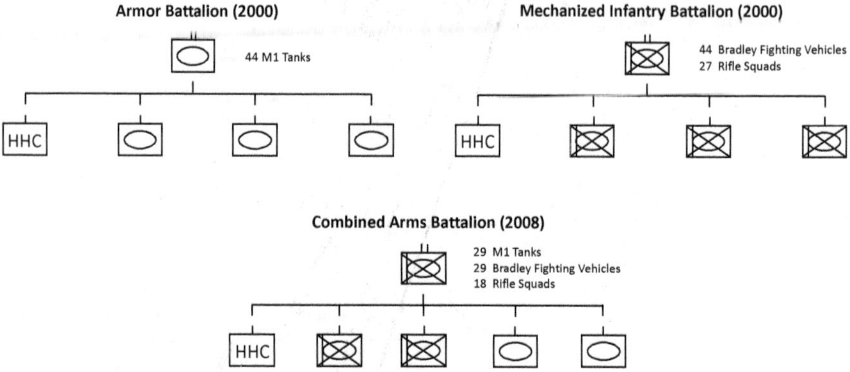

Figure 12.2. Combined Arms Battalion, 2008.
Source: Figure created by Liam Collins.

support company and an engineer company were also attached to the battalion. So, a combined arms battalion could pretty much support itself.

Our battalion consisted of two mechanized infantry companies (Alpha and Bravo), two tank companies (Charlie and Delta), an engineer company (Echo), a forward support company (Fox), and a Headquarters and Headquarters Company.[14] At the start of the battle of Sadr City, our Alpha Company was detached to forward operating base Prosperity. The base was home to our brigade headquarters and Alpha Company was tasked to provide entry control point and convoy escort duties. Alpha Company was relieved of this mission and returned to our battalion in May to support the battle of Sadr City.

For operations in Iraq, our battalion was augmented with four civil affairs teams, two tactical human intelligence teams, three explosive ordnance detachment teams, two tactical psychological operations teams, one multifunction (military intelligence) platoon, and one U.S. Air Force Joint Terminal Air Controller.

[14] Forward support companies "provide transportation, supply, maintenance, and food service support" to the unit they are assigned to support. They can provide immediate vehicle and weapons maintenance, dining, laundry, and other services usually maintained at higher levels. Thomas A. Knothe, "Forward support company operations in separate units," Army.com, May 5, 2014, https://www.army.mil/article/125019/forward_support_company_operations_in_separate_units. A battalion headquarters and headquarters company usually houses the battalion mortars, medics, snipers, and other specialty platoons that support the companies as needed. It is also where most of the staff and administrative personnel are assigned.

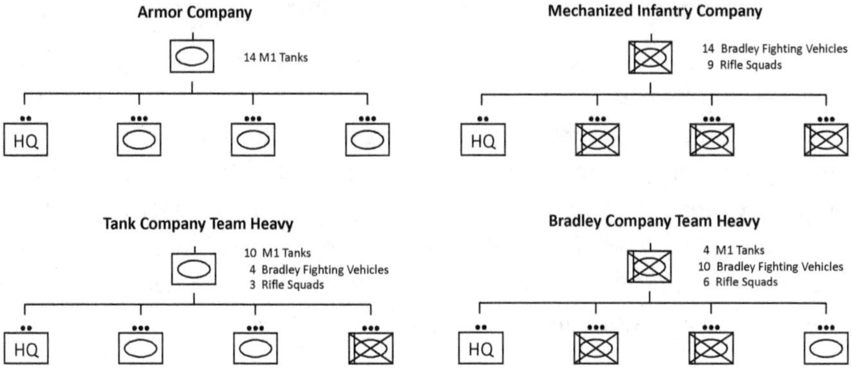

Figure 12.3. Company Teams.
Source: Figure created by Liam Collins.

Spencer: 1-68 Armor took the combined arms concept a step farther and swapped out infantry and armor platoons within each company to give each company a combined arms maneuver team. Is that correct?

MacMillan: Yes. One of the first things that we had to do was to establish the task organization for our assigned missions. We did not want infantry companies and armor companies to go into their battle spaces pure, because an infantry company had 2-3 times as many soldiers as an armor company.[15] At the time, there were strict minimum force requirements for patrols. These requirements specified the minimum number of vehicles and soldiers for a patrol.

A tank company, comprised of 16-person tank platoons, severely limited the number of patrols the company could create under those requirements, especially when considering the mandatory two-week rest and recuperation break that everyone had to take at some point during their 12- to 15-month deployment.

Thus, swapping tank and mechanized infantry platoons between the companies not only solved minimum force requirements, but it also allowed for a "cross-pollination" of skillsets within each company (see figure 12.3). The infantry soldiers were very skilled at dismounted patrolling while the armor soldiers were very skilled with the systems on their tanks and mounted tactics.

[15] At full strength, a mechanized infantry company had 135 soldiers and an armor company had 62 soldiers, Brandan Matsuyama, "U.S. Army Combined Arms Battalion (2016)," Battle Order, https://www.battleorder.org/usa-cab-2016.

While it proved to be an excellent construct for the fight in Sadr City, we did not establish this task organization for that mission. We established this task organization because it allowed us to have the necessary manpower to operate and run combat patrols at each combat outpost regardless of leave or casualties. This also allowed each company commander the necessary combat power to move about within his sector without having to task a subordinate platoon for the manpower to meet the minimum force requirements. Thus, we established true company teams within the battalion, with each armor company having an infantry platoon and each infantry company having an armor platoon. I would add that we deployed with our tanks and Bradley Fighting Vehicles (Bradleys), but we opted to use HMMWVs[16] (Humvees) for our patrolling vehicle, which I will explain later.

Spencer: If you think about it, while you may have reorganized your companies to meet minimum force requirements, it also provided the ideal task organization for urban warfare since studies have shown that the company teams that you created—ones that include a mix of armor and infantry—are often the most effective at urban warfare.[17] Urban combat requires a combined arms team comprised of infantry, armor, engineers, and fires. What you established allowed for a command relationship and a task organization that was ideal for a tough urban fight like the battle of Sadr City. I find that very interesting.

MacMillan: Yes, and it was very important that we established this task organization months prior to the battle. You do not want to create a new task organization at the last minute if you can avoid it. Elements need time to work together, to learn how they complement one another, and to establish the command relationship between the attached platoon leader and the company commander. It is important to do this early. The attached platoon should actually feel that it is an organic part of Alpha Company, for example, instead of feeling like a temporary attachment. They may

[16] HMMWV stands for high mobility multipurpose wheeled vehicle. It is a family of "lightweight, highly mobile, four-wheel drive tactical vehicles." See, "High Mobility Multipurpose Wheeled Vehicle (HMMWV)," Military.com, https://www.military.com/equipment/high-mobility-multipurpose-wheeled-vehicle-hmmwv.

[17] See, for example, Gian Gentile et al., *Reimagining the Character of Urban Operations for the U.S. Army: How the Past Can Inform the Present and Future* (Santa Monica, CA: RAND, 2017), https://www.rand.org/pubs/research_reports/RR1602.html.

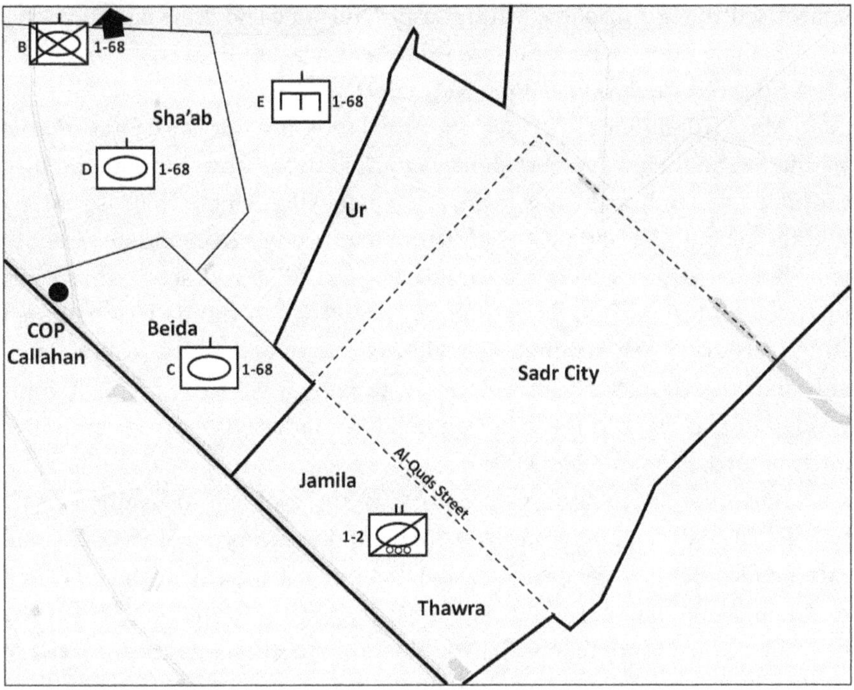

Figure 12.4. Forces around Sadr City prior to the battle, March 2008.
Source: Map created by Liam Collins.

look differently, act differently, and have different skills and vehicles, but whether they are armor or infantry soldiers, they all fall under the same blue infantry flag or red armor flag, depending on the company.

Spencer: Let us turn to the battle now. You said that your battalion assumed control of the area northwest of Sadr City in December 2007. Which neighborhoods were yours?

MacMillan: Bravo Company owned a mostly rural area north of the city that included the town of Hussaniyah. Charlie Company owned the Beida neighborhood, Delta Company had most of Sha'ab, and the engineer company had parts of Ur and Sha'ab and some of the rural area north of the city. The Headquarters and Headquarters Company owned Rashidiya which was north of Bravo Company's sector. The battalion's headquarters was located at combat outpost Callahan along the Sha'ab and Beida neighborhoods. Sadr City, Jamila, Thawra, and part of the Ur neighborhood fell within 1-2 Stryker Cavalry Regiment's area of responsibility (see figure 12.4). 1-2 Stryker was the second of three maneuver battalions that comprised the 3rd Brigade, 4th Infantry Division. The third battalion was 4-7 Cavalry, and it owned the area to the southeast of Beida.

Spencer: Can you describe Sadr City?

MacMillan: Sadr City was a very large, low-income, Shia district within Baghdad. It was a very densely packed urban area that contained an estimated two million people in only 22 square kilometers, which is roughly half the size of Manhattan.[18] The district had been called Saddam City but changed its name to Sadr City following the 2003 invasion, naming the district after Muqtada al-Sadr's father, Ayatollah Mohammad Mohammed Sadeq al-Sadr.[19] An estimated 1,000-2,000 active fighters, called Jaish al-Mahdi (JAM) or the Mahdi Army, who were loyal to Muqtada al-Sadr resided in the city, but the coalition believed Sadr could quickly mobilize an army of as many as 20,000.[20]

Since the start of the war, Sadr City represented an insurgent safe haven that coalition forces rarely visited with the exception of an occasional raid. In October 2007, a special operations raid into Sadr City encountered more militants than expected. The ensuing air strikes that supported the operation killed more than 40 JAM fighters, but Sadrists claimed dozens of civilians had been killed. Concerned with the civilian casualties and under pressure from the Sadrists, Prime Minister Maliki demanded his approval prior to any further coalition strikes into Sadr City. Concerned that the incident could push Maliki to oppose the renewal of the soon-to-be expiring United Nations Security Council Resolution that authorized the coalition's presence in the country, General Petraeus, the coalition commander, agreed to Maliki's demands. Thus, coalition forces encircled Sadr city with Iraqi checkpoints and joint security stations but would not enter it.[21]

Spencer: I know from my research that Sadr City was a particularly troubling area for the coalition dating back to at least early 2004. On April 4, 2004, a day known as Black Sunday, militants ambushed a 1st Cavalry Division patrol, killing eight soldiers.[22] Ever since that ambush, the coalition rarely entered Sadr City. I found a news release about the October

[18] Johnson et al., *The 2008 Battle of Sadr City*, xv.

[19] Stephen Farrell, "Sadr to split Iraq militia into 2 groups," *New York Times*, August 8, 2008, https://www.nytimes.com/2008/08/08/world/africa/08iht-iraq.4.15124374.html.

[20] Joel D. Rayburn and Frank K. Sobchak, eds. *The U.S. Army in the Iraq War, vol 2.* (Carlisle Barracks, PA: U.S. Army War College Press, 2019), 368.

[21] Ibid., 369-370.

[22] Ely Brown, "Inside the Ambush Known as Black Sunday," ABC News, November 7, 2017, https://abcnews.go.com/US/inside-ambush-black-sunday/story?id=50962302.

2007 airstrike that you discussed, and it was reported that the strike killed several children in addition to dozens of militants.[23] Following backlash from the strike, Maliki placed Sadr City off limits to U.S. ground forces.[24]

MacMillan: Yes, that is correct. The only units allowed to enter Sadr City were special operations forces for short duration raids and even they required Maliki's approval.

Spencer: If forces were forbidden from entering Sadr City, how did the battle start?

MacMillan: More than one-year into the surge, the surge was having its desired effect. With al Qaeda in Iraq largely subdued, Malaki now had the strength to focus on al-Sadr. Basra was a Shia dominated city 500 kilometers southeast of Baghdad. Like Sadr City, Basra was full of Sadr loyalists and its security situation had worsened through 2007 and early 2008. In March, Maliki finally had enough and started moving Iraqi forces against Basra.[25] In response, JAM militants launched a salvo of rockets from Sadr City into the Green Zone. Over the next two days, JAM overran a number of security checkpoints in and around Sadr City and this caused Maliki to authorize coalition and Iraqi forces to stop the rocket attacks and defeat JAM in Sadr City.[26]

Spencer: Did the JAM offensive come as a complete surprise?

MacMillan: No. Prior to the rocket attack on March 23, we knew that something was coming. Days prior to the attack, we noticed a decrease in civilian activity in our part of the city. The traffic on the roads was extremely light and we observed fewer people moving around the city. Additionally, the Iraqi company that normally picked up the trash from our dumpsters did not come that day. When we queried our brigade's support battalion about the trash pickup, they told us that all of their contacts were telling them that they were staying out of Baghdad. Many of the checkpoints in our sector, armed by Iraqi Army soldiers and Sons of Iraq, were left unmanned because they refused to come to work as well. We knew something was coming, so it was not a complete surprise when JAM started overrunning the Iraqi Army checkpoints in the Jamila, Beida, and Sha'ab neighborhoods.

---

23 Hurst, "US: Raid of Baghdad's Sadr City."
24 Johnson et al., *The 2008 Battle of Sadr City*, xiv.
25 Rayburn and Sobchak, *The U.S. Army in the Iraq War, vol 2.*, 356-9.
26 Ibid.

Spencer: I remember there was a day when the attacks spiked. Our battalion commander had a "SIGACT"[27] graphic that displayed the number of significant activities in our sector. It was basically a bar chart displaying the number of enemy attacks by day over the past three weeks. The bars for each day were pretty low and then on March 24th or 25th, the daily bar hit the top of the chart and stayed there. We called it "March Madness." It was obvious that the situation had exploded.

MacMillan: When JAM attacked the checkpoints, it forced us to deploy the battalion and company quick reaction forces in an attempt to reinforce the checkpoints.

Spencer: I think you recognized that something was about to happen, but you did not know when, where, or what it would be. And then the bottom dropped out. They overran the checkpoints. Al-Sadr's self-imposed ceasefire appeared to be over, and Malaki launched an offensive into Basra. How did this change your mission?

MacMillan: Previously, we were conducting fairly routine stability operations as part of a counterinsurgency campaign. Now, our immediate priority was to reinforce and reestablish the checkpoints so that we could maintain the cordon around Sadr City and prevent JAM forces from overrunning the Sha'ab neighborhood. We had to reinforce some of our companies using forces from our battalion Headquarters and Headquarters Company. We provided a patrol to Delta Company to reinforce the Sons of Iraq positions in their sector. We provided forces to Charlie Company to reinforce their positions. We also had to reinforce 1-2 Stryker positions in the Thawra and Jamila neighborhoods.

The exploding violence was not constrained to Sadr City, it was happening throughout the brigade's sector to include Hussaniyah, a densely populated area 20 kilometers north of Sadr City. It was almost a miniature-scaled Sadr City: a densely populated slum dominated by Shias with a strong JAM presence. They were experiencing on a micro level what was happening on a macro level in Sadr City. The battalion's Bravo Company cordoned off the city to prevent the movement of resources and

[27] There is no standard definition for SIGACTS in U.S. Army doctrine. For Operation Iraqi Freedom, SIGACTS included "attacks on Iraqi Security Forces, the civilian population, and infrastructure. It [did] not include criminal activity [or] attacks initiated by Coalition or Iraqi Security Forces." See, "OIF – Iraq Significant Activities (SIGACTS), Global Security, https://www.globalsecurity.org/military/ops/iraq_sigacts.htm; and David Chichetti, "Tactical-Level Multinational Interoperability During Operations Resolute Support and Freedom's Sentinel," *News From the Front*, December, 2019, https://usacac.army.mil/sites/default/files/publications/17970.pdf.

forces between Hussaniyah and Sadr City and to prevent militants inside the city from leaving to attack coalition convoys.

As we reinforced our Charlie and Delta companies in the Beida and Sha'ab neighborhoods to get the Iraqi checkpoints reestablished, the coalition was trying to figure out how to stop the rocket attacks that JAM was launching into the Green Zone.

Spencer: It is hard to describe the complexity of what happened in a single day. Coalition and Iraqi Security Forces had to immediately transition from stability operations to high intensity offensive and defensive operations. You mentioned the Sons of Iraq, which were locals that we hired as militia to stand guard at certain checkpoints. The daily threat against these militias and Iraqi Army soldiers at their checkpoints was relatively low and then the "shit hit the fan," and JAM militants attacked many of these checkpoints with AK-47 automatic rifles, rocket propelled grenades, and PKM machine guns.

MacMillan: The Iraqi Army was not prepared for an attack like this, so the checkpoints were easily overrun.

Spencer: Would you say you transitioned quickly from stability to offensive and defensive operations, or was it more of a counterattack to reclaim those checkpoints?

MacMillan: I would describe it more as a movement to contact which is an offensive operation.[28] We had to quickly transition from stability operations to offensive operations and had to put it in clear terms so that our soldiers could understand the transition. As I said, we conducted a movement to contact. We did not know where the enemy was or where the Iraqi Army and the Sons of Iraq had gone, yet we had to reestablish the checkpoints and get the Iraqi Army and Sons of Iraq to re-man the checkpoints.

Spencer: How and when did the new mission come about? Was it directed by a higher headquarters or was it something that you did on your own?

MacMillan: The answer is that it was a little bit of both. We knew that we had to reestablish checkpoints within our sector, we did not need our higher headquarters to tell us that. But we were focused on our battalion sector. The brigade had to consider the brigade's entire area, and as bad as

[28] Department of the Army, *Operations*, ADP 3-0 (Washington, DC: Headquarters, Department of the Army, 2019), 3-2, https://armypubs.army.mil/epubs/DR_pubs/DR_a/ARN18010-ADP_3-0-000-WEB-2.pdf.

our area was, the brigade realized that our battalion needed to support the other battalions in the brigade. So, the brigade directed us to support 1-2 Stryker in Thawra and Jamila.

In early April, our brigade headquarters directed us to seize and reestablish checkpoints and police stations around and inside Sadr City. This was a significant change in mission that would force us to move into parts of Sadr City that we had never been before. Other than a few short duration raids, U.S. forces had not been in Sadr City since the 1st Cavalry Division back in 2004. We—the division, brigade, 1-68 Armor, and our sister battalion the 1-2 Stryker Cavalry Regiment—conducted very detailed planning for the operation. Fortunately, two days before we were about to launch the operation, the Iraqi Army moved on their own to reestablish the checkpoints and police stations in and around Sadr City.

Spencer: So, it was not just the checkpoints that were overrun, but police stations and anything that was viewed as a national security element?

MacMillan: That is correct.

Spencer: Once the Iraqi Army took over the mission to reestablish checkpoints within Sadr City, what did your mission become?

MacMillan: Despite the Iraqi Army's advances, JAM was still firing rockets into the Green Zone, so we needed to stop the rocket attacks. Since we were still restricted from entering Sadr City, the mission to deny JAM the ability to launch rockets from within Sadr City was assigned to "Longknife Squadron," the regiment's aviation element that included Apache attack helicopters.[29]

Spencer: Basically, Sadr City was a "no-go" area for United States forces, so it was very challenging to stop the attacks when you are not allowed to enter. Can you discuss the rocket attacks a little more?

MacMillan: They launched some rockets from the safety of Sadr City, but they had to launch 107 mm rockets from Thawra and Jamila, the two neighborhoods immediately southwest of Sadr City, because the rockets lacked the range to hit the Green Zone from inside Sadr City. Militants would drive out of Sadr City and into the adjacent neighborhoods, launch the rockets, and then flee back to the safety of Sadr City. Thus, on April 1, we partnered with the Iraqi Army to seal off Jamila and Thawra from Sadr City to prevent JAM from bringing new rockets in and then searched within the two neighborhoods to recover any rockets that were already there.

[29] Longknife Squadron was the 3rd Armored Cavalry Regiment's 4th Squadron.

We conducted detailed planning for the operation which also included air support. We started the operation in the morning, and it lasted all day. It allowed us to demonstrate to the residents that they were safe and that we were getting rid of JAM militants from their neighborhood. This was also the first step toward reestablishing the Jamila market. The Jamila market was a major market that supported much of Baghdad with goods and services, but JAM was extorting the market and reaping the economic benefits. Pushing JAM out of Jamila and Thawra was critical to re-establishing the cordon around Sadr City and choking off resources to JAM.

Spencer: If you look at Thawra, Jamila, and Sadr City, on a map, you can see they are rectangular with grid layouts (see figure 12.4). Before you said that Jamila was outside of your sector, so this was a new area for you?

MacMillan: Yes, previously Jamila belonged to 1-2 Stryker. We assigned the mission to Charlie Company since they owned the adjacent Beida neighborhood. But the mission involved much more than operating in a new area, it also involved transitioning from stability operations to offensive operations, new vehicles, and new task organizations. As we were reestablishing checkpoints, we quickly learned that our Humvees lacked adequate armor protection. The company had been using up-armored Humvees, but the improvised explosive device (IED) threat from explosively formed penetrators (EFPs) in Thawra and Jamila was so great, that we had to bring in our armored vehicles.

EFPs are extremely sophisticated IEDs that could penetrate much thicker armor than a typical IED and they were the most lethal weapon that we faced in Iraq.[30] EFPs could easily penetrate the Humvees' armor, so we needed more heavily armored vehicles. Thus, we line hauled the armored vehicles that we had left at a base in Taji to the company outposts.

As a combined arms battalion, we actually had three sets of vehicles. We had our organic Bradleys and tanks, we had the Humvees, and we had Mine Resistant Ambush Protected (MRAP) vehicles. The army had procured the Humvees to support operations in Iraq and the vehicles remained in Iraq as units rotated in and out. The up-armored Humvees offered better protection against IEDs than the thin-skinned versions, but they remained vulnerable to IEDs, especially more sophisticated ones like the EFPs, so the army started procuring MRAPs. Its V-shaped hull, and

[30] John Ismay, "The Most Lethal Weapon Americans Faced in Iraq," *New York Times*, October 18, 2013, https://atwar.blogs.nytimes.com/2013/10/18/the-most-lethal-weapon-americans-faced-in-iraq/?mtrref=app.coschedule.com.

higher ground clearance offered much greater protection against IEDs. By the end of 2007, the U.S. military had procured nearly 3,500 MRAPs and they had started to arrive in Iraq and Afghanistan.[31]

Each vehicle platform offered its own advantages and disadvantages. When we arrived, the environment was what the military would consider an "uncertain" environment as opposed to being a "friendly" or "hostile" environment, so we felt that Humvees were the best vehicle platform for the situation and the mission at that point in time.[32] Their smaller frames allowed for quicker movement within Baghdad's dense urban terrain for daily patrols and movement to meetings with Iraqi leaders and Iraqi security forces. But they were extremely vulnerable to the EFPs which JAM used to line many of the roads leading to and around Sadr City.

The MRAPs offered better armored protection, but they were huge vehicles that were not designed as a fighting platform like the Bradley. They were designed to drive and transport troops along main supply routes, not as a mobile fighting platform. As the environment changed from "uncertain" to "hostile," we needed the mobility, firepower, and protection of the Bradleys and tanks. We needed a vehicle platform that was designed for high intensity combat. We needed its thicker armor, its state-of-the-art sight systems, and the firepower of the 120mm and 25mm guns to engage enemy fighters inside buildings.

One of the first challenges we faced was that our combat outposts lacked the space for the armored vehicles. During the surge, coalition units left the large forward operating bases and established smaller combat outposts throughout Baghdad. The unit that we had replaced, 2nd Battalion, 325th Infantry, was a light airborne infantry battalion. They did not need a lot of space in their combat outpost because all they had were Humvees. So, we had to increase the size of our combat outposts.

The transition from stability to offensive operations also forced us to increase our operational tempo. As a result, we had to change the composition of our combat patrols. Previously, a minimum of four vehicles

[31] Liam Collins, "Leading Innovation in War: The Vital Role of Leadership," (Unpublished paper, 2018), 200-16.

[32] The United States military defines an uncertain environment as an "Operational environment in which host government forces, whether opposed to or receptive to operations that a unit intends to conduct, do not have totally effective control of the territory and population in the intended operational area." See, Joint Chiefs of Staff, *Joint Operations*, JP 3-0 (Washington, DC: Joint Chiefs of Staff, 2018), GL-17, https://www.jcs.mil/Portals/36/Documents/Doctrine/pubs/jp3_0ch1.pdf.

were required for a patrol, but Charlie Company lacked the manpower to conduct combat patrols 24 hours a day with this requirement. Captain Todd Looney, the Charlie Company commander, figured out that he could conduct patrols around the clock if he decreased the size of the patrol to three vehicles—and with the armored vehicles replacing the Humvees, we thought it was an acceptable risk. Thus, he started running "hunter-killer teams," patrols that consisted of two Bradleys and one tank or two tanks and one Bradley. The tanks had a crew of four, the Bradleys a crew of three, and the infantry squad riding in the back of the Bradley usually totaled four to five. That decreased the number of personnel required for each patrol to 16 and it allowed Charlie Company the ability to conduct 24-hour operations.

The tanks were not immune to IEDs and EFPs, but when they hit a tank, they did not destroy it—a "catastrophic kill" in military vernacular—as they often did to a Humvee, and the casualties were significantly less as well. But it required the maintenance team to run 24-hour operations to quickly repair damaged vehicles to return them to the fight. This necessitated our maintenance company, Fox Company, to both pull 24-hour security for combat outpost Callahan while also running 24 operations to repair the armored vehicles. Ultimately, we had to modify the task organization within the battalion; we found that pushing some battalion level assets to the companies was more efficient than having the companies bring everything to the battalion combat outpost for repair.

The armored vehicles not only provided the patrols better armor protection and greater firepower, but they also really surprised JAM. JAM discovered that they could rarely inflict a mobility kill or a catastrophic kill against a tank or Bradley with a single EFP. In terms of firepower, the Humvees were armed with .50 caliber machine guns and 90 mm grenade launchers; by contrast, the main battle gun of a tank carries a 120 mm round and the Bradleys' main gun fires 25 mm rounds, nearly twice the diameter of a .50 caliber round.

Spencer: I find this battle fascinating because of its rapid transition from stability operations to offensive operations. I realize counterinsurgency operations are always comprised of offensive, defensive, and stability operations, but you went from 80-90 percent stability operations to 80-90 percent offensive operations overnight.[33] It necessitated so many changes: going into a new area, getting new vehicles, changing

[33] Joint Chiefs of Staff, *Counterinsurgency*, JP 3-24 (Washington, DC: Joint Chiefs of Staff, 2018), VII-11, https://www.jcs.mil/Portals/36/Documents/Doctrine/pubs/jp3_24.pdf.

the composition of a patrol, and then moving various assets around to run non-stop patrols throughout the sector. Luckily, as a part of a heavy brigade combat team, you knew how to fight using armored vehicles. If the light airborne infantry company would have been there, that would not have been an option. They did not have drivers and gunners nor training in the tactics to use the vehicles even if they had been present. So, how long did it take to convert the battalion from a Humvee based stability force into an offensive armored, mobile, protected, firepower force?

MacMillan: We transitioned from a light formation with Humvees back to our original armor formation within two to three days.

Spencer: That is crazy. I hope that we captured the lessons that you learned from that rapid transition somewhere. I am sure that there were hiccups—like you said, something as simple as parking space—that are recorded somewhere as I am certain that even with the wars in Afghanistan and Iraq at an end, it will not be the last time that a force has to quickly transition in a similar manner.

MacMillan: Our success stemmed from our ability to adapt quickly, but it also hinged on us having the vehicles in Iraq to move forward. That was not an option for 1-2 Stryker; they did not have the flexibility to use vehicles that offered more force protection.

Returning to the battle, we conducted non-stop combat operations for the next three to four weeks and it took a toll on all the companies. Delta Company, for example, had all of Sha'ab, which was a very large area (see figure 12.4). At the start of the battle, we had given them two additional patrols that we had established from the battalion's headquarters and headquarters company. They had to reinforce the Sons of Iraq checkpoints that JAM had overrun, and they had to conduct their own clearing operations.

JAM militants were driving around Sha'ab on the backs of scooters with rocket propelled grenades. They would ride up to a checkpoint, launch several rocket propelled grenades, and then take off. Thus, Delta company had to clear all of Sha'ab of guys with rocket propelled grenades on the backs of scooters. Talking about it now, it sounds kind of humorous, but it was a serious problem. They had the ability to rapidly move around the battlefield with significant firepower.

At this point of the battle, around late April, we were assigned a larger battlespace. Since 1-2 Stryker lacked armored vehicles, the brigade shrank their battle space and increased ours. The brigade assigned Jamila to us, which we assigned to Charlie Company. The battalion's total area was now too large for the forces that we currently had assigned, so we requested additional assets from the brigade, and they gave us a tank company and two Stryker companies. This was yet another huge transition for us in terms of the task organization change.

Spencer: So, you went from how many companies, four or five, up to eight?

MacMillan: We started with six—two tank companies, one mechanized infantry company, one engineer company, one forward support company, and one headquarters company—so we now had nine. It was a very large battalion task force with a large sector and a lot of responsibility. I mentioned earlier that we also had a number of attachments, but beyond our direct attachments, we had other assets supporting us. For example, we faced a fairly significant sniper threat within our battlespace, so we had a Navy SEAL counter-sniper team supporting us in Jamila.

Spencer: Were you the main effort in Baghdad at this point? Is that why you were getting all these assets? I can only imagine that given the urgency to stop the rocket attacks against the Green Zone that your battalion became the main effort for 3rd Brigade, which was the main effort for the division?

MacMillan: Yes, we were the main effort for all operations in Baghdad at that time.

Spencer: So, you were assigned a larger area of responsibility and received additional forces, but the attacks remained high. Were you starting to clear more areas and reestablish security in those areas?

MacMillan: Yes, at this point we had reestablished security in many of the areas surrounding Sadr City, so we transitioned to the next phase of the operation: cutting JAM resources to Sadr City to stop the rocket and IED attacks.

Spencer: At this point, you still have not entered Sadr City. It has been an enemy safe haven for the past four years, right?

MacMillan: That is correct. We had been ready to enter, but the Iraqi Army had entered instead. At this point in the battle, we had not gone northeast of Al-Quds Street, which is the boundary between Sadr City and Jamila and Thawra (see figure 12.4).

Spencer: Yet Prime Minister Maliki wanted you to stop the rockets. So, what did you do?

MacMillan: Sadr City contained an estimated two million inhabitants.[34] We had a very large battalion, but nowhere large enough to seize and clear Sadr City. So, we had to figure out how to stop the rocket attacks without clearing JAM from a densely packed urban area. We decided that the best way to prevent them from launching rockets was to cordon off Sadr City to prevent resources from reaching JAM and force al-Sadr back to the negotiating table to agree to a cease-fire.

Spencer: In military terms, you conducted a mission analysis to figure out how to create the desired effect on the enemy without physically entering their battlespace, because the dense urban terrain made direct engagement a very risky proposition.

MacMillan: That is correct. And this was at the end of the surge; we did not want to get bogged down in Sadr City and force a second surge of troops. The surge was working. Attacks in Baghdad and all across Iraq had decreased significantly in the past year.[35] The "March Madness," as you described it earlier, was a significant spike, but we wanted to ensure that this spike would be isolated and temporary. So, rather than entering Sadr City, we decided to seal it off and deny them access to the Jamila market and rocket firing points.

As you mentioned earlier, Jamila had a typical grid layout, so we attempted to use concrete Texas barriers to close off Sadr City from the market. We needed to prevent JAM trucks from getting to the market and loading up with supplies or from launching rockets, so we placed concrete barriers on the roads leading out of Sadr City and into Jamila along Al-Quds Street. Unfortunately, that technique proved to be relatively ineffective because, while stopping vehicles, individuals with small hand carts could easily bypass the barriers.

Thus, we determined that if we wanted to separate Sadr City from Jamila and Thawra, we would need to wall it off using giant Alaska barriers. These were prefabricated 20 feet tall concrete sections shaped like an inverted "T." The barriers could be moved into position and placed end to end by a crane to create a giant wall. Thus, we planned to wall off Thawra and Jamila from Sadr City along Al-Quds Street, a distance of nearly five kilometers. We named this Operation Gold Wall.

34 Johnson et al., *The 2008 Battle of Sadr City*, 21.
35 Ibid., 18.

Spencer: How did you plan this operation? Was it the brigade engineer simply doing the calculation to determine how many barriers you would need and then the battalion developing a plan on how to emplace them?

MacMillan: Although the wall was going into our battalion sector, it was a brigade operation, so a lot of the planning occurred at the brigade level. Critical to the operation was conducting route clearance operations to clear Al Quds Street of IEDs so that we could transport the concrete to their destination. Route clearance is very slow and methodical. Our organic engineers lacked the assets necessary to conduct the mission, so the explosive ordnance detachment teams that were attached to the battalion played a critical role in this mission. They would often encounter and clear 12-16 EFPs in a single kilometer.

Spencer: Obviously you cannot build a five-kilometer wall overnight. So, what would a typical night look like?

MacMillan: As I just mentioned, one of the first things we had to do was to clear the route. But this operation required us to modify some of our tactics, techniques, and procedures. Previously, when we encountered an EFP, the procedure was to disable it and then package it so it could be studied. We would send them off to be forensically analyzed, so that we could track the cell that was manufacturing the EFP. But it could take a couple hours to disarm and package a single EFP. With so many EFPs and the desire to get the wall built quickly, we lacked the time to disarm and package them, so instead, we destroyed them in place. This was one of the many changes that transpired as we transitioned from stability operations to offensive operations.

The engineers and explosive ordnance detachment teams became our route reconnaissance element. They not only had to destroy the EFPs that they discovered, but they also had to defeat any JAM fighters that were overwatching them. So, the engineers quickly transitioned from a route clearance mission to an armed reconnaissance mission.

Establishing the Gold Wall was a 24-hour operation, and it encompassed the battlespaces of both 1-68 Armor and 1-2 Stryker. We initially planned to start the wall in the middle—at the intersection of Jamila, Thawra, and Sadr City—where our sectors came together, and then simultaneously build the wall out to the west and east in each sector. But we quickly found this plan untenable. We determined that it would be more efficient to combine our efforts, so we started on the southeastern corner of Sadr City and built the wall to the northwest (see figure 12.5).

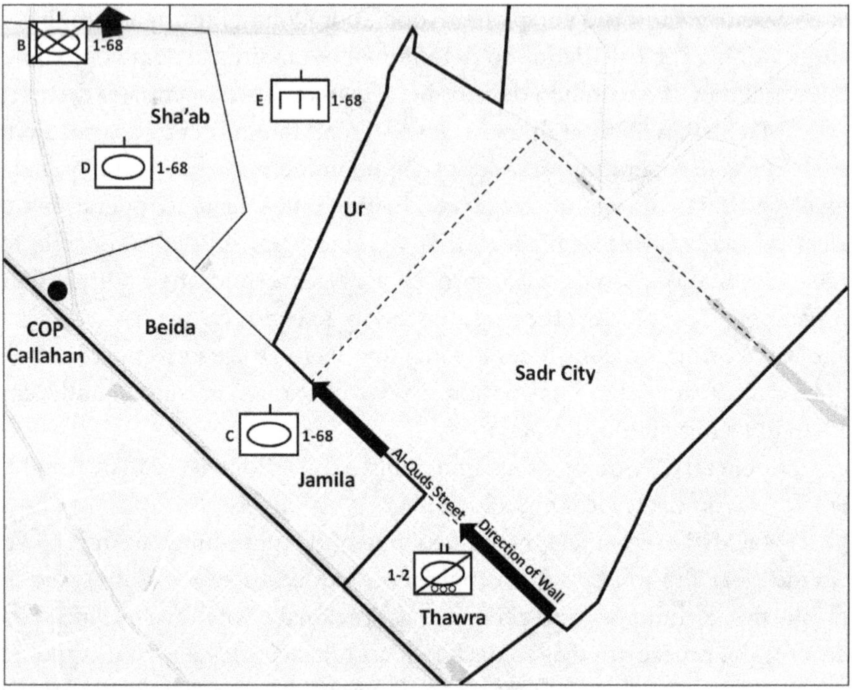

Figure 12.5. Operation Gold Wall, April 15 – May 15, 2008.
Source: Map created by Liam Collins.

Each battalion ran 12-hour shifts that initially rotated at 6 AM and 6 PM, since that was typically when we conducted shift changes for warfighting exercises in training. But it was the middle of April and temperatures were starting to rise, so we quickly moved the shift changes to 1 AM and 1 PM, so that one battalion was not stuck with the heat of the day or in periods of limited visibility for most of its shift.

Spencer: This is the part that fascinates me the most; the battle of Sadr City is really the battle of the Gold Wall at this point. You had to repeatedly conduct armed reconnaissance missions because the enemy was constantly emplacing new IEDs, then move in with massive amounts of armed forces to protect the semi-trucks carrying the barriers and the cranes at the construction site. I can imagine a semi-truck rolling up, a crane putting the wall section in place, and this process being repeated over and over again until the five-kilometer wall was completed.

MacMillan: That was it, and we called the construction of the wall a "defense in contact." Normally in field training, when you train for a defense, you might have a couple of days to build the defense before the enemy's main formation arrives. So, you could build the defense with little enemy contract. But when we were building the wall, we were constantly taking fire from the enemy, so we called it a "defense in contact."

The armored reconnaissance would clear 400 meters of the route with the tanks and Bradleys securing the cleared area. This security element would engage enemy fighters that attacked the wall-laying element. Another element would transport the concrete barriers from our concrete holding area located approximately one kilometer from the starting point of the wall to the current construction site on semi-trailers. A crane would then lift the barrier off the back of the semi-trailer and place it next to the adjacent wall section. Some poor soldier would then have to climb a ladder to physically unhook the crane from the top of the barrier, often under fire. The security element, a mix of combat vehicles and dismounted infantry, provided security for the wall-building force. After completing a large section, the process would start anew with the armed reconnaissance clearing the route and then combat forces securing the site and establishing a defense. This process continued 24-hours a day for more than 30 days. That is why we called it a defense in contact.

Spencer: You describe this as a defensive operation, but I view your use of concrete as being more offensive than defensive. According to U.S. Army doctrine, the purposes of the defense include creating conditions for a counteroffensive, retaining decisive terrain, or denying a vital area to the enemy, among others.[36] By contrast, the purpose of the offense is to "defeat, destroy, or neutralize the enemy force."[37] This operation was somewhat analogous to siege warfare in ancient times but rather than attacking a castle wall, you built a wall around them. You did not deprive them of food and water, but you did control the flows of weapons and fighters. By attempting to stop the construction of the wall, you also forced the enemy to expose itself to your tactical advantage.

Thus, the wall served defensive purposes. It clearly established a defensive line that denied terrain to the enemy: rocket firing locations and access to the Jamila market that they needed for monetary support. But at the same time, it neutralized JAM, which is an offensive purpose. The

---

36 Ibid., 10-11.

37 Department of the Army, *Offense and Defense*, ADP 3-90 (Washington, DC: Headquarters, Department of the Army, 2012), 7, https://www.benning.army.mil/mssp/PDF/adp3_90.pdf.

wall rendered [JAM] personnel and equipment incapable of interfering with a particular operation and drove them to the bargaining table.[38] Thus, the concrete wall was both offensive and defensive, but the mission of neutralizing the enemy was more important than the mission of denying terrain, so that is why I argue that it was an offensive operation. I realize that you did not launch concrete at the enemy, but I like to say you used concrete as a weapon in this battle to create multiple effects on the enemy.[39]

MacMillan: Yes. We established a defensive zone while we were in contact. JAM knew that we were trying to separate them from their source of financial support and resupply, but they were largely ineffective at disrupting our operation. The composition and the employment of our force resulted in their contact being limited and largely ineffectual. We eliminated threats to our force emanating from inside Sadr City, but never physically entered it. Our objective was not to seize Sadr City; our objective was more limited: to deny JAM access to firing positions and resources.

Spencer: What were some of the weapons that you commonly used and what were some new tactics, techniques, and procedures that you discovered? Were tanks firing rounds into buildings, for example?

MacMillan: Prior to Operation Gold Wall we had learned that JAM often camouflaged IEDs in trash. So, the route clearance element would use their Buffalo Mine-Protected Clearance Vehicle[40] to scoop out and sift through the trash just to find the IEDs. Sifting through all that garbage, which was purposely laid as camouflage, was time consuming but necessary. The tanks and Bradleys were fairly well protected from EFPs with their reactive armor, but the soldiers on the ground who were placing the concrete barriers, the trucks transporting the concrete, and the cranes, all remained vulnerable.

We had to develop a quicker way to clear the trash, so someone figured out that they could use the tank's 120 mm cannon to fire canister rounds to clear the IEDs. A canister round is like a huge shotgun shell. It is a 120 mm round that is packed with ball bearings, so it is a particularly

[38] The definition of neutralize comes from United States military doctrine. See, Joint Chiefs of Staff, *Joint Operations*, GL-13.

[39] John Spencer, "The Most Effective Weapon on the Modern Battlefield is Concrete," Modern War Institute, November 14, 2016, https://mwi.usma.edu/effective-weapon-modern-battlefield-concrete/.

[40] A Buffalo is a "six-wheel heavily armored [explosive ordnance disposal] vehicle." See, "Buffalo Mine-Protected Clearance Vehicle," *Army Technology*, April 26, 2021, https://www.army-technology.com/projects/buffalo/.

lethal anti-personnel weapon. We discovered that firing the canister rounds into the loose trash was an extremely expedient and effective method to disable and destroy IEDs. It was not one hundred percent effective, but it was effective enough to speed up the route clearance. So that was one useful technique that we discovered.

Spencer: What kind of collateral damage was caused by a canister round? How much damage would they do to the buildings on the left and right of the street?

MacMillan: It caused some collateral damage, but significantly less than other methods. For example, using a mine clearing line charge (MICLIC) is a common method to clear an obstacle belt. A MICLIC is a 110-meter-long line charge that contains more than 1,800 pounds of C4 plastic explosives. A common tactic, upon reaching an obstacle belt, is to fire the MICLIC to clear a lane that is one hundred meters long and 8 meters wide.[41]

We used them during Operation Desert Storm, and they were extremely effective. But that was an open desert void of civilians, this was a dense urban area with millions of people. We considered using MICLICs to clear the roads, but that would have caused significant damage to buildings in Sadr City on the northside of Al-Quds Street and to the Jamila market on the south side. Thus, the canister round seemed a better choice, given that we wanted to minimize civilian casualties and collateral damage.

Spencer: I think the use of IEDs will remain a common feature of contemporary and future warfare. It is a way to slow forces, especially more vulnerable vehicles.

MacMillan: It is simply a different way of building an obstacle belt. An obstacle belt in the desert might be a minefield with hundreds of mines, but in an urban area, it is dozens or hundreds of camouflaged IEDs.

Spencer: What other types of assets did you employ? Earlier you discussed using the Apache attack helicopters to overwatch JAM's rocket firing areas. What kind of intelligence, surveillance, and reconnaissance or fires platforms did you have available?

MacMillan: We maintained intelligence, surveillance, and reconnaissance 24 hours a day on the element that was building the wall as a force protection measure to identify any enemy that might threaten them.

[41] *M58 / M59 Mince Clearing Charge* (Philadelphia, PA: Day & Zimmerman Munitions Group, no date), https://cdn2.hubspot.net/hubfs/208537/assets/PDF/Product%20Sheets/NewWebsite_Munitions_CloseCombat_MICLIC.pdf.

The brigade controlled many of these close air support and unmanned aerial systems, but by 2008 the technology had advanced to the point that each company often had direct access to the video feeds so it could monitor what was going on its battle space. The company commanders could also talk directly to the Apache air weapons teams that were assigned to our area.

In terms of fires, we had Joint Terminal Attack Controllers in the brigade tactical operations center. They controlled air support, but luckily it was rarely needed. I think we only dropped a single 500-pound bomb during the operation and that was dropped by another unit that was operating within our sector. Once we fired artillery from a guided multiple launch rocket system on a structure in Sadr City that JAM was fighting from. But those were options of last resort. We started small, to minimize collateral damage, and only increased the munition when required.

In the particular case when we used artillery rockets, JAM fighters were moving to different positions throughout a particular building very quickly. Our small arms and larger vehicle weapon systems could not stop their attack, so we escalated to missile strikes from the Apache attack helicopters. But they also failed to impede the enemy. Because we could not physically enter Sadr City to clear the structure, we had to destroy it. So, we used laser-guided artillery which was extremely effective at destroying the target building while minimizing damage to the surrounding buildings.

Spencer: JAM clearly understood what you were doing by emplacing this wall, so one of the benefits of the operation was that it forced them to expose themselves. In counterinsurgencies, one of the biggest challenges is trying to discriminate the insurgent from the population. This problem is often even more challenging in an urban environment. But Operation Gold Wall forced the enemy to expose themselves to try to stop the wall construction, so it really helped to discriminate the enemy from the population.

MacMillan: We also used information operations. In advance of the operation, we air-dropped pamphlets in the areas immediately surrounding the planned wall location and told them that anyone seen with a weapon would be engaged. We followed our rules of engagement. We could not and would not indiscriminately fire tank rounds or 25mm rounds into the city, but we would engage anyone that was a threat. Lucky, JAM did not use human shields, so if they wanted to stop us from building the wall, they had to engage us with weapons directly, which exposed them. And when they did, our security force immediately returned fire and eliminated them so that the wall construction could continue.

Spencer: What was the highest number of barriers that you emplaced during one 12-hour shift?

MacMillan: We actually kept track of that because it was a competition between the two battalions to see who could lay the most. The most that we emplaced in one 12-hour period was 120 sections, and that was toward the end of the operation when we were really experienced. It was much slower at the start, not only because we were less experienced, but also because the enemy was much more aggressive. Once we reached the Jamila neighborhood, we noticed a significant decrease in JAM engagements. We assessed that they were running out of fighters, EFPs, IEDs, and ammunition, so that limited their ability to engage us. In total, we laid nearly 3,700 wall sections.

Spencer: What was the minimum number that you laid, on a bad night?

MacMillan: On the worst night, we probably laid only 10 or 15 barriers. It was early in the operation when JAM had plenty of ammunition, plenty of IEDs, plenty of rocket propelled grenade rounds, and stuff like that.

Spencer: Even taking the combat operations out of the equation, building a wall like that is an impressive logistical feat. It required 3,700 sections weighing eight tons each.[42] They had to be loaded on trailers, driven to the site, and lifted and put into place by a large crane. After being put into place, a soldier had to climb up a 20-foot ladder and expose himself to the enemy to unhook it from the crane. In my dream world, we would find a way to emplace a concrete wall that does not require a soldier to unhook it, and ideally, a way to build it much faster.

MacMillan: As I said, this was a brigade operation, and the brigade support battalion was responsible for coordinating the movement of the concrete wall sections to a concrete holding area. Our battalion never had to worry about getting them there, we simply pulled them from the stockpile in the holding area. The operation definitely required a significant engineering effort. For example, early in the operation we contracted Iraqi cranes to place the wall sections, but they were too susceptible to enemy fire, so we brought in a U.S. Army engineer crane.

[42] Mark Thompson, "The Great Wall of Qatar," *Time*, April 24, 2013, https://nation.time.com/2013/04/24/the-great-wall-of-qatar/.

We also required corps-level heavy equipment transporters—armored semi tractor-trailers—to transport the concrete from the holding area to the construction site. While a 20-foot wall offers significant protection, we still had to maintain overwatch of the entire wall as it was being built. If we did not, someone could destroy a section of the concrete, and then we would have to clear the area again and reestablish that part of the wall.

Spencer: As we in the military know, an obstacle is only an obstacle if you observe it. If not, the enemy can easily breach or bypass it. How did the wall end? I know that there were walls along Sadr City's northwestern and the southeastern sides that had been emplaced as early as 2006, maybe even earlier, it is hard to trace their exact emplacement. Did the Gold Wall connect with these other walls?

MacMillan: The wall on the northwestern side did not extend all the way to the Gold Wall. We would extend it later; that was another operation. Operation Gold Wall ended on May 22 when the wall was completed, but the fighting ended days earlier on May 11 when al-Sadr asked for a ceasefire. With JAM's supplies exhausted and many of its fighters dead, a cease fire became al-Sadr's best option. Additionally, many JAM leaders had left Sadr City and fled to Iran, leaving JAM largely depleted in Sadr City.[43]

Spencer: By the end of Operation Gold Wall, I can imagine that you viewed the operation as a success. You had built the wall, stopped rockets from being fired into the Green Zone, cut off JAM's logistical support, attritted their forces who had tried to prevent it, and had reestablished checkpoints that controlled passage in and out of Sadr City.

MacMillan: That is correct. The Iraqi Army established checkpoints along this wall to control the flow in and out of Sadr City. We did not completely close off Sadr City; we left a few openings that the Iraqi Army could control. We also started aggressively providing economic support to the businesses within Jamila and Thawra to increase the economic growth of the area. It was not as simple as building the wall and we were done. Once we completed the wall, we transitioned back to conducting predominantly stability operations and that involved trying to spur economic growth in the local area.

[43] Johnson et al., *The 2008 Battle of Sadr City*, xvii, 77.

Spencer: This reminds me of the "three-block war" made famous by General Krulak.[44] In one part of the neighborhood you were fighting high-intensity combat to emplace this wall. In another part of the neighborhood, you were conducting reconstruction and stability operations in an attempt to revive the economic vitality of the urban area.

MacMillan: That is correct. Attacks decreased throughout May and June. After the wall was completed, for a period of time they stopped altogether.

Spencer: I think that there are multiple lessons that can be learned from your experience and many lessons that we did not have time to cover, ranging from the tactical to the strategic level. What are some lessons that the army should take forward as they prepare for dense urban operations in the future?

MacMillan: A key factor to our success was our ability to rapidly adapt to the enemy. Rather than simply fighting with what we had, the brigade and division changed boundaries and changed task organizations to provide the right force for the mission. Equally important was conceptually understanding the change of the strategic environment. While we quickly transitioned from stability to offensive operations, we had to understand that we were still engaged in a counterinsurgency campaign.

For example, as discussed, it might have been more expedient in the short term to use a MICLIC to clear Al-Quds Street, but it would have made it more challenging when we transitioned back to stability operations. It would have cost millions of dollars to reconstruct portions of the Jamila market that we would have destroyed. Down to the individual soldier, everyone understood that they needed to use an escalation of force approach to protect the Jamila market because we understood that revitalizing the city's economic capability and stability of the area was what was needed in the long run.

At the tactical level, our success was undergirded by flexibility, adaptability, and understanding the operation's next phase. At the battalion level, we had to support the companies' current operations while at the same time planning the next. We had to plan two or three steps ahead to

[44] Krulak described how within three city blocks forces might be required to conduct full-scale combat operations, peacekeeping operations, and humanitarian aid. General Charles Krulak, the Commandant of the United States Marine Corps, published the idea of the "three-block-war" in a 1999 article. See, Charles C. Krulak, "The Strategic Corporal: Leadership in the Three Block War," *Marines Corps Gazette* 82, no. 1 (1999), 14-17, https://mca-marines.org/wp-content/uploads/1999-Jan-The-strategic-corporal-Leadership-in-the-three-block-war.pdf.

facilitate the rapid transition from offensive operations back to stability operations.

To reiterate, I think the biggest lesson that can be learned by the battle of Sadr City is that combat, and especially urban combat, often requires rapid change. People are often resistant to change, but you have to anticipate the next phase, or the next transition, and make the changes that are required to be effective. That is what will lead to success in the long term.

One final note. Anyone that has ever been in the army knows that this mission was ultimately successful because of the soldiers. Soldiers from 1-68 Armor, along with all our attachments, were extremely hard workers and true professionals. They put in long hours, in dangerous neighborhoods to make this mission a success and six of these heroes paid the ultimate price.

13

The Battle of Mosul

The Islamic State of Iraq and al Sham (ISIS) emerged from the remnants of al Qaeda in Iraq which had been founded by Abu Musab al-Zarqawi in 2004.[1] U.S. special operations forces killed Zarqawi in 2006, and in the spring of 2010, the special operations task force killed most of the group's leadership, leaving the group decimated.[2] However, the Islamic State of Iraq, as ISIS was called at the time, was able to reemerge because the group was no longer being actively pursued after American troops withdrew in 2011.[3] At the same time, Sunni frustration with the Iraqi government began to grow, so some turned to the Islamic State of Iraq to break away or resist against the Shia-led government in Baghdad.[4]

Concurrently, the Arab Spring movement was spreading across the region and helped spark a civil war in neighboring Syria. This civil unrest provided a safe haven for the Islamic State of Iraq and other terrorist groups to grow and on April 9, 2013, the group rebranded itself as the Islamic State in Iraq and al Sham.[5] Taking advantage of the degraded security environment, ISIS consolidated power and seized control of large swaths of territory in Syria and Iraq including Raqqa, Aleppo, Mosul, and Tikrit.[6]

---

[1] Al Sham generally refers to "greater Syria" or the Levant.

[2] Liam Collins, "Dismantling al-Qaida in Iraq," in *Routledge Handbook of U.S. Counterterrorism and Irregular Warfare Operations*, eds. Michael A. Sheehan, Erich Marquardt, and Liam Collins (New York: Routledge, 2021), 275-7, DOI: 10.4324/9781003164500-21.

[3] Joseph Logan, "Last U.S. troops leave Iraq, ending war," *Reuters*, December 17, 2011, https://www.reuters.com/article/us-iraq-withdrawal/last-u-s-troops-leave-iraq-ending-war-idUSTRE7BH03320111218.

[4] Brain Fishman, "The History of the Islamic State," in *Routledge Handbook of U.S. Counterterrorism and Irregular Warfare Operations*, eds. by Michael A. Sheehan, Erich Marquardt, and Liam Collins (New York: Routledge, 2021), 61, DOI: 10.4324/9781003164500-5.

[5] For the ISI's statement, see Abu Bakr al-Baghdadi, "Declaring the Establishment of the Islamic State of Iraq and al-Sham," al-Furqan Media Establishment, April 9, 2013.

[6] "Timeline: the Rise, Spread, and Fall of the Islamic State," Wilson Center, October 28, 2019, https://www.wilsoncenter.org/article/timeline-the-rise-spread-and-fall-the-islamic-state.

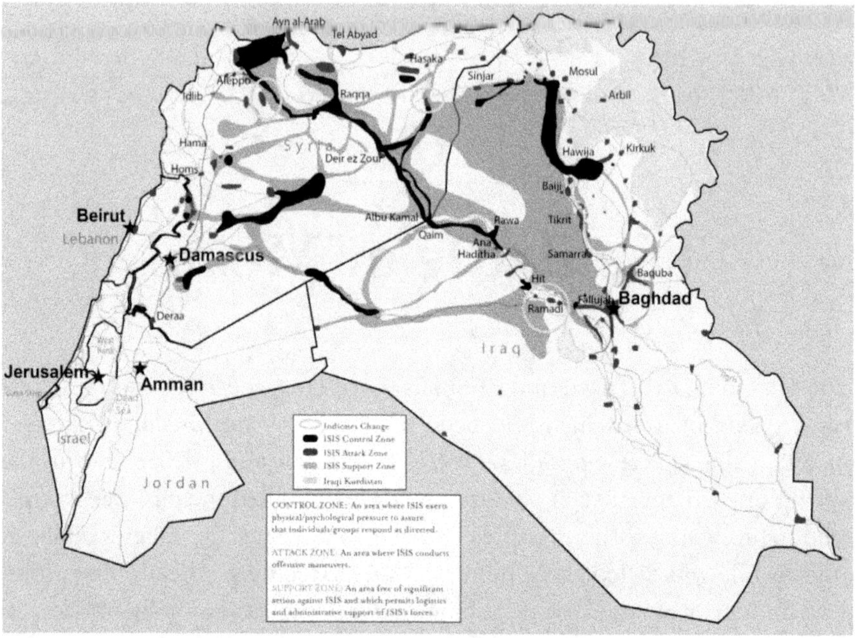

Figure 13.1. ISIS sanctuary in Syria and Iraq, March 2016.
Source: "ISIS Sanctuary: March 3, 2016," Institute for the Study of War, March 3, 2016, https://www.understandingwar.org/sites/default/files/ISIS%20Sanctuary%2003%20MAR-01_12.png

In June 2014, ISIS declared a caliphate with its capital in Raqqa.[7] Mosul fell in a matter of days when the Iraqi military fled (see figure 13.1).

Fearing Baghdad would be next to fall, the United States sought to stem the ISIS advance, but there was little appetite by U.S. policymakers, military leaders, or its public to send in large troop formations. Instead, the United States opted for a bombing campaign and began airstrikes against ISIS in Iraq in August 2014 and expanded airstrikes into Syria the following month. In October 2014, the U.S.-led coalition named the campaign Operation Inherent Resolve and enlarged the campaign to include ground troops with the establishment of a combined joint land component command.

Elements from the 1st Infantry Division formed the core of the land component command with the mission to "advise and assist the Iraqi Security Forces to help them go on the offensive against [ISIS] and conduct

[7] Carla E. Humud, *Syria Conflict Overview: 2011-2021* (Washington, DC: Congressional Research Service, 2021), 1, https://crsreports.congress.gov/product/pdf/download/IF/IF11080/IF11080.pdf/.

intelligence, surveillance and reconnaissance flights."[8] Nine months later, they were replaced with troops from the 82nd Airborne Division. In October 2015, the U.S. deployed special operations forces into Syria to support the Syrian Democratic Forces in their fight against ISIS.[9]

In this chapter, John Spencer interviews Major General Roger Noble.[10] At the time of the interview in 2020, Noble was serving as the head of the Australian Army's Military Strategic Commitments. He currently serves as the Ambassador for Counter-Terrorism within Australia's Department of Foreign Affairs and Trade. In February 2016, Noble deployed to Iraq for his third operational tour to serve as the deputy commander for the combined joint forces land component command. He served in that position until he redeployed in late November 2016. Much of his time was spent supporting the Iraqis in their battle for Fallujah and the start of their battle for Mosul.

John Spencer: Sir, I would like to start by discussing your time in Iraq, especially any observations you had dealing with major urban combat operations during the fight against ISIS. Can you tell me when you were in Iraq, what was your position, and what were your primary missions?

Roger Noble: This was my third deployment to Iraq. I arrived in late February 2016 to serve as the deputy commanding general of the combined joint forces land component command. Major General Gary Volesky's 101st Airborne Division assumed command of the land component command from the 82nd Airborne Division on March 8.[11] The Iraqis had recently retaken Ramadi, which was the first setback for ISIS.[12] Up until that time, ISIS had been advancing quickly towards Baghdad. During the deployment, I spent most of my time on two major battles: the battle of Fallujah and the beginning of the battle of Mosul. They were part of a larger campaign, Operation Inherent Resolve, that involved clearing ISIS from

[8] Wilson Center, "Timeline;" and "'Big Red One' Leaders return to Fort Riley," KMAN News Radio, June 30, 2015, https://1350kman.com/big-red-one-leaders-return-to-fort-riley/.

[9] Humud, *Syria Conflict*, 2.

[10] The views and the opinions expressed within this chapter are the author's and do not necessarily reflect those of the Australian Army or Australian government.

[11] "101st Airborne Division assumes command of CJFLCC-OIR from 82nd Airborne Division in Iraq," *Clarksville Online*, March 8, 2016, https://www.clarksvilleonline.com/2016/03/08/101st-airborne-division-assumes-command-cjflcc-oir-82nd-airborne-division-iraq/ .

[12] W.G. Dunlop, "Iraq's Ramadi retaken, but rebuilding it a huge task," *Agence France-Press*, February 8, 2016, https://news.yahoo.com/iraqs-ramadi-retaken-rebuilding-huge-task-172924072.html .

Syria and Iraq. I found myself in Iraq at the right time to gain substantial urban warfighting experience.

Spencer: That must have been an exciting time to get there, given the number of cities that ISIS controlled at that time (see figure 13.1).

Noble: Yes. At that time there was little confidence in the Iraqi government and military. The Iraqi population, the United Nations, and the international community were all pretty negative. They considered ISIS almost unbeatable. Retaking Ramadi was the first positive step forward. Before that, it had been quite bleak. But Ramadi was just the beginning; it was a long way to Mosul.

Spencer: From your vantage point, what was the character of the fighting when you got there?

Noble: I think the biggest shock for me was that it was largely a conventional fight with brigade, division, and even corps-level operations. I had been to Iraq twice before and I was expecting more of a counterinsurgency fight, which is typically fought at the platoon, company, and battalion level. But this time I found an enemy that was fighting a conventional defense in urban terrain. I was not expecting that. ISIS was using unconventional weapons and unconventional means to generate a conventional defense in the urban terrain. They were forced into the urban terrain because out in the open they were too vulnerable. Probably the biggest lesson I learned, and I learn it every time I go to war, is to pay close attention to the enemy: understand exactly how they are thinking and how they operate. These guys were clever, ruthless, and used everything they had to generate a pretty conventional defensive effect.

Spencer: I think a lot of scholars have said this time period in Iraq and in Syria was almost a return to siege warfare, where the enemy was embedding themselves in cities, building big defensive plans, and basically saying, "come and get me."[13]

Noble: That is correct. They were not a conventional army, although they were armed with some conventional army weapons and equipment that they had seized from the Iraqi Army. More often, they improvised. Lacking conventional mines, they used hundreds or thousands of improvised explosive devices (IEDs) to create minefields. They improvised

[13] See, for example, Amos Fox, "The Reemergence of the Siege: An Assessment of Trends in Modern Land Warfare," Institute of Land Warfare, Landpower Essay No. 18-2 (Washington, DC: Association of the United States Army, 2018), https://www.ausa.org/sites/default/files/LPE-18-2-The-Reemergence-of-the-Siege-An-Assessment-of-Trends-in-Modern-Land-Warfare.pdf .

to create an equivalent obstacle effect. If you lack tanks or large caliber direct fire weapons, what do you do? You use vehicle-borne IEDs and suicide bombers to blow a hole where you want to blow a hole. What we would do with a tank round they could do with a human. They were pretty sophisticated in putting it all together to create a conventional defense. I never thought I would find myself conducting a division attack against prepared positions in urban terrain, but that is what I found.

Spencer: That is amazing! I call that the city attack. You might call it a deliberate attack with a penetration, an encirclement, or an envelopment. We have a bunch of fancy words for it, but you are basically attacking a city of defenders.[14]

Noble: That is exactly right. And in a city like Mosul, that is more than one million people. So, we would sometimes ask ourselves: are really going to participate in an attack against a city with one million people? The big lesson is you do not do that overnight. You have to plan months in advance, and it requires a whole-of-Iraqi-government, inter-agency, and international effort to be successful. It is not just the tactical problem of clearing the enemy from buildings. It is the whole setup that makes the execution successful in the longer term.

Spencer: So, is that what happened? It took months just to plan for the attack on Mosul?

Noble: Yes. Prior to deploying to Iraq, I went to the home of the 101st Infantry Division at Fort Campbell, Kentucky to start planning the operation. We were trying to divine the strategic intent and the objectives for our tour in Iraq. It all kind of distilled down to getting Mosul. That was it. But we did not think we would be able to liberate Mosul during our tour, given how long it had taken to liberate Ramadi. Nonetheless, Major General Volesky directed us to develop a plan to retake Mosul. So, we adopted a campaign approach to determine what was required to advance to and sustain an attack to clear Mosul, a million-person city.

The big lesson for me, in a big struggle like that, is that you have to think well ahead. It is classic campaign planning, sequencing what you

[14] The US Army has five forms of maneuver: envelopment, frontal assault, infiltration, penetration, and turning movement. See Department of the Army, *Offense and Defense*, ADP 3-90 (Washington, DC: Headquarters, Department of the Army, 2019), 2-17, https://armypubs.army.mil/epubs/DR_pubs/DR_a/pdf/web/ARN18377_ADP%203-90%20FINAL%20WEB.pdf.

want to do and understanding the steps to get there.[15] I will be frank, at the beginning, we did not think we would get to Mosul during our tour. We thought we would get somewhere towards it. Yet by the time I left Iraq, the attack was underway and pretty well advanced. I think the key was that we had thought it through intellectually, developed an initial plan, and then constantly adjusted the plan based on the changing conditions. Of course, the plan did not unfold as we thought it would, but we had branches, sequels, and were constantly adjusting it if that makes sense to you.

Spencer: Yes, for sure. If we look back at the American-led attacks on Fallujah in 2004, the tactical and operational thinking was there but most people forgot about the political objective and the political will required for the operation. In the first battle of Fallujah in April 2004, the American forces lost the political will of the Iraqi government and members of the coalition so much so that the military operation had to be halted.[16] In the second battle of Fallujah in November of 2004, much more planning was given to maintaining political will throughout the duration of the operation.[17] Was that a major concern for your time there, maintaining political will?

Noble: Absolutely. The political objective is the most important thing. The biggest difference between Fallujah in 2004 and my tour was that in 2016 it would be the Iraqis, not coalition troops, doing the fighting. I remember being a battle group commander during an earlier deployment to Iraq. I would attempt to get the Iraqis to lead, but if operations did not go well, we would find a way to do it ourselves.

This time we had to rely on the Iraqis to conduct the operations because we were not permitted or resourced to do them. And guess what? The Iraqis were ultimately in charge of the plan, and they really did lead the fight from the outset. And we learned—for me, Fallujah 2016 was the battle where I probably learned more lessons than I ever learned in my life—from them because the Iraqi approach to defeating the enemy was about winning a cognitive fight against them—convincing them that they had lost—rather than clearing the city block by block and killing them all. That shaped their entire plan.

---

[15] For U.S. doctrine on campaign planning, see Joint Chiefs of Staff, JP 5-0, *Joint Planning* (Washington, DC: Joint Chiefs of Staff, 2020), https://www.jcs.mil/Portals/36/Documents/Doctrine/pubs/jp5_0.pdf?ver=us_fQ_pGS_u65ateysmAng%3d%3d .

[16] Timothy S. McWilliams and Nicholas J. Schlosser, *U.S. Marines in Battle: Fallujah, November-December 2004* (Quantico, VA: U.S. Marine Corps History Division, 2014), 3, https://www.usmcu.edu/Portals/218/FALLUJAH.pdf .

[17] Ibid., 12.

Their mindset was that they had to get their opponent to understand that they were defeated and convince them to leave, surrender, or lose the will to fight. So they would tell us, "We are going to liberate Fallujah, not clear it." In the West, the maneuver tactical plan is typically focused on the enemy and the principal method is destroy or clear.[18] Their maneuver plan was really part of a larger information operations campaign, which is the reverse of us in some ways. Yet they were extremely successful in adopting that approach and applying it to Fallujah. They tried the same approach in Mosul, but for a variety of reasons, it did not work quite as well. There was a lot of frustration between the coalition and their Iraqi counterparts because we have two different worldviews about how to approach a fight: we wanted them to clear the city and they wanted to liberate it. In the end, I think their model proved very successful.

Spencer: I did see news in late November 2014 about some frustration with the pace of the Iraqi Army.[19] The Iraqis' siege of Fallujah took about three months. Was a lengthy siege done on purpose, to place effects into the city without having to clear it?

Noble: One of the salutary lessons from urban combat is that it is hard, difficult, and lethal. There was no shortage of people telling the Iraqi Army to go faster, to clear this, and do that. But if you are the soldier standing on the Tatafar Bridge at Objective Peach south of Fallujah and you are trying to attack a prepared position, that is not an easy thing to do. So, the Iraqis basically surrounded the city and managed the stakeholders before they entered the city. And for a city like Fallujah, the stakeholders numbered in the hundreds.

The most important stakeholders included the coalition; the Iranians; the Popular Mobilization Forces, which include approximately 340 different groups; and the Iraqi Security Forces, a heterogeneous conglomerate that included the Iraqi Army, the Iraqi Counter Terrorism Service, and the Federal Police, each with different allegiances, objectives, and controls. The really clever thing that the Iraqi government did was leveraging Iraqi society. They messaged Fallujah's population using local tribes, Islamic charities, and the United Nations which was there to mitigate the war's

[18] For a definition of destroy, clear, or other tactical mission tasks, see Department of the Army, *The Army Universal Task List*, ADRP 1-03 (Washington, DC: Headquarters, Department of the Army, 2015), https://armypubs.army.mil/epubs/DR_pubs/DR_a/pdf/web/adrp1_03.pdf.

[19] Eric Schmitt, "Obstacles Limit Targets and Pace of Strikes on ISIS," *New York Times*, November 9, 2014, https://www.nytimes.com/2014/11/10/world/middleeast/trouble-pinning-down-isis-targets-impedes-airstrikes.html.

impact on civilians. The Iraqis occupied positions around the city and then sent a clear message to ISIS fighters within the city: "We have won, and you have lost, it is just a matter of time before you are going to die."

They put all that pressure on ISIS fighters before starting an extremely slow tactical advance. However, they advanced on multiple axes, which we did not like. We thought if they massed their combat power, they could finish sooner. But it was not a race for them, it was an effort to get the enemy to conform to what they wanted the enemy to do, which was to leave. So, we were frustrated. We thought they should move quicker and use fewer axes. But they were executing the operation in their very Iraqi way. In the end, and much to the coalition's surprise—certainly I was surprised—it worked pretty well. ISIS reached a breaking point and eventually attempted a breakout from the city rather than staying and fighting to the death.

Spencer: Fascinating. Did they attempt to empty the city of its people?

Noble: No, a lot of people remained in the city throughout the siege. ISIS tried to hold many of them in and made it difficult for them to leave. But ISIS's ability to leverage the population to ISIS's benefit was fairly crude in Fallujah but became more sophisticated over time. ISIS learned from its experience in Ramadi and Fallujah. In Fallujah, ISIS tried to control the civilian population. They tried to prevent the civilians from leaving the city, but they failed to close it off. The United Nations, the Iraqi government, and other Iraqi stakeholders managed to create a humanitarian corridor where civilians could, to a limited extent, leave and get away. While some escaped, many remained. One of the big challenges for the Iraqis and the coalition was the civilians in the city. We had to obey humanitarian law and the law of armed conflict.[20] That was non-negotiable.

The more civilians that remained in the city, the closer and more entangled they became in the fight. That slowed down operations and increased the risks that we would inadvertently kill them. Given the purpose was to liberate them, there was no point in liberating a city if you killed everyone in it. So, it was a very difficult problem. But I think the Iraqis, particularly in Fallujah, were successful in mitigating ISIS's capacity to exploit the civilian population. Although it was not a good place for a civilian to be, I can tell you that much. In Mosul, they became more effective

[20] International Committee of the Red Cross, *What is International Humanitarian Law?* (Geneva: International Committee of the Red Cross, 2004), https://www.icrc.org/en/doc/assets/files/other/what_is_ihl.pdf; and Krystyna Blokhina, "Geneva Conventions and their Additional Protocols," Legal Information Institute, June 10, 2019, https://www.law.cornell.edu/wex/geneva_conventions_and_their_additional_protocols# .

at using civilians as part of their maneuver plan, by moving them around to create a human shield around their forces.

Spencer: You just mentioned something that reminded me about risk. I was reading an interview you did back in 2016.[21] In that interview you characterized the fighting as consisting of conventional ground combat with high casualties and high risk. How do you define risk, especially in major urban fights?

Noble: Underneath all these campaigns the information operations and the narrative is the close combat piece. What I saw time and time again was that success basically rested on the capacity of the Iraqi Army to win in a fight, because that is the currency of the narrative: "We will beat you. We can defeat you in close combat in an urban situation where you have prepared for us to come." It was critical for the Iraqi Army to be successful, so we put significant effort into ensuring their combat forces could fight and win. But in the end, the tremendous risks fell on the individuals and the units involved.

An infantry company with tanks and combat engineers in support had to attack a prepared defensive position. That was the critical requirement and the Iraqi Army had to be able to do that. Plenty of people criticized them but the Iraqis were better than the enemy when they were effectively armed, equipped, and enabled. They could win in that situation, but often at tremendous cost. There were many fallen Iraqi soldiers—Iraqis would call them heroes and Americans would too—whose commitment was unquestionably as high as it gets. So that is it. You have to be good enough at the tactical level to let the rest happen. There is high risk to the people involved and there is risk to the mission if you cannot win the tactical fight. Because as soon as the enemy can stop you, your narrative falls apart. Does that make sense?

Spencer: Yes, it makes a lot of sense. Was there ever a risk of losing political will throughout the country or with other key stakeholders?

Noble: One of the beauties of ISIS as an enemy is that they were so hated that they unified enemies against them. Iraqis would often comment "I never thought I'd see those five people in a room together" because they were united to defeat ISIS. Across the many Iraqi stakeholders, there was a strong commitment to defeat ISIS, but initially low levels of confidence

---

[21] Siobhan Heanue, "The Australian heading up efforts against Islamic state," ABC News Australia, September 25, 2016, https://www.abc.net.au/news/2016-09-25/iraq-australian-army-second-command-islamic-state/7874072?nw=0&r=HtmlFragment .

in their capacity to do it. I talked to Prime Minister Haider al-Abadi fairly frequently in the strike cell and it was clear as to what he thought he had to do politically. He was focused on defeating ISIS and that required the Iraqi Army and the population having enough confidence to enable them to do it. The Iraqi Army had to both be tactically successful and have an effective information operations campaign that targeted both the enemy and the civilian population.

Effective information operations required understanding what the population was thinking. And anyone who has been in Iraq for ten minutes knows that is not an easy thing to do, because Iraqis are anything but homogenous. It depends on who they are, where they are from, what their connections are, and the list goes on. It is no different than us, just in their own context. Thus, anything that threatened their national will became a major threat for us. The enemy knew that Baghdad was our center of gravity, so they were always trying to attack the city.

In 2003, al Qaeda in Iraq leader al-Zarqawi orchestrated a suicide bombing campaign that prompted United Nations Secretary General Kofi Annan to pull all but a skeletal staff out of Iraq, effectively ending the United Nations presence there.[22] In 2016, ISIS was once again trying to use large bomb attacks in Baghdad to undermine confidence in the government and its narrative that it could actually defeat ISIS.[23] The political struggle for confidence and purpose was constant and because it was ISIS's method to attack the Iraqis' political will.

I do not lose a day's sleep thinking about fighting ISIS. There is nothing good about them, but one thing they taught me was the power of aggression. Any opportunity ISIS had, they attacked. Occasionally ISIS was successful, and a tactical success of no real tactical significance could have a strategic impact because they started to believe "we really can win, we can stop them." That was just a rolling part of that narrative clash with them.

Spencer: You mention capacity, what did it take to liberate a large city like Mosul? What were some of the surprises you encountered between planning and execution?

---

[22] Kirk Semple, "Truck Bombing; Panel Faults U.N. on Lax Security for Iraq Office," *New York Times*, October 23, 2003, https://www.nytimes.com/2003/10/23/world/struggle-for-iraq-truck-bombing-panel-faults-un-lax-security-for-iraq-office.html.

[23] For example, the Karrada bombing in Baghdad on July 3, 2003, killed 382 civilians. See, Moni Basu, "In Iraq, thousands of terrorism's victims go unnamed," CNN, January 12, 2017, https://edition.cnn.com/2017/01/12/world/iraq-terrorism-faceless-victims/index.html.

Noble: You have to take the long view for effective campaign planning and think about sequencing what you need. During planning, we identified the need to establish a major forward operating base close to Mosul to support the attack. We selected Qayyarah Airfield West (Q-West), an airbase 70 kilometers south of Mosul that the Americans had used during Operation Iraqi Freedom. But first we had to liberate Q-West from ISIS which had captured the airbase in 2014.[24] After recapturing the base we would have to clear and restore it, which would not be easy. The base was completely destroyed, and ISIS had seeded it with IEDs. To fortify the base, we required T-walls—large concrete barriers shaped like an upside-down T. I asked my engineer "How many T-walls do we need?" He replied, "5,928." I asked him, "How long will it take to get that many?" He said, "About six months." So, we ordered the T-walls before the Iraqi Army took its first step towards Q-West. That is but one of those millions of details that you must consider during planning or suffer during execution.

A second example of what was needed to liberate Mosul was a floating bridge. The Iraqi Army only had a single floating bridge.[25] For a number of reasons, our plan hinged on using this single bridge to cross the Tigris River. If you want to lose sleep at night, just have one bridge. So, a massive amount of effort went into protecting this bridge, from its movement to the river, to protecting it after it was built. During planning, about two months before the bridge went in, a smart engineer major remarked, "You better have a way of protecting the bridge from the water." I, along with the rest of the staff, thought "ISIS doesn't have any frogmen or Navy SEALs." So, we all laughed. But after the bridge went in and we started crossing the bridge, we saw dozens of refrigerators floating down the river towards the bridge. Since refrigerators are watertight, ISIS used them to float IEDs toward the bridge.[26]

When you are advised your single bridge is under attack by floating IEDs you are glad for expert engineer advice that meant a defensive barrier/net was already in place. The Iraqi Army was able to destroy some of the

---

[24] Charles Rivezzo, "Air Force engineers resort Q-West to support Mosul offensive," U.S. Air Force Central, October 25, 2016, https://www.afcent.af.mil/News/Article/985022/air-force-engineers-restore-q-west-to-support-mosul-offensive/.

[25] Yeganeh Torbati and Idrees Ali, "U.S. forces help Iraqis build bridge on way to Mosul," *Reuters*, July 27, 2016, https://www.reuters.com/article/us-mideast-crisis-usa-islamicstate/u-s-forces-help-iraqis-build-bridge-on-way-to-mosul-idUSKCN10722Q.

[26] Jim Garamone, "Iraqi Military Demonstrates New Skills in Mosul Campaign," DoD News, August 2, 2016, https://www.defense.gov/News/News-Stories/Article/Article/885558/iraqi-military-demonstrates-new-skills-in-mosul-campaign/.

floating bombs by shooting every high caliber weapon they had at them. The rest were stopped by the net that the smart engineer major had placed upstream to protect the bridge. What can be learned from this short vignette is to pay attention to detail, think it through, and never underestimate the enemy. We did not lose the bridge, which was good for everybody. To this day, I will never underestimate the enemy.

Spencer: That is fascinating, especially the T-wall story. I wrote my very first urban warfare article about concrete and about how I, as a company commander in Iraq, became a concrete expert. The article is titled, "The Most Effective Weapon on the Modern Battlefield Is Concrete."[27]

Noble: That and a garden spade. I used to have a slide, mainly for air force officers, of a picture of a joint strike fighter aircraft, or something similar, with its cost—millions of dollars—under it. Next to it, I had a picture of a spade with its cost—five dollars—under it. And that is what ISIS attempted to do to defeat us. They dug holes under the ground. It is not new; soldiers have been digging for a long time. Entrenching tools or E-Tools became standard issue in World War I.[28] Soldiers had to dig a good hole to protect themselves. In the 21st century, building a concrete wall can offer similar protection to what digging did in the past. Yet despite our technological advancements, it often ends up an infantryman's job to enter a trench, enter a tunnel, or scale a wall. That is a salutary lesson.

Spencer: Yes, for sure. Was the subterranean environment a major obstacle to the forward momentum?

Noble: In the urban areas the subterranean environment posed a major challenge, and ISIS was good at leveraging it in the defense. In Fallujah, they tunneled under the advance route. When the Iraqi Army advanced through and penetrated ISIS's forward defensive line, ISIS fighters went underneath and attacked from the rear. In a single day, south of Fallujah, ISIS used this technique to destroy four of the Iraqi Army's M1 Abrams tanks. ISIS was unable to execute this tactic on an industrial scale, but it was a clever tactical tool that they used. ISIS also fortified and dug inside the city for protection, because they lacked air power and possessed few intelligence, surveillance, and reconnaissance assets.

---

[27] John Spencer, "The Most Effective Weapon on the Modern Battlefield Is Concrete," Modern War Institute, November 14, 2016, https://mwi.usma.edu/effective-weapon-modern-battlefield-concrete/.

[28] See, for example, "Model 1910 Entrenching Tool," *Smithsonian*, https://amhistory.si.edu/militaryhistory/collection/object.asp?ID=220.

By contrast ISIS faced an Iraqi Army, supported by the coalition, with enough intelligence, surveillance, and reconnaissance platforms to create an "unblinking eye," where death could come 24/7 from a whole range of platforms. I thought surely this would significantly constrain their freedom of maneuver, but that was the world that they lived in, and they learned to hide, camouflage, disguise, move underground, or move with civilians in a way that taught me lessons about what was possible. So, if I ever find myself in a scenario without air support and an enemy pursuing me with an "unblinking eye," I will not give up, because I have seen what is possible if you are creative, aggressive, and determined.

Spencer: I agree. I think we undervalue the need to train on defense operations. Most of the urban training events that I have observed involve a unit in the attack. For some reason we, the military, do not envision future scenarios where we would be defending. Despite, like you just said, the fact that if we are facing a near peer threat, we likely will not have air supremacy and may need to defend from within a city.

Noble: Defense in an urban area is tough and it is hard to prevail if the opponent has sustainable combined arms systems. An effective defense still requires an effective maneuver or offensive component. Earlier in my tour, ISIS was effective because they maintained an offensive capability. Watching an unmanned aerial vehicle feed, I witnessed ISIS conduct what could best be described as a conventional company attack through Peshmerga—the Kurdish branch of the Iraqi Armed Forces—lines. ISIS used vehicle-borne IEDs for improvised breaching, unmanned aerial vehicles, rockets, mortars, and infantry to conduct its company-sized attack under conditions of limited visibility. They deliberately waited to conduct the attack during bad weather at night to limit the employment and effectiveness of our intelligence, surveillance, or reconnaissance assets. When attacking in this manner, ISIS was often successful. These early tactical defeats put the Iraqi Army mentally on their back foot.

To defeat ISIS and take away their offensive capability, we had to target ISIS in depth or in all domains as it is called in current doctrine.[29] We eventually degraded their offensive capacity, so they could do nothing but dig in and fight from prepared positions. This meant a long, hard fight, but it took away their ability to regain the initiative. The key to winning an

[29] Congressional Research Service, *Defense Primer: Army Multi-Domain Operations (MDO)* (Washington, DC: Congressional Research Service, 2021), https://sgp.fas.org/crs/natsec/IF11409.pdf.

urban fight is taking the enemy's maneuver capacity and making it a worst-case scenario for them where their only choice is to hold their ground or surrender. And then if you mass your force and execute combined arms operations properly, you will defeat them. Which is what the Iraqis did, but it took a while.

Spencer: Did the Iraqi Army have urban warfare doctrine?

Noble: No. We spent a lot of time training the Iraqi Army, but we did not teach them offensive urban warfare tactics; instead, we focused on basic conventional operations. I did it myself when I trained an Iraqi brigade in 2005. I focused on stability and counterinsurgency operations. They lacked experience and tradition with urban operations. The Iraqis do not have doctrine, at least not the way that we conceive of it, no volumes of documents and a training system that implies it. One thing that the coalition did really well was training the Iraqi combat brigades that were going to be doing the fighting. We sequenced brigades through the coalition-led training effort, where they were also outfitted with mission essential equipment that we take for granted but they often lacked: body armor, proper helmets, and small arms, for example. We called it "just-in-time training," where they would cycle through the training at a base in Taji and then drive north to attack ISIS. The most important task was getting them to be proficient at conducting offensive operations in complex terrain.

How did it work out? The results were mixed. Some units were good, others were not. Success often depended on leadership. Units with leaders who were heavily involved in the training often performed well. Units with poor leadership, where the company and battalion commanders were not involved, often fell apart when tested. In 2016, the Iraqi Army was no place for the faint of heart, it was a difficult fight. So, army fundamentals, such as leadership and discipline, were absolutely critical. And like all armies, the quality within the Iraqi Army was mixed. One of their best units was their Counter Terrorism Service which had been mentored and advised continuously since 2003.[30] But as a whole, it was a mixed bag across the force.

Spencer: So, we trained the Iraqi military in fundamentals, but combined arms maneuver into dense urban terrain was not one of the training objectives?

---

[30] David Witty, *The Iraqi Counter Terrorism Service* (Washington, DC: Brookings Institution, 2015), https://www.brookings.edu/wp-content/uploads/2016/06/david-witty-paper_final_web.pdf.

Noble: Counterinsurgency had been the training focus for many years, but in 2016 the training effort deliberately focused more precisely on the skills needed for success in offensive urban operations—but often at a basic level.

Spencer: I realize they lacked a formal lessons learned process like most Western armies, but do you think they learned a lot and will carry those lessons forward?

Noble: I am not sure. Within the coalition and the Iraqi government, there was a feeling of desperation at the time: "We have to get ISIS out of this country." So, we invested very little time in formal processes—after action review processes, big organizational training, or systems learning curves—so that when a brigade learned the hard way, the next brigade forward did not have to learn the same lesson. There was simply no time to introduce this type of thing; we were focused on preparing the operational army for combat as opposed to building the institutional army. And the Iraqis were Iraqi, so we should not have tried to turn them into the Australian Army or the United States Army. It had to be the Iraqi Army—they were leading the operations. It was their plan. If we did not like their plan, we would tell them, but they owned it.

Over time, they became more proficient. Their institutional confidence grew as they advanced towards Mosul; they started to believe that they could liberate a city and conduct complex attacks because they had. They saw the value of the training and experience in a way that they had not in the decade when the focus was on counterinsurgency operations.

I am not sure what they will carry forward, but what they did was incredibly difficult, and they really did it themselves. Without us, they probably would have been successful, but it would have taken them years longer. The coalition's support was critical in speeding them up, moving them around, magnifying their combat power tenfold, and giving them great confidence. Simply by being there, they knew that if it got really grim, the coalition would bail them out. Ultimately, it worked! I was prouder of the Iraqi Army at the end of that tour than at any time previously, and I had two previous tours in Iraq.

It is also important to note that the orders to attack Mosul were given by Iraqis not the coalition. The attack order came from the Prime Minister and the Chief of Staff of the Army and was given in Arabic. I was in the room with the combined joint task force and land component command commanders when the order was issued. We sat quietly along the wall and did not speak. That is how far they had progressed. It took some heartache to get there, but they developed the plan, issued the order, and executed it. In some ways that was one of my proudest professional days because we did not speak.

One Iraqi officer did have a question about fire support that he directed to the combined joint task force commander. But it was a pretty good question and given the coalition commander was providing the fires, it was appropriate that he, as opposed to one of the Iraqi officers, provide the response. While there was a lot of criticism of the Iraqi Army and they had a long list of failures, they performed pretty well in 2016.[31]

Spencer: I would like to highlight one of the criticisms, because I get it a lot.[32] I get requests from organizations wanting to discuss the protection of civilians in urban combat. In your 2016 ABC interview, you foreshadowed the risk to civilians, and we just talked about how ISIS purposefully moved them around, using them as human shields to restrain the Iraqi Army's use of force.[33] Did you see any best practices that were developed to counter that strategy?

Noble: The thing about ISIS is that their strategy is population centric. You cannot have a caliphate without people or territory. It is where they started. Their organizational focus was on population control and perception management. They were always deeply invested in the population. As the fight progressed, they deliberately planned and used the population as a maneuver tool. By the battle of Mosul, they were moving and shifting the population fairly quickly and effectively in an attempt to constrain our fire and maneuver to slow us down.

They realized our moral and legal obligation to minimize civilian casualties and leveraged it to their advantage because they did not adhere to international humanitarian law. I thought our targeting engagement protocols were one of our best practices and were as good as I have ever

---

[31] See, for example, Eric Schmitt and Michael R. Gordon, "The Iraq Army Was Crumbling Long Before Its Collapse, U.S. Officials Say," *New York Times*, June 13, 2014, https://www.nytimes.com/2014/06/13/world/middleeast/american-intelligence-officials-said-iraqi-military-had-been-in-decline.html.

[32] See, for example, Suadad Al-Salhy and Tim Arango, "Sunni Militants Drive Iraqi Army Out of Mosul," *New York Times*, June 10, 2014, https://www.nytimes.com/2014/06/11/world/middleeast/militants-in-mosul.html; and Loveday Morris and Missy Ryan, "After more than $1.6 billion in U.S. aid, Iraq's army still struggles," *Washington Post*, June 10, 2016, https://www.washingtonpost.com/world/middle_east/iraqs-army-is-still-a-mess-two-years-after-a-stunning-defeat/2016/06/09/0867f334-1868-11e6-971a-dadf9ab18869_story.html.

[33] Heanue, "The Australian heading up efforts."

seen. We were ruthless on ourselves. The commanding general directed us to have no civilian casualties, a standard much stricter than the law of armed conflict.[34] If we had any doubt, we would not shoot. That was our mantra.

We leveraged the coalition's intelligence resources to understand the civilian pattern of life, not five minutes before a strike, but three weeks prior. It was much different from my previous tours, where you might look at a building or target area only minutes prior to a strike. We studied an area to understand its pattern of life, so we understood not only the specific place where we were shooting, but we also understood who had been moving in and out of the area. We would only launch fires when we were confident that no civilians were present. That was our model, and we had a set of protocols. If we suspected civilians might be present, we would not shoot.

The challenge was that ISIS pulled the civilians into the close fight and deliberately obscured their own movement and positions. This slowed us down and forced us to conduct more intelligence, surveillance, and reconnaissance and take more precautions, which exposed our troops that were in close proximity to the enemy. Our enemy fully understood this calculus. I was a target engagement authority, meaning I was the one that had to approve fires into populated areas, but I had a lot of confidence in our system. We had lawyers, weapons experts, pilots, Iraqis, people on the ground, and databanks to help us quickly make decisions about firing or not firing against a target. It required a really rigorous self-assessment and lessons learned process, particularly when it involved complex terrain with civilians everywhere. That did not mean we were able to completely eliminate civilian casualties. Even with our robust system we occasionally missed seeing some civilians, but we did our best and stayed well within international humanitarian law.

Spencer: Earlier I asked you what the Iraqi Army learned from this experience. I read the Mosul Study Group report, which was a quick U.S. Army research project.[35] I would like to flip the question and ask you

---

[34] Ryan Dowdy et al., *Law of Armed Conflict Deskbook*, 16th ed. (Charlottesville, VA: U.S. Army JAG's Legal Center and School, 2016), https://www.jagcnet.army.mil/Sites%5C%5Cio.nsf/0/EEF9422EB2C293B68525805A0063ABED/%24File/LOAC%20Deskbook%20final%20with%202016%20index%20(20%20Sep).pdf.

[35] Mosul Study Group, *What the Battle for Mosul Teaches the Force* (Ft. Eustis, VA: TRADOC, 2017), https://www.armyupress.army.mil/Portals/7/Primer-on-Urban-Operation/Documents/Mosul-Public-Release1.pdf.

what you think the Australian Army learned from these battles about future combat?

Noble: I read the Mosul Study Group report and it is largely focused on the tactical level. I think our doctrine is pretty sound, so I do not think there were any revolutions in our approach to the tactical close fight. I do not think there was anything particularly revolutionary at the tactical level; the micro-tactical level of battle is important, but it is an evolution we already knew.

I think the biggest lesson was adopting an all domain approach to targeting. All domain targeting involves employing both lethal and non-lethal means to target an adversary in all domains to put him into a position where he cannot be successful; and for us, that was an urban defense.[36] That is the thing that I learned. That not only includes targeting his physical capacity, which is what we tend to do: for example, shoot at his tanks. But it also includes targeting the enemy's mental capacity—his mind—getting the enemy into a position where they will not fight, or fight in a way that suits you. I never thought I would be involved in corps-sized attacks or conduct large deception operations, but I was.

We used fire, maneuver, and various means of information operations to deceive the enemy: to get ISIS to not understand what we were doing or to get them to conform to what we wanted them to do. As an example, we planned a brigade attack towards the end of my tour at Shirqat, a Sunni Arab town 55 km south of Mosul along the Tigris River. It was a supporting effort and we had time to plan the operation with our Iraqi Army partners. It was probably the first all-domain brigade attack against an enemy, where we used every domain to deceive and influence them. Ultimately, the enemy brigade left the city before we closed with them, so it was extremely effective.

Spencer: Your comments about influence fascinate me. There were reports that in Tal Afar, the Iraqi Army encircled the city, used an information operation to influence the enemy within the city, and left a backdoor out of the city, because their objective was to retake the terrain, not necessarily kill all the enemy.[37]

Noble: When you think about it, it comes back to your purpose. Are you trying to kill all the enemy? Is the way to a man's heart killing the guy

[36] The domains include air, land, maritime, space, and cyberspace. See, Congressional Research Service, *Defense Primer: Army MDO.*

[37] Rukmini Callimachi, "Tal Afar After Liberation From ISIS: Battered but Still Standing," *New York Times*, September 1, 2017, https://www.nytimes.com/2017/09/01/world/middleeast/tal-afar-mosul-islamic-state.html.

next to him? There is some strength in that thinking. If you are a dominant tactical force, you are very difficult to resist both mentally and physically. The Iraqis said we are going to try to defeat this enemy and clear them out of our cities. Towards the end of the battle of Fallujah, the Iraqis had the city surrounded. In late June 2016, two large convoys of ISIS vehicles fled the city, driving across the desert in different directions.[38] The initial coalition response was, "Who let them out?" To this day, I do not know.

I have no knowledge that the Iraqis intentionally let them out, but it would not surprise me if they did because they conducted some information operations that we did not know about. It is entirely plausible that the Iraqis left them a back door out of the city knowing the coalition would kill them after they took it. And that is pretty much what happened since convoys on an open desert road proved to be easy targets for coalition airstrikes.[39]

There's a sophisticated difference between us and the Iraqis. Our goal was to defeat ISIS; their goal was to defend Iraq, defeat ISIS ideology, get ISIS out of their cities, and minimize collateral damage in the process. Thus, the back door idea, having a way out for the enemy, whether it was planned or unintentional, is not a stupid thing to consider. The alternative, leaving a highly determined person in a prepared position with no way out, would have likely resulted in a long, costly drawn-out fight that may have destroyed the city that they were trying to save. This alternative has happened many times before.[40]

Spencer: Yes, I am pretty sure Sun Tzu said something similar thousands of years ago.[41]

---

[38] Mustafa Salim and Thomas Gibbons-Neff, "Iraqi, U.S. aircraft bomb convoy of Islamic State fighters fleeing Fallujah with their families," *Washington Post*, June 30, 2016, https://www.washingtonpost.com/news/checkpoint/wp/2016/06/30/iraqi-gunships-u-s-jets-target-islamic-state-convoys-outside-fallujah-and-ramadi/.

[39] Ralph Ellis, Barbara Starr, and Hamdi Alkshali, "Airstrikes hit convoys carrying ISIS militants out of Fallujah, U.S. officials say," CNN, June 30, 2016, https://www.cnn.com/2016/06/30/middleeast/iraq-militants-airstrike/index.html.

[40] Examples of a military sealing off a city completely and not leaving the enemy an escape route and destroying most of a city to liberate it include the 1942-43 Battle of Stalingrad and the 2004 Second Battle of Fallujah. See John Spencer and Jayson Geroux, "Urban Warfare Project Case Study Series: Case Study #1 — Stalingrad," Modern War Institute, June 26, 2021, https://mwi.usma.edu/urban-warfare-project-case-study-1-battle-of-stalingrad/; and McWilliams and Schlosser, *U.S. Marines in Battle*.

[41] Sun Tzu said, "When you surround an army, leave an outlet free. Do not press a desperate foe too hard. When a foe is cornered, they must fight for their lives and will do so with the energy of final fear. If you force them to go down in a blaze of glory they will do so, taking more of your troops than you might otherwise expend." See Sun Tzu, *The Art of War* translated by Lionel Giles, http://classics.mit.edu/Tzu/artwar.html.

Noble: That is the thing; none of this is new. It is just context driven. It is about the objective and the enemy. People often claim that they will fight to the death. We have had scenarios in our armies where those orders have been given.[42] Occasionally it happens, but most often it does not. In most armies, people do not generally do that. In many cases it may be advantageous to provide the defender an option to run, an option to stop fighting. If not, then a long, deadly fight will often result.

This partly explains why the Japanese were such tenacious fighters on Pacific islands during World War II. Culturally, surrender was not an option, and since there was no easy escape off an island, it should not be surprising that they fought so hard. The Australian Army confronted this firsthand across the Pacific. Similarly, ISIS fighters trapped in a city who felt a fight to the death was their only option, were equally tenacious.

You cannot underestimate the potential to influence the enemy's thinking, and then use all available tools to build a scenario that makes them do what you want them to do. And if they do not, then you have to go in and kill them until they comply.

Spencer: Yes, it is pretty straightforward; compel them to do your will.

Noble: People were universally surprised to see the Iraqi army clear the Euphrates River Valley in roughly two months after seizing Fallujah; nobody thought they could do it that quickly. A decrease in the enemy's confidence was one factor which contributed to the army's rapid advance. After losing Fallujah, their confidence wavered. They started to think: "The Iraqi Army's methods are working. We cannot hold the caliphate. We cannot hold the ground. We do not want to fight to the death." We tracked the narrative from Fallujah's survivors and could see their confidence fall. There is a cleverness in spreading a lack of resolve across the force, which the Iraqis understood because they understood ISIS very well. They did not view the group as a homogenous set of humans. They viewed ISIS as different groups from different countries—albeit a majority were Iraqis—with different purposes. They spent a great effort messaging Iraqis within ISIS about how they should behave and what their options were. There was a cleverness to it. Often, we fail to consider if there are better options than simply trying to kill the enemy.

[42] Examples include Joshua Chamberlain at the Battle of Gettysburg, and Mustafa Kemal Atatürk at the Battle of Gallipoli in World War 1 who is famously quoted as telling his men: "I do not order you to fight, I order you to die." See, "Remembering Gallipoli, the grinding WWI battle to take the Dardanelles," *Euronews*, April 21, 2015, https://www.euronews.com/2015/04/21/remembering-gallipoli-the-grinding-wwi-battle-to-take-the-dardanelles.

Spencer: Yes, there are so many lessons here. And many of them point to the earliest stage of planning, to the operational design. Gathering information on the enemy and knowing the end result you are trying to achieve.

Noble: And it is that depth of understanding and the context. So, if you are fighting in somebody else's country, like Iraq, only the Iraqis truly understand their own environment. We do not know what is really happening. We do not know the cultural connections. The better we can understand them, and the better we understand the levers that work, the more successful we will be. And it is often the really simple things that work. In Iraq, we dropped leaflets, and to our surprise, they worked. The Iraqis loved leaflets. You cannot jam a leaflet or prevent access to it by shutting off the internet; it floats down from the sky. To counter, ISIS made it a death penalty to pick up a leaflet in Mosul.

Leaflets provided us practical lessons in culture and context. We created one leaflet that we had printed in another regional country, using a version of Arabic and stating things in a way that was clearly not Iraqi. To the local reader, it was clearly from another country. Having the incorrect word or phrase on the leaflet turned it from a positive message into a reinforcing narrative for ISIS to use back at us: "you see who's really behind this leaflet." So having that kind of depth of understanding in terms of the context is something we all know from counterinsurgency, but it actually applied in this kind of pretty conventional fight as well. You have to know your enemy, you have to know the ground, and you have to know the big context in which it's all happening.

Spencer: Yes, and you have to know your partner. One of the other questions I often get asked is about the nastiness of these types of fights and if it creates an environment that is ripe for ethical challenges?

Noble: The nature of war is rooted in violence, enmity, and hatred. So, atrocities are always a concern and one that we were worried about, especially given the lack of discipline among some of the combatants. ISIS produced messages demonstrating atrocities, often when they did not actually occur. They would put people in Iraqi Army uniforms and then kill them for messaging purposes: to demonstrate their superiority in combat. Other times, ISIS would dress in Iraqi uniforms, kill civilians, and then use the video as evidence that the Iraqi Army was murdering people. Sometimes it was true and sometimes it was not. In close combat, death, violence, enmity, and hatred rule. So, the discipline of the Iraqi military and the framework in which they operated was always under pressure and they were not perfect. But neither were Western armies when experiencing the same close combat challenges. If you want to go to a place where the soldier is under the most pressure, go to the old city in Mosul in the last days of the fight. Ethical challenges are an enduring feature of war, and the Iraqi security forces, while not perfect, largely performed well under intense pressure.

Spencer: Since we know this type of environment is ripe for ethical challenges, do you think some of the things we know help prevent atrocities—good leadership, training, and things like that—need to be strongly reinforced in partners?

Noble: We could talk forever about what makes a good army, but two things are discipline and leadership. They go together. It is imperative to understand that there are a set of rules that you must operate within, and it requires a system, discipline, training, and culture that reinforces it at every level. Often, it is not a general who is making an ethical decision; instead, it is four soldiers in a room. They must often make a split-second decision with a single noncommissioned officer. Do they make the right one? That is when an army is really tested. And when they make the decision, they do not have a senior officer looking over their shoulder telling them what to do. Armies that consistently make the right choice have good leaders who instill a system of training, discipline, and culture that allows the soldiers to make the correct ethical choice. All armies must be on constant guard against deviation and drift—the logical human result of sustained exposure to violence and stress.

Spencer: Definitely. This has been a fascinating talk and I know that the readers will love to read about your first-hand experience.

Noble: Good! I am very proud of what we did. It was a seminal professional experience. I love the U.S. Army, and all the armies I have served with, but particularly the U.S. Army. And of all of the units in the U.S. Army I shall always be a Screaming Eagle.[43]

[43] The symbol of the 101st Airborne Division that Noble served with is a Screaming Eagle.

14

The Battle of Marawi

The Battle of Marawi occurred against the backdrop of a 400-year struggle of the Moro Muslim population on the Philippines' second largest island of Mindanao against invaders, Christian settlers, and for self-determination. The Muslim Sultanates there resisted Spanish colonization so effectively that imperial sovereignty only ever extended to guarded settlements, with the Marano ancestors of today's inhabitants of Marawi remaining unconquered until the arrival of U.S. troops in 1901.[1]

Likewise, the United States and the Japanese occupiers, as well as Filipino Government forces, have always been violently resisted. The Moros have never viewed themselves as a part of an integrated Catholic Philippine state.[2] On the West coast of Mindanao and adjacent islands, Muslims make up 90 percent of the population despite only representing 5.6 percent of a Filipino population of over 110 million, of whom 80 percent adhere to Catholicism.[3]

In the 1960's, the Ferdinand Marcos administration encouraged Christian settlers to move to Mindanao from other areas. Abuses by their *Ilaga* militia triggered ethnic tensions, leading initially to the formation of Moro political secessionist movements. The introduction of martial law in 1972 and the prohibition of political parties led to the growth and eventual dominance of various insurgent groups including the Moro National Liberation Front (MNLF) and the Moro Islamic Liberation Front (MILF).

---

[1] Mamitua Saber, "Maranao Resistance to Foreign Invasions," *Philippine Sociological Review* 27, no. 4 (1979), 273-282, https://www.jstor.org/stable/23892402.

[2] Peter Chalk, "The History of Terrorism in Southeast Asia," in *Routledge Handbook of U.S. Counterterrorism and Irregular Warfare Operations*, eds. Michael A. Sheehan, Erich Marquardt, and Liam Collins (New York: Routledge, 2021), 119, DOI: 10.4324/9781003164500-10.

[3] Philippine Statistics Authority, *Fact Sheet– Autonomous Region in Muslim Mindanao* (Cotabato City, Philippines: Philippine Statistics Authority, 2017); and Central Intelligence Agency, The World Factbook 2021 (Langley, VA: Central Intelligence Agency, 2021), https://www.cia.gov/the-world-factbook/countries/philippines/#people-and-society.

When the relatively secular MNLF accepted the Philippine government's offer of semi-autonomy in 1987, the MILF and other rebel groups, especially the Abu Sayyaf Group and the Bangsamoro Islamic Freedom Fighters, rejected the agreement and continued fighting.[4]

Due to the Abu Sayyaf group aligning itself with al Qaeda after 9/11, the U.S. government began to support the Philippines' counterterrorism efforts with Operation Enduring Freedom – Philippines.[5] In 2014, the MILF signed the *Bangsamoro* peace deal in return for further autonomy. Importantly this agreement included a reduced military presence in Muslim areas and the integration of some former MILF members into the military and police. The government continued to fight the Abu Sayyaf Group and the Bangsamoro Islamic Freedom Fighters who did not accept the deal, as well as battling the 50-year insurgency of the communist New People's Army.[6]

The 2014 *Bangsamoro* peace deal coincided with the Islamic State of Iraq and al Sham (ISIS) eclipsing al Qaeda in reputation and power, and the Islamic State in the Philippines started to emerge. Many of the separatist groups in the Philippines pledged *bay'a,* or an oath of allegiance, to the ISIS including the Abu Sayyaf Group, Ansar Khalifah Philippines, Bangsamoro Islamic Freedom Fighters, and Maute Group, named after the Marawi clan family that led it.[7]

In 2015, these formerly distinct groups united as Daulah Islamiyah Wilayatul Mashriq (Islamic State – Eastern Region), more commonly referred to as the Islamic State in the Philippines (IS-P).[8] In 2016, the ISIS appointed Isnilon Hapilon, the former head of the Abu Sayyaf Group, as the emir of the IS-P whose goal was to establish a *wilayah,* or Islamic governorate, in Southeast Asia.[9]

---

[4] Francisco Lara, "Transition-induced Violent Conflict and the Bangsamoro State-building Project," *Philippine Sociological Review* 67 (2019), 5-34, https://www.jstor.org/stable/26933203.

[5] Edwin Amadar and Bobby Tuttle, "The Rise of ISIS in the Philippines and the Battle of Marawi," *CTX* 9, no. 2 (2019), 31-2, https://nps.edu/documents/110773463/120088589/CTX+Vol+9+No+2+WEB+FINAL-2.pdf.

[6] Cristopher C. Harmon, "The Philippines face the New People's Army: Fifty years in the field," in *Routledge Handbook of Democracy and Security*, ed. Leonard Weinberg et al. (New York: Routledge, 2020), 265-281, https://www.taylorfrancis.com/chapters/edit/10.4324/9781315755724-17/philippines-face-new-people-army-christopher-harmon.

[7] Amadar and Tuttle, "The Rise of ISIS," 31.

[8] International Crisis Group, *The Philippines: Militancy and the New Bangsamoro* (Brussels: International Crisis Group, 2019), 3, https://d2071andvip0wj.cloudfront.net/301-the-new-bangsamoro.pdf.

[9] Caleb Weiss, "Islamic State details activity in the Philippines," *Long War Journal,* June 12, 2016, https://www.longwarjournal.org/archives/2016/06/islamic-state-details-activity-in-the-philippines.php; and Amadar and Tuttle, "The Rise of ISIS," 31.

Two circumstances allowed the IS-P to grow from 2014-2017. First, the United States ended its counterterrorism operations in the Philippines in 2015, withdrawing most of its material support from Mindanao.[10] Second, in 2016, newly elected President Rodrigo Duterte made fighting illegal narcotics a priority which took attention away from the separatists on Mindanao. Together this allowed the Islamic State in the Philippines to grow "into a fighting force with surprising staying power."[11]

The battle of Marawi started on May 23, 2017, when a Philippine police arrest operation fortuitously triggered the premature initiation of a plan by Islamic State affiliated fighters to seize the city of 200,000. IS-P failed to secure Marawi and confine its population, yet the battle lasted until October 23, 2017.[12] The government ultimately "freed" the city but in the process destroyed much of it. The fighting resulted in the deaths of at least 920 militants, 165 soldiers, and 47 civilians.[13]

In this chapter, John Spencer interviews Dr. Charles Knight, a senior lecturer in terrorism, asymmetric conflict, and urban operations at Charles Stuart University; and a senior researcher at the University of New South Wales. Knight is both an academic and a practitioner, having served in several militaries, including more than 30 years in the Australian Army.

Knight discusses how and why the battle started and the various phases of the battle. He describes how the Armed Forces of the Philippines (AFP) ultimately prevailed, despite having to clear the entire city of the IS-P from strongpoint defenses, and how the AFP effectively leveraged an information campaign to win over the local populace. Finally, he offers some lessons that can be learned from this battle.

John Spencer: What are some lessons that can be learned from the Battle of Marawi?

---

[10] Linda Robinson et al., *U.S. Special Operations Forces in the Philippines, 2001-2014* (Santa Monica, CA: RAND, 2016), xiii, https://www.rand.org/pubs/research_reports/RR1236.html.

[11] Sidney Jones, "How ISIS Got a Foothold in the Philippines," *New York Times*, June 4, 2017, https://www.nytimes.com/2017/06/04/opinion/isis-philippines-rodrigo-duterte.html.

[12] PCIJ data team, "Quick Stats: Marawi City," Philippine Center for Investigative Journalism, May 25, 2017, https://pcij.org/data/116/quick-stats-marawi-city; and Ashley Westerman, "Over 120,000 People Remain Displaced 3 Years After Philippines' Marawi Battle," NPR, October 23, 2020, https://www.npr.org/2020/10/23/925316298/over-120-000-people-remain-displaced-3-years-after-philippines-marawi-battle.

[13] Institute for Policy Analysis of Conflict, *The Jolo Bombing and the Legacy of ISIS in the Philippines* (Jakarta: Institute for Policy Analysis of Conflict, 2019), 9, https://www.jstor.org/stable/resrep24206?refreqid=excelsior%3A0fcc243564a27120501b05f1b65ba93c.

Charles Knight: As Otto von Bismarck said, "Only a fool learns from his own mistakes. The wise man learns from the mistakes of others." So, it is important to study this battle, to actually learn from it, and to incorporate those lessons into training and education in meaningful ways. That might seem obvious, but history suggests armies more often observe lessons than actually learn and change their approaches and methods.

I would highlight that point by posing a puzzle about preparedness for urban conflict more generally. Only 2-3 days prior to the Battle of Marawi, the Philippine Army's Scout Rangers were fighting communist rebels in the jungle. They were a competent, professional force from a nation with a long history of capable fighters. American General Douglass MacArthur famously said, "Give me ten thousand Filipinos and I shall conquer the world!"[14] Yet, the Scout Rangers, like many before them, were not prepared for an urban fight. Why?

Often the same urban combat lessons are learned, forgotten, re-learned, and forgotten again. During the Mexican-American War's Battle of Monterrey in 1846, U.S. troops had to knock holes in the walls to progress against the Mexicans. "Battering rams, sledgehammers, pickaxes, and even small cannons were assembled for the job" to move from house to house that often shared a common wall.[15] In 1848, during uprisings all across Europe, professional soldiers being shot down by rebels with rifle muskets, were forced to knock holes in walls to progress.[16]

During the Franco-Prussian War, the Bavarians sustained heavy casualties from the enemy fighting from stone houses until the Bavarians used their artillery to knock holes in the stone walls.[17] Likewise, the British used their artillery to blast holes in walls during the Easter Uprising in Dublin in 1916.[18] Again and again, armies have learned to create holes in walls to reduce their vulnerability in urban combat.

[14] Colin Fraser, "The Fighting Filipinos: Give me ten thousand Filipinos and I shall conquer the world! —MacArthur," War History Online, March 17, 2016, https://www.warhistoryonline.com/world-war-ii/the-fighting-filipinos.html.

[15] Martin Dugard, *The Training Ground: Grant, Lee, Sherman, and Davis in the Mexican War, 1846-1848* (New York: Little, Brown and Company, 2008).

[16] Mike Rapport, *1848: Year of Revolution* (New York: Basic Books, 2009).

[17] Geoffrey Wawro, *The Franco-Prussian War: The German Conquest of France in 1870-1871* (New York: Cambridge University Press, 2003), 172-7, DOI:10.1017/CBO9780511511820.

[18] Jerome Reilly, "1916 row explodes as An Post claims 'there are no bullet holes in GPO wall'," *Independent*, August 28, 2005, https://www.independent.ie/irish-news/1916-row-explodes-as-an-post-claims-there-are-no-bullet-holes-in-gpo-wall-26212121.html.

In their early battle reports, the Scout Rangers, reported discovering something new: ratholes. It was their name for what other armies have called mouseholes: "openings that are made from the interior or exterior of a structure—walls, floors, ceilings, roofs—to facilitate inter- and intra-building communications and movement."[19] The use of mouseholes, like most of the lessons from Marawi, was not new at all, rather the method can be retrieved from the history books reporting conflicts back to ancient times—and indeed old U.S. Army doctrine manuals that the Filipino military once used.

Returning to the battle itself, in 2017, IS-P militants planned to seize Marawi and declare it a *wilayah*, a province of the Islamic State. It was an intentional act of great symbolism with the potential to inspire support across the region, and particularly on the island of Mindanao, where the Muslim Moro people have a history of fierce resistance to outsiders. The Moro people have fought the Spanish, the Americans, and, for many decades now, the Philippines government.

The people of Marawi, the *Maranaos*, are a subgroup of the Muslim Moros (see figure 14.1). They are a mercantile trading community, and they have their own distinct subculture of violent clan warfare called *Rido*.[20] The tradition of fighting amongst the clans in Marawi is so endemic that in some areas nearly everybody has a concrete reinforced cellar, or a *buho* shelter as they called them, where they keep firearms and store food for protection and subsistence when the clashes break out. Thus, prior to the start of the Battle of Marawi, readymade defenses existed throughout the city. Furthermore, for the Moro population, and the *Maranaos* in particular, there exists a cultural obligation to fight for one's blood relatives. It is similar to the *Pashtunwali* in Afghanistan.[21]

On March 27, 2014, the MILF signed a final peace agreement with the Government of the Philippines, yet some of the more radical members of the group wanted to reject the peace deal and to continue fighting for an

[19] Department of the Army, *Combined Arms Operations in Urban Terrain*, ATTP 3-06.11 (Washington, DC: Headquarters, Department of the Army, 2011), B-1, https://www.globalsecurity.org/military/library/policy/army/attp/attp3-06-11.pdf.

[20] Wilfredo Torres III, ed., *Rido: Clan Feuding and Conflict Management in Mindanao* (Manilla: Manila University Press, 2014).

[21] Andrea B. Coulson et al., "Hospitality codes and Social Exchange Theory: The Pashtunwali and tourism in Afghanistan," *Tourism Management* 45 (2014): 134-141, https://doi.org/10.1016/j.tourman.2014.03.019.

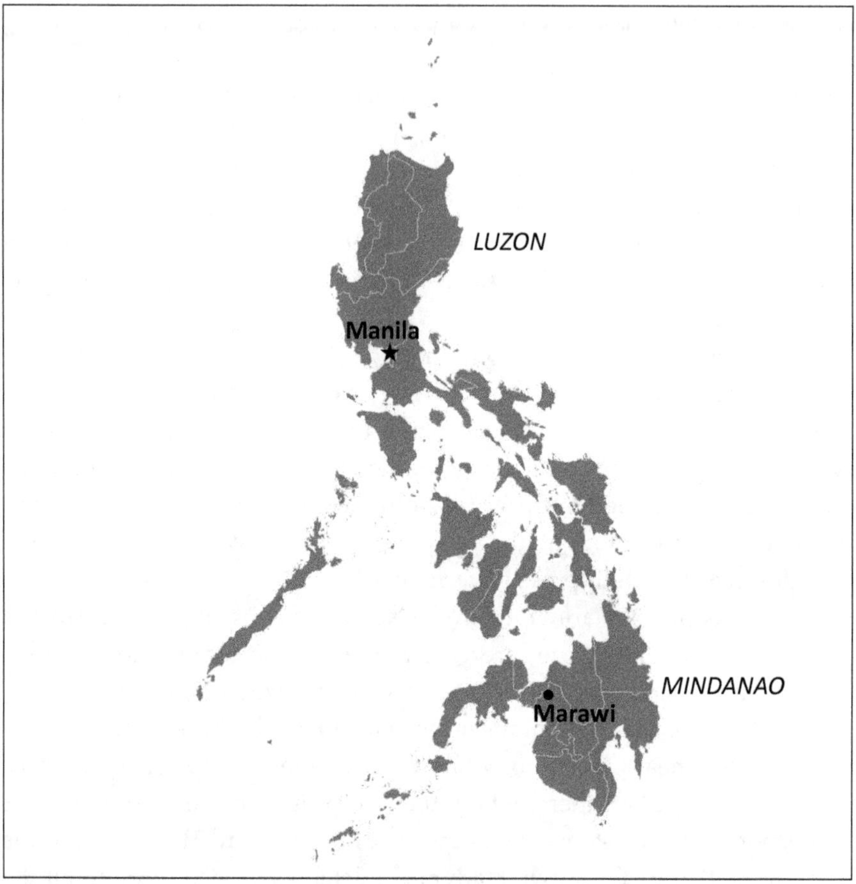

Figure 14.1. The location of Marawi on Mindanao.
Source: Map adapted by Liam Collins from Image 1538208875 / Map of Philippines © Shutterstock.com, http://shutterstock.com.

independent state with *Sharia* law.[22] At that time, ISIS was on the rise in the Middle East after seizing Raqqa earlier in the year, and only months later, it would declare a caliphate.[23] Previously unaffiliated separatist groups started uniting under the Islamic State banner. Isnilon Hapilon, the head of the Abu Sayyaf Group, was the first to swear his oath of allegiance to ISIS

[22] "Philippines, Muslim rebels sign final peace deal to end conflict," *Reuters*, March 27, 2014, https://www.reuters.com/article/us-philippines-rebels/philippines-muslim-rebels-sign-final-peace-deal-to-end-conflict-idUSBREA2Q1W220140327.

[23] "Timeline: the Rise, Spread, and Fall of the Islamic State," Wilson Center, October 28, 2019, https://www.wilsoncenter.org/article/timeline-the-rise-spread-and-fall-the-islamic-state.

in 2014. The Maute Group, one of the more powerful clans within Marawi, swore *bay'a* in 2015, and Ansar Khilafah Philippines swore *bay'a* in 2016.[24]

The power of these previously unaffiliated groups grew exponentially with this union. The Maute Group possessed local knowledge of the area while the Abu Sayyaf Group possessed greater capacity and experience, having conducted attacks against the government for years. Together, the whole was greater than the sum of the parts. The military had information that there was some level of cooperation between the groups and some level of infiltration into Marawi. They even mounted a few operations to try to learn more, but they simply lacked clarity on the level of integration between the groups. Their efforts to learn more proved problematic as they tried to avoid antagonizing an always suspicious community. Furthermore, the warnings they did provide to political elites were largely disbelieved.[25]

What seemed to have happened in 2017 was that the *Maranao* within the Marawi community started to accept the insurgent narrative and believe that the insurgents were the protectors of the Muslim population. The evidence for this is the fact that hundreds of insurgents, emanating from all over the country and even outside of the Philippines, were able to infiltrate into the city in early May with the military remaining unaware.[26]

Fighting in a city is always challenging, but in Marawi it was especially so. The city is bounded to the south by Lake Lanao; and the Agus River, which is fairly fast flowing, steeply banked, and splits the city in half. Only three bridges span the river, limiting the number of routes into the city (see figure 14.2). The 103rd Brigade of the Philippine Army was stationed at Camp Ranao, located just outside the city to the northwest.[27] There were a few police stations inside the city, but the 2014 peace deal ensured that the military presence in the city was kept to an absolute minimum.

The Battle of Marawi lasted five months and was confusingly referred to as a siege by the press, but it is most easily understood as consisting of six

[24] Amadar and Tuttle, "The Rise of ISIS," 33.

[25] Jasminder Singh, "Insights for Practitioners and Policymakers from the Marawi Siege, May–October 2017," in *Learning From Violent Extremist Attacks: Behavioural Sciences Insights for Practitioners And Policymakers* (Singapore: World Scientific Publishing, 2018), 35.

[26] Criselda Yabes, "Factors and Forces that Led to the Marawi Debacle," *Kasarinlan: Philippine Journal of Third World Studies* 34, no. 1-2 (2019), 83-104, https://www.journals.upd.edu.ph/index.php/kasarinlan/article/view/7075/pdf_1.

[27] Ver Marcelo, "Gov't forces, Maute group clash in Marawi City," CNN, May 23, 2017, https://cnnphilippines.com/news/2017/05/23/marawi-city-clash.html.

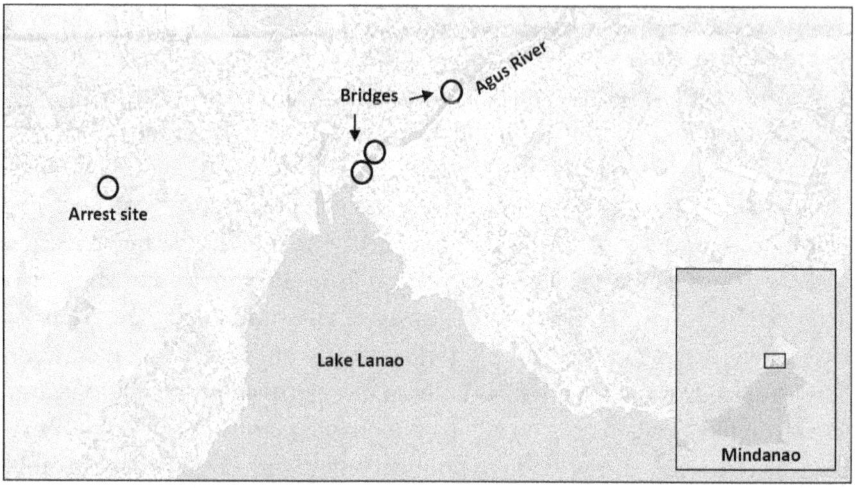

Figure 14.2. Map of Marawi.
Source: Map created by Liam Collins and Charles Knight.

stages: (1) insurgent infiltration, (2) military raid, (3) anarchy, (4) reaction, (5) consolidation, and (6) deliberate recovery operations.[28]

The ISIS-affiliated Maute Group planned to seize Marawi on the first day of Ramadan (May 26).[29] While that date has religious symbolism, it also has a practical effect because citizens typically stock up on extra food supplies for Ramadan. They planned to infiltrate the city, isolate it, and then attack Camp Ranao. They knew that most of the army brigade was located 30 kilometers away fighting communist insurgents in the jungle, so they placed ambushes on the three roads leading into the city along which they expected Philippine reinforcements to come. The Philippine military knew that something was happening but lacked the clarity to truly understand.[30]

The second stage was the raid, which set in motion the subsequent events. The police and the military received intelligence that Isnilon Hapilon would be meeting with Maute Group leaders, brothers Abdullah and Omar Maute, in a safe house in western Marawi (the arrest site on

[28] See, for example, Shibani Mahtani and Regine Cabato, "Philippine forces cleared this city of Islamist militants in 2017. It's still a ghost town," *Washington Post*, February 1, 2019, https://www.washingtonpost.com/news/world/wp/2019/02/01/feature/philippine-forces-cleared-this-city-of-islamist-militants-in-2017-its-still-a-ghost-town/.
[29] Jim Gomez and Todd Pitman, "AP Exclusive: Video shows militants in Philippine siege plot," *Associated Press*, June 6, 2017, https://apnews.com/article/manila-philippines-islamic-state-group-ap-top-news-international-news-8f6649e404964ab4a2d280537c86c83f.
[30] Singh, "Insights for Practitioners," 35.

figure 14.2).[31] The government forces mounted an immediate arrest operation. Although they expected minimal resistance, anticipating maybe a dozen IS-P fighters, the raid included special operations forces, rangers, reconnaissance troops, and an armored fighting vehicle unit. As the military convoy approached the safehouse, a firefight broke out. To the Filipinos' surprise, approximately one hundred militants emerged from buildings all around. As the firefight continued, the insurgent leaders narrowly escaped through a hole in the wall.[32]

The sound of that firing brought on the third stage, which could be characterized as anarchy: anarchy, assaults, atrocities, and ambushes. The fighting triggered a rampage and hundreds of militants from all across Marawi came out from their hiding places. They burned a Catholic church, two schools, and the city jail; beheaded a police officer; and took churchgoers and residents hostage.[33] In some cases they executed civilians who could not recite the *Shahada,* which is an expression of Muslim faith.[34] More than 500 fighters attacked Camp Ranao.[35]

What ensued was pure chaos. Troops that were not actively defending against the attack on Camp Ranao moved to relieve the pinned-down raiding force, only to be ambushed enroute. Some of the forces, including the initial raid force, became pinned down and trapped for several days. The militants ambushed police and army teams responding in armored vehicles.[36] President Duterte, who was in Moscow at the time, declared martial law, the cessation of military operations elsewhere in the country, and sent additional forces to Marawi.[37] Within Marawi, several hundred residents joined the rebellion but most of the population simply bunkered down—some very heroically sheltered their Christian neighbors, as there were a small number of Christians living in Marawi.[38]

---

[31] "What happened in Marawi?" *Al Jazeera,* October 29, 2017, https://www.aljazeera.com/features/2017/10/29/what-happened-in-marawi.

[32] Briefing to Charles Knight and Katja Theodorakis at the Philippine Army Operations Research Center on October 10, 2018.

[33] *Al Jazeera,* "What happened in Marawi?"

[34] Amnesty International, *'The Battle of Marawi': Death and Destruction in the Philippines* (London: Amnesty International, 2017), 13, https://www.refworld.org/pdfid/5a0e99724.pdf.

[35] Marcelo, "Gov't forces."

[36] Knight and Theodorakis, *The Marawi crisis,* 12.

[37] "Duterte declares martial law after Mindanao attack," *Al Jazeera,* May 24, 2017, https://www.aljazeera.com/news/2017/5/24/duterte-declares-martial-law-after-mindanao-attack.

[38] Jonathan Kaiman, "A Muslim landowner in the Philippines took in Christians to protect them from Islamist extremists," *Los Angeles Times,* September 7, 2017, https://www.latimes.com/world/asia/la-fg-philippines-islamic-state-20170907-story.html.

The fourth stage was the reaction: the reaction, the rescues, the exodus, and the containment. Across the city small groups of civilians hid and used mobile phones and social media to plead to the government forces for rescue.[39] Soon, rescue became a top priority. Ad hoc groups of police, soldiers, and volunteers led by civil community leaders attempted to find ways to bring trapped civilians out of the city. This was also the stage when the narrative started to shift. Until this time, most of the city's population believed that the IS-P were their liberators and protectors, but the IS-P started to lose the population's support when the residents saw them setting fire to the church and college and attacking the hospital. It was their fellow residents who were destroying their own city and the community started to worry.[40]

As the uprising continued, the residents' fears increased. While uprisings in Philippine cities were not unusual, they typically lasted only a day or two. With some of the inhabitants fearfully recalling the days of martial law under President Ferdinand Marcos in the 1970s and the brutal tactics used by the military at that time, the population started to flee the city.[41] At the same time that the residents were fleeing, police and military reinforcements started arriving from across the country. The armed forces were trying to fight the insurgents and prevent more fighters from infiltrating while at the same time conducting rescue operations. It was chaotic.

The fifth stage was consolidation: reinforcing the security forces, reframing the narrative, and realigning. By early June, government forces had rescued 1,500 people but 500 hostages remained.[42] Over several weeks, the armed forces and police slowly began to secure the perimeter of the city as a task force was assembled from across the country. From the start, the AFP emphasized both symbolic and very practical actions to support its soft power message and frame the government response.[43] As an example, one of the combat engineer regiments that was sent to support combat operations was re-tasked to construct shelters for the homeless residents

[39] Knight and Theodorakis, *The Marawi crisis*.

[40] Ibid., 23-25.

[41] John Malvar, "Forty-eight years since Marcos declared martial law in the Philippines, World Socialist Web Site, September 23, 2020, https://www.wsws.org/en/articles/2020/09/23/mart-s23.html.

[42] Lt. Gen. Emmanuel Bautista briefing to Knight and Theodorakis at the Philippine Army headquarters on October 12, 2018.

[43] Knight and Theodorakis, *The Marawi crisis*, 20-7.

who had fled.[44] The government believed it was extremely important to be doing something to support the displaced population, even if it was only a small gesture.

Similarly, the government raised what they called the Hijab troopers, 105 women from police and military from around the country that deployed to Marawi as part of an all-female civil relations company wearing modest Muslim dress. In addition to assisting in the implementation of rehabilitation and recovery programs, they helped man checkpoints to conduct the searches of women because that was a potential cultural flashpoint.[45]

Surprisingly, the Philippine government gave significant and unprecedented political authority to local leaders for all non-military decisions to further the narrative that the government was working with the local community.[46] The government also negotiated a short-lived cease fire to facilitate the evacuation of trapped residents. The June 4 ceasefire lasted only four hours, but it allowed the government forces to free 134 residents.[47]

At that time, the government was building the joint task force for subsequent combat operations. The task force consisted of three brigade-sized combined arms task groups—one marine, and two army—each with embedded police units. In addition to the joint task force, the Filipinos created two additional joint task groups, each commanded by brigadier generals. One joint task group was responsible for the cordon and everything inside the city that was not kinetic. The second joint task group was responsible for information operations, managing internally displaced persons, and supporting the civil-military operations. Having the commanders of these

[44] John Unson, "2 model houses for Marawi rehabilitation being built," *Philstar*, July 31, 2017, https://www.philstar.com/nation/2017/07/31/1723249/2-model-houses-marawi-rehabilitation-being-built.

[45] Dempsey Reyes, "AFP, police deploy female troops to Marawi," *Manilla Times*, August 30, 2017, https://www.manilatimes.net/2017/08/30/todays-headline-photos/top-stories/afp-police-deploy-female-troops-marawi/347553; and Lourdes Veneracion-Rallonza, "Sustaining 'Firsts': Women, Peace and Security in the Philippines," Manila University, April 12, 2018, http://ateneo.edu/ls/soss/political-science/news/research/substantiating-%E2%80%98firsts%E2%80%99-women-peace-and-security-philippines.

[46] Discussions between Knight and Theodorakis and the Governor of Marawi City and an assemblyman in Marawi on October 11, 2018.

[47] "Gunfire during ceasefire stops civilians fleeing fighting in Philippines," *Reuters*, June 4, 2017, https://www.theguardian.com/world/2017/jun/04/gunfire-ceasefire-civilians-fleeing-fighting-philippines-marawi-isis.

joint task groups commanded by brigadier generals sent the message that the non-lethal operations were just as important as the lethal operations.[48]

Once that was underway, the Philippine forces moved to the final stage: deliberate recovery operations. The AFP described the approach that it developed for the systematic recovery of the city as "SLICEing": strategize, locate, isolate, constrict, and eliminate.[49] The task force had a challenging mission. They had to conduct a systemic offensive operation, but it was complicated by the fact that they had to limit their offensive firepower, because they had to consider the hostages and a prohibition against damaging mosques.

Spencer: You discussed how this started with militant infiltration and the initial raid by the Philippine military. How much time was there between the martial law being declared and the complete evacuation of the city? The United Nations reported that 98 percent of the city's population was displaced.[50] How long did the evacuation take and was getting them out of the combat area a planned operation?

Knight: The military did not have a deliberate plan for the evacuation. But first, it is important to return to the initial raid. Many have characterized it as a "botched raid," but when you examine it, the plan was competent based on the number of fighters that they expected to find.[51] In fact, despite the initial surprise and the large number of militants that they faced, most of the raiding force survived despite having been ambushed by one hundred fighters. So, their performance during the raid was fairly effective. It just was not the situation, much like the American's experience in Mogadishu in 1993, that the troops expected.[52] While they failed to make the arrest, the raid triggered the insurgents to execute their plan prematurely. This prevented them from establishing checkpoints to keep Marawi's citizens from fleeing. Had that actually happened, the problem would have been many times worse, because the AFP would have had to attack the city with most of its population still in place.[53]

---

48 Briefing at the Philippine Army Operations Research Center on October 10, 2018.

49 Knight and Theodorakis, *The Marawi crisis*, 13.

50 "Marawi Crisis," UNHCR, https://www.unhcr.org/ph/marawi-crisis.

51 See, for example, Romeo Ranoco, "Thousands flee Philippine city after rebel rampage claimed by Islamic State," *Reuters*, May 23, 2017, https://www.reuters.com/article/us-philippines-militants/thousands-flee-philippine-city-after-rebel-rampage-claimed-by-islamic-state-idUSKBN18K08M.

52 Mark Bowden, *Black Hawk Down: A Story of Modern War* (New York: Penguin Books, 1999).

53 Gomez and Pitman, "AP Exclusive."

To return to your question, President Duterte swiftly declared martial law on May 24, the day after the raid. It took the citizens about three days to realize this was not a typical uprising; it was something more. As they witnessed the militants burn buildings and distribute videos executing Christians, the community started to view the IS-P as a threat rather than a protector.[54] So after three days of chaos, small groups started to leave. The mass exodus started soon thereafter and lasted about a week.[55] Opinions differ as to what prompted people to leave but it seems to have been a combination of the declaration of martial law, the realization that a cease fire seemed unlikely, and advice from community leaders.

Spencer: Where did they go?

Knight: Those who could, fled to relatives and friends, the others became the internally displaced persons that we saw on our television screens huddling in makeshift shelters. They moved away from the fighting as best they could, but the terrain around Marawi is fairly rugged and offers only two major evacuation routes out of the city, so most people moved along these routes. Within the city community leaders played a key role in guiding inhabitants away from immediate fighting as well as negotiating with the more amenable groups of fighters to allow safe passage of civilians under white flags. An equally important role in the evacuation was played by former insurgents who were party to the 2014 *Bangsamoro* ceasefire and worked with community leaders to put checkpoints and representatives out along the evacuation routes to guide the fleeing residents.[56]

The internally displaced persons reception areas that were run by the military were further away. This ensured that the military was not conducting these soft power missions inside and close to the city where they were concurrently attempting to both locate and contain IS-P while mounting the occasional rescue patrol into rebel-controlled areas. But what is really significant, as mentioned before, was the prioritization of soft power during this stage. The government gave religious and local leaders unprecedented authority to decide what would happen and then a surprising portion of the limited military capability that was available was

[54] Rohan Gunaratna, "Ending the Fight in Marawi," *Counter Terrorist Trends and Analyses* 9, no. 10 (2017), 1, https://www.jstor.org/stable/pdf/26351557.pdf?refreqid=excelsior%3A4b2ff59294cab160508ac722582e1c8d.

[55] Ranoco, "Thousands flee Philippine city."

[56] Briefing at the Philippine Army Operations Research Center on October 10, 2018.

deployed according to the wishes of the local leaders.[57] Nevertheless, this was all very, very messy.

Spencer: Can you describe the joint task force in a little more detail? What comprised it? What did they marshal together? Was it part of the Philippine military already or did they cobble it together?

Knight: Camp Ranao housed an infantry division headquarters, but there was never more than a single brigade stationed there which consisted of two infantry battalions and an armored battalion. At the start of the battle, most of the brigade was fighting insurgents in the jungle, but they joined the fight within the first few days. For the first week of the battle, only the military and the police that were normally based at Camp Ranao and Marawi were involved, so it was a local fight.[58]

Within four or five days, additional forces started to arrive on the island by ship and it soon became a national fight. The first element to arrive was of regimental-size and it was a composite of different types of units. Over the following weeks, additional army and police units arrived as well as a marine team consisting of several battalions, and they were assembled into composites. The police, artillery, and medical support were sliced to support the three combat task groups (two army and one Marine). The joint task force and task groups that were created, were task organized for the mission at hand and the task organization constantly changed as more resources arrived from elsewhere in the Philippines. Does that make sense?

Spencer: Yes, that makes absolute sense. They had armored units, so did they have main battle tanks?

Knight: No, the AFP did not have main battle tanks. Their armored units contained various types of M113 armored personnel carriers, Simba armored personnel carriers, and a variety of wheeled armored personnel carriers.[59] This offered protection against small arms up to 7.62mm and artillery shell splinters, but the vehicles were vulnerable against more powerful weapon systems.

Spencer: Why did it take five months to clear the city?

Knight: It is useful to break the militants' tactics, techniques, and procedures into two phases. During the initial phase, for the first month or

[57] Discussions with the Governor of Marawi City and an assemblyman in Marawi on October 11, 2018.

[58] Briefing at the Philippine Army Operations Research Center on October 10, 2018.

[59] For more on Simbas and M113s, see, "Simba," Military Today, http://www.military-today.com/apc/simba.htm; and "M113 Armored personnel carrier," Military Today, http://www.military-today.com/apc/m113.htm.

so, the IS-P kept the AFP at bay while that chaotic situation continued and while the AFP built up its combat power. During the second phase, which was during the deliberate recovery operations stage that I described earlier, the IS-P resisted the AFP for four months by using the buildings within the city as strongpoint defenses. Trying to clear a well-prepared enemy from an urban area is almost always a slow fight. So, four months for a city of this size is by no means slow.

For example, it took a marine battalion several weeks to cross one of the bridges and secure the far side while suffering very heavy losses in the process. Why was that? One significant factor was situational dominance. The militants not only had the advantage of local knowledge, but they also employed hobby drones pretty effectively.[60] So, they knew the AFP positions and could deploy to engage them. The AFP, by contrast, completely misjudged the size of the enemy force and, at least in the early stages, lacked good intelligence about their composition and disposition.[61]

The militants also conducted extremely effective anti-vehicle ambushes in the first few days, deterring further mounted attacks. The police and army sent in pairs of armored vehicles to relieve their trapped colleagues. The militants set up chicanes, an artificial narrowing or turn on a road, or roadblocks at ambush sites to slow the vehicles, and then once the vehicles had slowed or stopped, the militants engaged them with rocket-propelled grenades (RPGs).[62] While armed with older RPG-2s, which were fielded by the Soviets in the 1950s, they were still capable of penetrating up to 20 centimeters of rolled homogeneous armor, so they could easily penetrate any of the armored vehicles in the AFP inventory.[63]

After disabling vehicles with RPGs, the militants often threw petrol bombs and charges onto the vehicles from the buildings above. At the same time, other militants shot out the vehicle's optics.[64] Often, the IS-P fighters filmed the attack for propaganda purposes, capturing the panicked crew rotating the vehicle's turret and firing wildly. The police teams were all killed. One of the army teams lost several people to RPGs in the first strike,

---

[60] Tom Allard, "One week to cross a street: how IS pinned down Filipino soldiers in Marawi," *Reuters*, September 25, 2017, https://www.yahoo.com/news/one-week-cross-street-pinned-down-filipino-soldiers-060001842.html.

[61] Singh, "Insights for Practitioners," 35.

[62] Knight and Theodorakis, *The Marawi crisis*, 14.

[63] For more on RPG-2s, see, "RPG-2," Military Today, http://www.military-today.com/firearms/rpg_2.htm.

[64] Briefing by veteran junior officers to Knight and Theodorakis 'on the ground' in the Marawi main battle area on October 11, 2018.

but the platoon commander managed to evacuate the rest of the platoon into a nearby building. But even with air and artillery support, it still took five days for the AFP to rescue the stranded element.[65] So, early in the battle, the armored response was knocked out.

The AFP quickly recalled its forces from the jungle. The unit force marched from operations 30 kilometers away, briefly paused at Camp Ranao to pick up ammunition, and then went straight into battle. As they pushed out on foot, they encountered sharpshooters positioned to provide enfilading fire down the narrow streets, which soon stopped their advance. The better trained insurgents could provide effective fire from 100 meters or more. Less skilled fighters, to include prisoners who the militants had freed and armed at the start of the battle, were assigned to the flanks to provide defilade fire on the Filipinos as they attempted to evade the enfilading fire.[66]

Often, the militants were unable to place particularly effective fire on the AFP, but the combination of being pinned in the street and the roving militant groups often overwhelmed the AFP troops. In one particular incident, a Marine company attempting to secure the Maipandi bridge was ambushed by a group of four militants who initiated the attack with an improvised explosive device (IED) and then threw hand-launched bombs amongst the Marines as they took cover. In the ambush and subsequent firefight the Marines suffered 53 casualties including 11 dead, effectively wiping out the company.[67]

Despite the setbacks and the heavy casualties, the AFP continued to push. But the AFP was fighting a sophisticated opponent, one who was also employing information operations to support their lethal operations. The militants collected videos, transmitted messages with the videos to the Middle East where ISIS processed, edited, and released the videos in a matter of hours. Some videos showed hostages pleading with President Duterte to halt the attack.[68] At the same time, local officials protected only by white helmets began to go into IS-P areas to rescue trapped civilians while religious leaders at different times attempted to negotiate with the

[65] Knight and Theodorakis, *The Marawi crisis*, 14.

[66] Ibid., 12.

[67] Briefing by veteran junior officers 'on the ground' in the Marawi main battle area on October 11, 2018.

[68] See, for example, Paterno Esmaquel II, "Propaganda shows Marawi priest echoing kidnappers' demand," *Rappler*, May 30, 2017, https://www.rappler.com/nation/171413-propaganda-shows-marawi-priest-echoing-kidnappers-demand/.

militants.[69] So, it was an extremely challenging environment; not only was the tactical fight difficult, but at the same time, the AFP did not want to compromise the ongoing rescue operations.

Early in the battle, the AFP simply lacked the resources necessary to defeat or dislodge IS-P from the city. Even after recalling the troops from the jungle, a brigade sized element was not enough for a city the size of Marawi. The limited capability, the tactical uncertainty, and the political ambiguity created a tactical pause which allowed the military to assemble more forces. During this time, the military isolated the main battle area, the population fled, and rescues continued. The importance of the rescue operations is enormous because this was where the military won the battle of the narratives. It changed the way the community perceived the armed forces and the police, because they witnessed the Filipino forces trying to rescue their friends and relatives. The bulk of the rescue operations occurred during the first month of the battle.

In addition to using ambushes effectively, the IS-P also leveraged the subterranean environment to their tactical advantage. Once again, this was not unique to Marawi and has occurred throughout the history of urban conflict. Prior to the battle, in addition to *buho* shelters, Marawi already had a fairly extensive tunnel network, dating back to at least the 1970s when tunnels were dug to allow religious leaders to move around the city and avoid arrest by the government forces.[70] But the IS-P expanded the tunnel network, often forcing their hostages to dig tunnels to connect to existing tunnels and bunkers.[71] They used the cellars of the mosques to store munitions and hold hostages, because they correctly calculated that the AFP would not target religious buildings.[72]

The militants also used the mosques for protection since the AFP would not fire on them. When I was in Marawi, young officers showed me a number of enemy positions. They explained that every time that they tried to attack a mosque, they would suffer heavy casualties because

[69] "'White Helmet' team a ray of hope for trapped Marawi residents," ABS-CBN News, May 30, 2017, https://news.abs-cbn.com/news/05/30/17/white-helmet-team-a-ray-of-hope-for-trapped-marawi-residents.

[70] Discussions with the Governor of Marawi City and an assemblyman in Marawi on October 11, 2018.

[71] "Lessons from Marawi," Indo-Pacific Defense Forum, March 23, 2020, https://ipdefenseforum.com/2020/03/lessons-from-marawi/.

[72] Decemay P. Padilla, "300 hostages still held in Marawi mosque," *Philstar*, July 21, 2017, https://www.philstar.com/headlines/2017/07/21/1720323/300-hostages-still-held-marawi-mosque; and Indo-Pacific Defense Forum, "Lessons from Marawi."

they were forbidden from using high explosives or anything that would cause considerable damage. They were only successful when employing CS gas, a type of riot control gas, to incapacitate the militants.[73] In contrast, the AFP had an easier time seizing structurally similar buildings where they could use their full inventory of weapons and munitions.

Snipers posed another tactical challenge for the AFP in Marawi. Once again, snipers are fairly common in urban combat, yet they came as a complete surprise to the AFP. The term sniper gets bandied around quite a lot to mean anybody who is shooting from any distance. To be clear, a sniper is a highly trained marksman, who can consistently hit a target from several hundred meters away and, working with an observer, fire one or two shots from a well concealed position and then move.[74]

When I was there, I examined some of the sniper positions, including some in mosques, and some of the firing ports were extremely small, no larger than the end of a paperback book. Even from inside the room it was difficult to see some of the small firing ports if you were not looking for them. The likelihood of spotting a firing port from hundreds of meters away, from the street, or from the window of another building, was extremely small. That was what the militants did, but the AFP did not expect it. Yet, the sniper counter-sniper fight was a familiar aspect in Stalingrad as well as other urban battles, so they should have expected it.[75]

To bolster their defenses, the militants often placed command detonated IEDs outside of buildings. As troops clustered outside to assault or as armored vehicles moved in to protect the assault force, the militants would detonate the IED. The IED and RPG threat relegated the armored vehicles to providing only limited support, often serving primarily in a casualty evacuation role.[76] It is important to note that although they had only "thin-skinned" armored vehicles, even a main battle tank is vulnerable in this type of urban combat, because an enemy equipped with RPG-7s could easily penetrate a tank when engaging a tank from above, such as from a second story window, where the tank's armor is thinnest. Furthermore,

---

[73] CS, or chlorobenzylidenemalononitrile, is a riot control agent used by the military. Knight and Theodorakis, *The Marawi crisis*, 16; and briefing by veteran junior officers on October 11, 2018.

[74] "Sniper," Military History, https://military-history.fandom.com/wiki/Sniper.

[75] Anthony Beevor, *Stalingrad: The Fateful Siege: 1942-1943* (New York: Penguin Publishing Group, 1999), 203.

[76] Briefing at the Philippine Army Operations Research Center on October 10, 2018.

many of the streets in the main battle area were too narrow to allow the use of modern tanks.

Inside the buildings, the militants placed IEDs in doorways, stairwells, and ceilings. That made any normal entry too risky. Troops entering through the existing apertures were risking death and in the early stages of the fight, many soldiers died this way, before they learned to blow holes through walls and ceilings and to enter from another direction.[77]

The militants also employed the tactic of "hugging the enemy." This tactic, dating back to at least Stalingrad, involves staying in close proximity to the enemy to prevent them from employing their indirect fires or airdropped munitions.[78] However, the militants modified this tactic slightly to counter the AFP's methods. For example, the AFP conducted pre-assault fires on a building prior to entering. Upon hearing the aircraft overhead or seeing artillery tubes moved into position for direct fire engagement, the militants would withdraw to the neighboring building through mouseholes. Once the pre-assault fires had ceased, the militants would return through the mouseholes, staying "one wall away" from the likely AFP assault, and then fire or throw grenades on the unsuspecting AFP troops through loopholes once the assault began.[79] So, it was often an extremely close proximity fight.

At the operational level, the three combat task forces slowly squeezed the IS-P militants into a smaller and smaller space within the city. The Marine task force group approached from the river to the northeast and the two army task force groups approached from the west and northwest, with all three pushing south towards the lake (see figure 14.3).

Clearing the city was a very slow process and it took the Filipinos at least a month to learn best practices of how to fight in an urban environment. By the later stages of the fight, they not only had their organic assets, but they also had foreign air support, foreign special operations forces advisors, and they were employing the latest urban tactics, to include ones that had been refined in the Middle East only two weeks prior.[80]

[77] Knight and Theodorakis, *The Marawi crisis*, 16.

[78] Russell W. Glenn et al., *Denying the Widow-Maker: Summary of Proceedings of the RAND-DBBL Conference on Military Operations on Urbanized Terrain* (Santa Monica, CA: RAND, 1998), 12, https://www.rand.org/pubs/conf_proceedings/CF143.html.

[79] Information comes from inspection of IS-P positions while being briefed by junior officer veterans in Marawi on October 11, 201811 Oct 2018.

[80] Neil Jerome Morales and Simon Lewis, "U.S. joins battle as Philippines takes losses in besieged city," *Reuters* June 10, 2017, https://www.reuters.com/article/uk-philippines-militants-usa-idUKKBN191066.

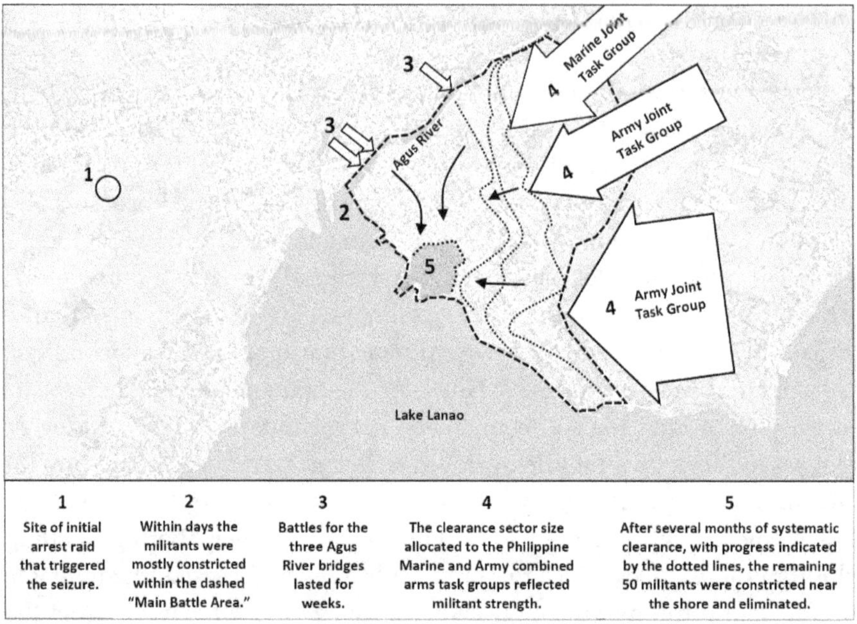

Figure 14.3. The Battle of Marawi, May – November 2017.
Source: Map created by Liam Collins and Charles Knight.

By the latter stage of the battle, the AFP were basically conducting repeated "deliberate stronghold assaults." This label was popularized by British and Australian special forces to describe the careful planning, preparation, and close quarters tactics used in rare one-off recovery operation against holed-up terrorists. But in Marawi, the AFP were eventually conducting one stronghold assault after another, at a rate of one building per unit, per day.[81] Deliberate assaults of strongholds take time.

The first phase of a deliberate stronghold assault—reconnaissance and planning—was usually the longest. The AFP conducted detailed reconnaissance. They observed the target building from every possible angle, making extensive use of drones when they became available. They attempted to identify enemy positions, booby traps, and the withdrawal routes (because the route was likely to be free of booby traps), and the best options for assault entry. After conducting the necessary reconnaissance, they developed the assault plan.

[81] Briefing by veteran junior officers 'on the ground' in the Marawi main battle area on October 11, 2018.

The second phase was establishing overwatch of the target to observe the target prior to the assault and to provide supporting fires during the assault. In training this is often treated quite glibly because the focus is often on the assault phase. Special forces tend to be reasonably good at urban overwatch, because they have snipers who are regularly trained to perform this task in a counterterrorist context, but conventional forces are rarely familiar with the techniques required, especially the coordination required to continue firing after the assault teams have entered the building.

The urban battle is often fluid with the enemy fighting from a neighboring room or on a floor above or below, and the overwatch element needs to be able to provide support to protect the assault force once it has entered. This requires detailed fire control methods which take time to plan. Commanders may need to spend an hour or two talking to the supporting fire and overwatch elements to ensure each sniper, machine gunner, or grenadier understands all their fire control measures: where to shoot, where not to shoot, when to start, when to stop, and so on. Failing to do this can easily result in friendly casualties.

The third phase was the assault which was preceded by high explosive munitions into the building. As the battle continued, the Filipino's ideal method was to hit each room with high explosive direct fire prior to entering. They had learned that a 105mm artillery round detonating in a room would trigger most IEDs, typically without collapsing the building. Yet much of the time the obstruction of other buildings and walls meant that they could not position artillery tubes for direct fire into every room, so they would often have to use aircraft to drop bombs, which had been their initial approach. However, the aircraft were less preferred because the enemy could hear the aircraft approach, flee the building to avoid the bombing, and then return once the bombing had stopped.[82] Also, aircraft bombs caused more friendly fire casualties, through misguidance or mistargeting, than direct fire artillery, and significantly much more destruction to the building.

The other, less obvious, problem with the bombs was that the soldiers would sometimes get too complacent. They would see the destruction from aerial delivered bombs and think, "no one could survive that." It is tough to blame them. They showed me photos from one objective, where two 250-kilogram bombs hit the middle of a five-story building and detonated on the second or third floor. The holes in the concrete were massive,

[82] Briefing at the Philippine Army Operations Research Center on October 10, 2018.

everything was shattered, and rubble had fallen onto the lower floors. Yet somehow, at least one insurgent survived the bomb blasts and was able to kill a couple of assaulters.[83] Thus, firing a 105mm artillery round or detonating a significant amount of high explosive into every space before entering should certainly not be considered overkill.

During or after the pre-assault fires or bombings, the AFP would create an entry breach. In the early months it was often by hand, or with sledgehammers.[84] Breaching walls is something that some in the military talk about but rarely practice, and it is a lot slower and more difficult than many think. As the Filipinos discovered, an alert enemy had time to move and engage through mouseholes with small arms or explosives. Consequently, as the means became available, AFP used engineers to place explosive breaching charges to breach walls, but this still required someone to expose themselves as they moved to the wall or to the contact point to place the charge. To reduce that vulnerability, the AFP learned to use a combined arms approach and used their armored vehicles to provide ballistic breaching and intimate fire support.

To exploit their armored vehicles the Filipinos developed some fascinating improvisations and tactics. Discovering that the M113s quickly became stuck in rubble and realizing that the vehicles lacked the tractive power to clear rubble when fitted with their own dozer blades, the AFP installed home-made armor, often including hardwood timber, on bulldozers to push a path through the rubble. This allowed the M113s to move forward, engage quickly, and then withdraw before the insurgents could redeploy anti-armor weapons against them. They took the turrets from unserviceable Scorpion tanks to create 76mm armed fire support vehicles.[85]

Perhaps most remarkably, they reshaped the urban battlefield so that armored vehicles could fight from inside buildings. To achieve this, the Filipino engineers constructed substantial ramps leading into the upper stories of buildings. This allowed the M113 and other armored vehicles to maneuver up into and within suitable buildings to engage the enemy

[83] Briefing by veteran junior officers 'on the ground' in the Marawi main battle area on October 11, 2018.

[84] Briefing to Knight and Theodorakis at 103 Infantry Division headquarters on October 11, 2018.

[85] The FV1010 Scorpion is a "light tank, or armored reconnaissance vehicle by British classification" that "was developed in the late 1960s." Its main gain was 76.2 mm. See, "FV101 Scorpion," Military Today, http://www.military-today.com/tanks/fv101_scorpion.htm.

in adjacent structures and streets while relatively well protected from RPG fire.[86]

Then, and only then, would they attempt the assault. After years of combat in Afghanistan and Iraq, most Western armies know how to conduct close quarters combat within buildings. Certainly, Australia training terms sent to the Philippines focused on room and corridor combat as well as first aid and light demolitions.[87] However, it is important to note that the Filipinos combined close quarters combat with an increasing role of close infantry firepower, especially light, medium, and heavy machine guns firing through walls, rifle launched grenades and, towards the end when they became available, 40mm "long" high explosive grenades fired from Chinese versions of the RPG-7.[88] The Filipinos essentially relearned the World War II lesson that if everything preceding the assault is done well, then you should be picking your way through rubble and enemy corpses rather than having an attritional fight inside the building.

Of course, the assault is not over when the building has been cleared, it must be reinforced. Again, this lesson dates back to at least World War II. Today in training, the units from most militaries will capture a house, quickly reorganize, and then move to clear the next one. That was the tactic that the AFP started with, but they realized that it was a deadly mistake against insurgents who were happy to lose their lives—well not happy, but you understand—sought to take as many people with them as possible.

Given the opportunity, the enemy would rapidly counterattack inside buildings. So, the AFP realized that immediately after the first wave assaulted, another wave should be sent in to reinforce. That second group brought in sacks of rubble and soil to fortify the secured area by blocking loopholes and mouseholes before the unit moved to clear the next part of the building. After the building was cleared, they would bring in defense stores and many more prefilled sacks and prepare the newly captured

---

86 Briefing to Knight and Theodorakis at 103 Infantry Division headquarters on October 11, 2018; and Modern War Institute, "MWI War Council – Fighting Terrorism in the City, The Battle of Marawi," YouTube video, 51:07-51:36, November 19, 2019, https://youtu.be/txUdnu1shf8?t=3041.

87 James Lewis, 'The battle of Marawi: Lessons for developing urban capabilities," *Australian Army Journal* 15, no. 1 (2019), 31-51, https://search.informit.org/doi/epdf/10.3316/informit.528977125865626.

88 "Philippines starts transition to RPG-7 infantry rocket launchers," *Asia-Pacific Defence Journal*, January 31, 2019, https://www.asiapacificdefensejournal.com/2019/01/philippines-starts-transition-to-rpg-7.html.

building as a protected fighting position before continuing the advance.[89] This is why it would typically take about 24 hours to clear a single enemy strongpoint.

What I have described was the general AFP tactic, but each unit executed it slightly differently based on their equipment, experience, and capability. For example, units with good night vision goggles often conducted their assaults at night to gain an advantage over the militants.[90]

As well as the tactical destructive battle, we should consider the soft power battle of narratives that was concurrently being waged. Lieutenant General Emmanuel Bautista, who commanded the operation, told me that he considered it equally important.[91] No matter the tactical outcome of the lethal battle, if the manner in which it was waged inspired the resumption of a wider Moro insurgency, then it would have been considered a strategic loss.

Here it might be useful to compare the slow strategic success of Marawi to the strategic disaster of the first battle of Fallujah in Iraq in 2004. The core issue was clearly not destruction. If you compare pictures of Marawi to pictures of Fallujah after the 2004 battle, it is immediately clear that the level of destruction in Marawi was much worse. In both the first and second Battles of Fallujah, while the insurgents committed similar or worse abuses against the civilian population, the U.S. forces retained a reputation amongst the Iraqi population of being heavy handed or for committing abuses, notwithstanding greater efforts to limit collateral damage than in Marawi.

Both battles involved a small minority of hardcore militants within the city: IS-P in Marawi and al Qaeda in Iraq in Fallujah. In both cities the attacking forces were initially viewed as alien invaders and their arrival dreaded. But in Fallujah, the United States actions generated public support for the resistance locally and across the Middle East while in Marawi, IS-P appears to have suddenly lost the sympathy and support amongst most of their potential constituency. Why was Marawi different? Why did IS-P not achieve its strategic objective of inspiring the thousands of armed veteran Moro insurgents to join them in "liberating" Mindanao when Fallujah appears to have encouraged insurgent recruitment fighters across Iraq?

---

[89] Briefing by veteran junior officers 'on the ground' in the Marawi main battle area on October 11, 2018.

[90] Ibid.

[91] Bautista briefing on October 12, 2018.

Maranao community leaders and politicians told us that the sudden outburst of brutality, vandalism, and hostage taking by the IS-P militants was shocking and alienating. They witnessed the Maute brothers destroy the college where they were educated, which contradicted the brothers' claims to be building an Islamic society. The influx of militants who were outsiders, especially from Abu Sayyaf Group, was not seen as Islamic solidarity but as a malign influence. Turning against the militants, however, did not mean they would embrace the AFP whom they had long feared and distrusted.[92] These leaders said that their attitudes changed slowly as a result of watching the behavior of the troops risking their lives to save hostages and as well as a very different approach from AFP officers as they implemented a policy of empowering established community leaders as far as possible.

The question of whether the failure of IS-P to trigger a wider Moro uprising was a Filipino success or a militant failure is debatable. However, it seems clear that the insurgents created a window of opportunity in the first "battle" in a communications campaign that allowed the Philippine government to prevail over the militants while rehabilitating the reputation of their own army.[93] Yet, it is important to remember that this battle never ends. If the government does not rebuild the city, then the population may turn to the insurgents in the future, just as some of Iraq's population started to support ISIS in 2014 after it rejected al Qaeda in Iraq years earlier.

It seems that the relationship between the conduct of the tactical kinetic battle and the information campaign was key to the AFP's success. Counterintuitively, it is possible that AFP tactical setbacks may have provided the opportunity for Filipino domestic information operations to shift the narrative of the battle. While the length of time taken to recapture Marawi has been widely criticized, the first hesitant and then cautious way in which the Philippine government conducted the operation was consistent with a declared policy of minimizing risk to hostages and giving militants scope to surrender.

Very visibly, the AFP used traditional influence methods like loudspeakers and banners to communicate with the insurgents and hostages as well as actions like dropping messages in plastic bottles from helicopters. This messaging, which was echoed and amplified in the

---

[92] Discussions with the Governor of Marawi City and an assemblyman in Marawi on October 11, 2018.

[93] Knight and Theodorakis, *The Marawi crisis*, 20-7.

mainstream media, was initially designed to demoralize and defeat the insurgents, and to give hope to the hostages. However, it also reinforced a broad narrative emphasizing themes of hostage plight, AFP heroic rescues, and AFP reform.[94]

The delay before the systematic recapture of the city could begin was caused by several factors including initial tactical setbacks, the need to wait for reinforcements, and the initial necessity of involving local and community leaders in the rescue and evacuation efforts. This delay serendipitously created opportunity. It allowed for an increasing empowerment of local and traditional leaders to lead welfare efforts for the internally displaced persons. The political signaling this achieved seems to have shifted *Maranoan* attitudes. It was arguably a successful operational level information operations effort, whether by design or accident.

At the strategic level, the government attempted to influence the international media. The Filipinos absolutely understood that their influence operations were just as important as their lethal operations. They applied martial law, but differently than they had in the past. The task force commander who was responsible for the welfare effects made it a priority to get on television to explain what they were doing and to get out of the headquarters to personally engage with relevant stakeholders.[95]

Fortuitously the AFP had created a social media operators' course just as the battle was unfolding, so they were able to leverage 80 newly trained social media operators in their information support organization to influence the enemy and their hostages using wider social media. What these operators did, and the fact that they were allowed to do it, was impressive. It seems that the young Filipino soldiers were trusted and given virtual carte blanche to get on the internet, hack, change the messaging, and do whatever their country needed.[96] Unlike practice in Western armies, they did not need a senior officer to sign off before releasing each message.

To counter the ISIS message that the soldiers were engaging in brutal tactics, the Filipinos put helmet cameras on their own soldiers, and pieced together combat footage showing their soldiers killing the IS-P militants. Interestingly, and in contrast to most Western practice, they took a robust approach to countering the IS-P messaging portraying the AFP as engaged in wanton destruction. Rather than emphasizing good intentions with their

94 Ibid.

95 Briefing at the Philippine Army Operations Research Center on October 10, 2018.

96 Bautista briefing on October 12, 2018.

targeting and lamenting collateral unavoidable error, they sought instead to portray "reluctant determination to do whatever it takes."[97]

Notably, they provided exciting first-person videos to the media of the military in close combat, aggressively but surgically engaging the militants. The overall message was one of resolve, but for the civilian population and international audiences, it was intended to convey the difficulties of urban combat and create empathy for the soldiers. For the IS-P, the message was, "We are not intimidated by your brutal execution videos, rather you have made us determined, and we can make equally compelling footage showing how we will kill you." This suggests that Filipinos understood that narrative messaging showing very human soldiers unapologetically killing might better counter anti-AFP messaging than spokespeople expressing regret about collateral deaths or damage.[98]

Spencer: I have studied many urban battles and Marawi seems unique. Battles in other urban areas might have periodic deliberate assaults on fortified structures, but in Marawi, it was constant. The AFP was forced to conduct a very deliberate clearing from the outside-in, almost squeezing the enemy. I do not know if it was the construction of the buildings, how the enemy built its defenses to create a fortified structure in every building, how the AFP conducted the attack, or a combination of the three, but it resulted in a massive amount of damage. Correct me if I am wrong, but I seem to remember seeing that 95 percent of the buildings in the main battle area were damaged, with more than 50 percent destroyed.[99] Has the city been repopulated yet?

Knight: No, the city has yet to be repopulated and that remains contentious. Three years after the battle, more than 120,000 residents remained displaced.[100] The politics surrounding rebuilding are complicated, but if this is not managed, then it seems likely that the AFP could find itself fighting more insurgents in Mindanao in the near future. It is not encouraging.

---

[97] Quote is paraphrase from briefing by veteran junior officers 'on the ground' in the Marawi main battle area on October 11, 2018.

[98] Knight and Theodorakis, *The Marawi crisis*, 20-7.

[99] Ervin Malicdem, "Aftermath of the Battle of Marawi," *Shadow Expeditions*, November 30, 2017, https://www.s1expeditions.com/2017/11/223-marawi-battle-structures.html.

[100] Ashley Westerman, "Over 120,000 People Remain Displaced 3 Years After Philippines' Marawi Battle," NPR, October 23, 2020, https://www.npr.org/2020/10/23/925316298/over-120-000-people-remain-displaced-3-years-after-philippines-marawi-battle.

Spencer: What about the tactics, techniques, and procedures of the fight itself? Could the AFP have employed different tools—such as flamethrowers, specialized vehicles, and specialized weapons—to accomplish the mission without creating the same level of destruction? It seems like the AFP fired high explosive munitions into nearly every building.

Knight: The Filipinos learned that they needed to try to put high explosive munitions in every room prior to clearing it or place their troops at great risk during the assault. Of the weapons they had available, the most suitable was their 105 mm towed howitzers which they used in a direct fire mode to fire into buildings. Some did not have direct fire sights, so they created improvised aiming devices taking 105mm diameter sized plastic bottles, cutting the end off and taping crossed threads over the end and attaching them to the artillery tube—aiming this way was remarkably accurate, much more accurate than indirect fire from afar.[101] It is worth mentioning that their most flexible howitzer was the old OTO Melara Mod. 56 pack howitzer which was designed to be broken down into parts that can be carried by a mule. Thus, it could be easily disassembled and carried around buildings by soldiers.[102]

But the Australian Army, as well as many other Western armies, no longer have light artillery like this. So, from that perspective, the AFP were actually better armed for this particular fight than many other armies. Whether by accident or deliberate design, they were able to use direct artillery firepower successfully as part of their combined-arms system. As a result of this battle, the AFP decided to pursue a new armored vehicle. There was some discussion about acquiring more larger artillery guns as well, but they are going to stick with the 105mm because of its utility in urban fighting.

I think their use of bulldozers was pretty ingenious as well. The armored personnel carriers could not progress through the rubble-filled streets. They discovered that they needed bulldozers to clear the debris, but they lacked armored bulldozers, so they installed improvised armor on regular bulldozers to create the capability.[103] The AFP also employed

---

[101] Briefing to Knight and Theodorakis at 103 Infantry Division headquarters on October 11, 2018.

[102] For more on the OTO Melara Mod. 56, see, "OTO Melara Mod. 56," Military Today, http://www.military-today.com/artillery/oto_melara_mod_56.htm.

[103] Briefing at the Philippine Army Operations Research Center on October 10, 2018.

drones and electronic warfare assets effectively, but specific details remain classified.

Returning to Bismarck's quote, "Only a fool learns from his own mistakes. The wise man learns from the mistakes of others." There are many other aspects of the battle that one can learn from: the difficulties of medical evacuation in an urban environment, the huge demand on ammunition, the use of sappers to blow pathways and to clear obstacles, the use of armored vehicles inside buildings, and the use of improvised armor. The Filipinos even tried installing wooden armor on their thin-skinned vehicles for added protection, although the benefit was probably more psychological than physical.[104]

The key point is that there was little that was truly new in Marawi; many of the lessons that the AFP learned could have been learned by studying previous urban battles. But the AFP, like many militaries, was unprepared for urban combat. So, the biggest lesson to learn from Marawi is that militaries must be ready for urban warfare.

Spencer: I have one final question. You mentioned that they used tear gas liberally and with fairly good success to clear enemy-held buildings. But the United States largely failed to employ tear gas in its counterinsurgency campaigns in Afghanistan and Iraq. Why are riot control gasses not used more often? It seems to put the attacker, defender, and civilians at less risk.

Knight: That is a vexing question. Western nations are more likely to employ tear gas and other riot control agents for internal policing operations than they are to employ them in war due to fears of violating the Chemical Weapons Convention. Certainly, the situation in Iraq was ambiguous enough to prevent Western nations from using irritant gasses. Unfortunately, that often forced them to use lethal means when nonlethal means might have been better, so that is what I mean by a vexing question.

When I last visited the International Committee of the Red Cross in Geneva to discuss this very issue, it was clear that they do not want to backtrack on the prohibitions despite the dilemma. While it is often a more humane and better tactical option, there is a real fear that it would create a slippery slope where more deadly chemical weapons might be employed. That said, smoke weapons will, in many situations, have the same effect, and certainly that is the Russian's understanding.

[104] Kyle Mizokami, "These ISIS-Fighting Philippine Tanks are Clad in DIY Wooden 'Armor'," *Popular Mechanics*, June 7, 2017, https://www.popularmechanics.com/military/weapons/a26804/wooden-armor-tank-rockets/.

The Russians have the shoulder launched RPO-series of weapons. While the RPO-A, which has a thermobaric warhead gets the most attention, they also have an incendiary warhead, the RPO-Z, and a smoke-producing warhead, the RPO-D, both of which are quite nasty.[105] The Russian contemporary battle drill is to shell the hell out of the target, drive up in an MT-LB multi-purpose armored vehicle, exit out the back, and then fire a number of these shoulder-launch weapons into the target to eliminate the threat prior to entering the building.[106]

Spencer: Thank you, I know that is a vexing question. You know where I stand on the issue, I have always felt we should use these non-lethal munitions more often, and when I heard that the AFP used them to great effect, that only enhanced my belief. It seemed like they were able to employ riot control agents in a manner that protected their force, protected hostages or civilians that were present, minimized collateral damage, while at the same time offered a less lethal, but just as effective, way to subdue the enemy. Well, Charles, this has been a fascinating conversation. I know people are going to love it, and I really appreciate your time and expertise in giving it to us.

Knight: Thank you! I really enjoyed chatting. We really should share the lessons that the Filipinos offer us. Thanks, John.

[105] For more on RPOs, see, "RPO-A Shmel," Military Today, http://www.military-today.com/firearms/rpo_a_shmel.htm; "RPO PDM-A Shmel-M," Military Today, http://www.military-today.com/firearms/rpo_shmel_m.htm; and "RPO Rys," Military Today, http://www.military-today.com/firearms/rpo_rys.htm.

[106] For more on the MT-LB, see, "MT-LB," Military Today, MT-LB Multi-Purpose Armored Vehicle | Military-Today.com; and M. A. Kharitonov et al., "Improvement of flamethrower Incendiary weapons of the troops," *Journal of NBC Protection Corps* 2, no. 2 (2018).

15

The Battle of Shusha

The Nagorno-Karabakh region has been contested since Armenia and Azerbaijan first declared their short-lived independence in 1918. The Soviet Union kept tensions subdued during the Cold War, but the First Nagorno-Karabakh War broke out in 1989 as the Soviet Union was dissolving. The war lasted until 1994 and resulted in a decisive victory for the Karabakh Armenians. The Armenians of Nagorno-Karabakh declared an independent state, the Republic of Artsakh, though it remains unrecognized by the rest of the world. Also, as a result of the conflict, Armenia gained control of seven Azerbaijan regions that surrounded Nagorno-Karabakh (see figure 15.1).[1]

Over the next 25 years, the contested region experienced low level conflict that occasionally flared up, but it never escalated to large scale war and control over Nagorno-Karabakh and the seven surrounding territories remained unchanged until the 2020 Nagorno-Karabakh War. Leading up to the war, Armenia, the poorer of the two nations, failed to modernize its military and was still equipped with "legacy Soviet-era weapons, some obsolete."[2] Azerbaijan, by contrast, was able to use some of the wealth that it had gained from oil and gas exports to modernize its military. It improved its special forces and officer corps and heavily invested in modern weaponry ranging from cyber capabilities to anti-tank guided munitions, to unmanned aerial vehicles.[3]

Azerbaijan initiated the war on September 27, 2020, and by early November, Azerbaijani forces had reached the outskirts of Shusha, a small city in Nagorno-Karabakh's west. Shusha not only controlled the road

---

[1] Svante E. Cornell, *The Nagorno-Karabakh Conflict* (Uppsala, Sweden: Department of East European Studies, 1999), 3-23.

[2] Uzi Rubin, *The Second Nagorno-Karabakh War: A Milestone in Military Affairs* (Ramat Gan, Israel: Begin-Sadat Center for Strategic Studies, 2020), 6, https://besacenter.org/nagorno-karabakh-war-milestone/.

[3] Ibid.

Figure 15.1. Map of Nagorno-Karabakh and the surrounding area, 2016.
Source: © Wikimediacommons, https://commons.wikimedia.org/wiki/File:Nagorno-Karabakh_Map2.png.

through the Lachin corridor, a pass that connects Armenia with Nagorno-Karabakh, but its higher elevation offered it a commanding position over Stepanakert, the capital of the self-proclaimed Republic of Artsakh.

The two nations had fought over the city in 1992, but this time found themselves in opposite roles. By November 9, Azerbaijan forces had seized Shusha. With Shusha lost, Armenia sued for peace and the war ended the following day. As a result of the six-week war, Azerbaijan regained control of its seven regions and portions of Nagorno-Karabakh, but at a heavy cost of life. The Armenians lost an estimated 2,317 soldiers during the war, while the Azerbaijanis lost 2,783.[4]

In this chapter, John Amble, the editorial director for the Modern War Institute at West Point, interviews Colonel (California) John Spencer. Spencer is the chair of urban warfare studies at the Modern War Institute, co-director of the Urban Warfare Project, host of the *Urban Warfare Project Podcast*, and director of urban warfare training for the 40th Infantry Division of the California Army National Guard. In July 2021, he co-authored a report titled "The Battle of Shusha City and the Missed Lessons of the 2020 Nagorno-Karabakh War." Because of that report, he was invited to

[4] Paul Iddon, "The Last Azerbaijan-Armenia War Changed How Small Nations Fight Modern Battles," *Forbes*, May 25, 2021, https://www.forbes.com/sites/pauliddon/2021/03/25/the-last-azerbaijan-armenia-war-redefined-how-small-nations-fight-modern-battles/?sh=66691b037dd3.

Azerbaijan and Shusha, Nagorno-Karabakh and in November 2021 he traveled to Nagorno-Karabakh to study the war.

John Amble: You have studied this particular battle of Shusha that occurred during the 2020 Nagorno-Karabakh War very closely. You were arguably the first American military–affiliated researcher to visit and walk the ground where the fighting occurred. Before the trip, you had written an article arguing that this fight for a city was a decisive battle in the war.[5] You have also written that it holds lessons about urban warfare in the future. This war seems to have generated several sweeping and sometimes categorical declarations about the future of war, especially things like unmanned systems and other technologies, as well as the relevance of legacy platforms. For example, some have argued that the war demonstrated that the age of armor is over, while others have argued that the war demonstrated that armor is more relevant than ever.[6] So, I would like to start by asking what led you to the conclusion that there are some urban lessons that can be learned in this war that largely took place in fairly sparsely populated, mountainous areas.

John Spencer: This was only a six-week war, but it generated much discussion. I watched the war as it unfolded and the debates that followed but I did not want to weigh in until I had time to really research the conflict. I was, however, very skeptical of claims that armor is not important in contemporary conflict. In every battle examined in this book, to include battles as recent as 2016-2017's battle of Mosul and 2017's battle of Marawi, armor was found to be extremely important. I had planned to travel to the region in the summer of 2021, but the trip was canceled as both countries were closed to international visitors due to COVID. I was finally able to get into Azerbaijan in November 2021. I was excited to get on the ground to see and study the war. After my field research, I concluded that the battle of Shusha was a decisive battle with strategic implications. Other researchers

[5] John Spencer and Harshana Ghoorhoo, "The Battle of Shusha City and the Missed Lessons," Modern War Institute, July 14, 2021, https://mwi.usma.edu/the-battle-of-shusha-city-and-the-missed-lessons-of-the-2020-nagorno-karabakh-war/.

[6] See, for example, Benjamin Brimelow, "A brief, bloody war in a corner of Asia is warning about why the tank's days of dominance may be over," *Business Insider*, November 24, 2020, https://www.businessinsider.com/drones-in-armenia-azerbaijan-war-raises-doubt-about-tanks-future-2020-11; and Jon Hawkes, Sam Cranny-Evans, and Mark Cazalet, "The tank is dead. Long live the tank," *Wavell Room*, October 1, 2020, https://wavellroom.com/2020/10/01/a-critical-analysis-of-the-future-of-the-tank/.

had focused on the role that unmanned aerial vehicles played in the war, but the strategic role that Shusha played was largely ignored.[7]

Amble: Your article does a good job of contextualizing the battle. Can you provide some of that context to the readers and explain why you believe Shusha to be key terrain in this region?[8]

Spencer: The greater Shusha region, consisting of the city of Shusha and ten surrounding villages, sits within the heart of Nagorno-Karabakh. The city has been at the center of Azerbaijani and Armenian fighting for centuries and it holds cultural significance to both groups. Shusha is often described as the "cradle of Azeri culture," but it also holds Armenian religious and cultural history, serving as a fortress city in the Middle Ages and still home to famous Armenian sites such as the Ghazanchetsots Cathedral, the church of Kanach Zham, and a monastic convent.[9]

The city has changed hands multiple times, most recently in 1992 when Armenia seized the city on its way to a larger victory. At that time, the Shusha area was estimated to have contained 1,600 Armenians and 23,000 Azerbaijanis.[10] During the battle, most of the city was destroyed and much of the Azerbaijani population was killed, expelled, or fled. At the start of the 2020 war, the city's population, almost all ethnic Armenians, totaled roughly 5,000.[11]

---

[7] See, for example, John Antal, "The First War Won Primarily with Unmanned Systems: Ten Lessons from the Second Nagorno-Karabakh War," John Antal, 2021, https://higherlogicdownload.s3.amazonaws.com/AUVSI/7bc57aaa-ae26-4c7a-93f9-512dc4a1bca0/UploadedImages/Ten_Lessons_from_the_2d_Nagorno-Karabakh_War_by_John_Antal_2021-03-08F.pdf. For example, Uzi Rubin calls Shusha a "strategic city" in his assessment of the war, but his analysis focuses largely on the role of unmanned aerial vehicles. Rubin, *The Second Nagorno-Karabakh War*, 8.

[8] The United States Army defines key terrain as "an identifiable characteristic whose seizure or retention affords a marked advantage to either combatant." Department of the Army, *Offense and Defense*, ADP 3-90 (Washington, DC: Headquarters, Department of the Army, 2019), 3-10, https://armypubs.army.mil/epubs/DR_pubs/DR_a/ARN18010-ADP_3-0-000-WEB-2.pdf.

[9] Francis Farrell, "The battle for Shusha: the cauldron of generational pain at the heart of the Nagorno-Karabakh war," *New Eastern Europe*, November 12, 2020, https://neweasterneurope.eu/2020/11/12/the-battle-for-shusha-the-cauldron-of-generational-pain-at-the-heart-of-the-nagorno-karabakh-war/.

[10] Elchin Amirbayov, *Shusha's Pivotal Role in a Nagorno-Karabakh Settlement* (Cambridge, MA: Belfer Center, 2001) https://www.belfercenter.org/publication/shushas-pivotal-role-nagorno-karabagh-settlement#session.content_root#documents/CSP6.pdf.

[11] Thomas de Waal, "Shusha Armenians Recall Their Bittersweet Victory," Institute for War & Peace Reporting, May 10, 2002, https://iwpr.net/global-voices/shusha-armenians-recall-their-bittersweet-victory; and Colin Freeman, "War tips favour Azerbaijan as Army captures key town in Nagorno-Karabakh," *The Telegraph*, November 8 2020, https://www.telegraph.co.uk/news/2020/11/08/war-tips-favour-azerbaijan-army-captures-key-town-nagorno-karabakh/.

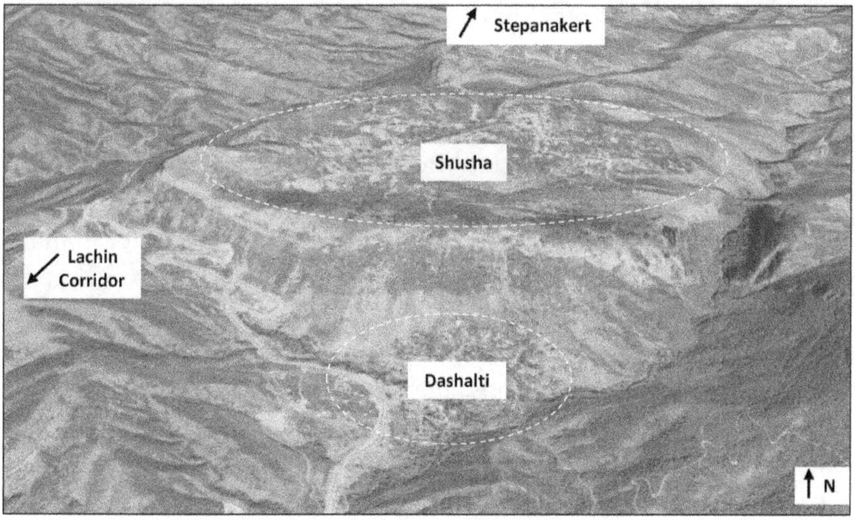

Figure 15.2. Three-dimensional view of Shusha, 2022.
Source: Map created by Liam Collins.

The city is a formidable natural fortress. Steep cliffs drop off from the edges of the city on three sides and a single road passes north to south through the city (see figure 15.2). Stepanakert, the largest city and capital of the self-proclaimed Republic of Artsakh, is fifteen kilometers to the north along this winding road. But the straight-line distance between the two is just a few kilometers, and Shusha has the advantage of sitting several hundred meters above Stepanakert in elevation. Shusha's location serves as an ideal defensive buffer zone for Stepanakert.[12]

Shusha also serves as a critical defensive node along the Lachin corridor, the primary mountain pass for the road that links Armenia to Nagorno-Karabakh. Given its cultural significance and key geographic advantage to both Azerbaijan and Armenia, controlling Shusha was a major objective of both parties during the 2020 conflict.[13] Thus, Shusha is a key and strategic terrain for the region.[14] It is geographically advantageous terrain that controls the primary route—or line of communication in military vernacular—into the region, and it was politically and culturally vital to the objectives of both parties in the war.

[12] Spencer and Ghoorhoo, "The Battle of Shusha City."
[13] Ibid.
[14] Tural Hamid, "Free Shusha: The road to victory – Part I," Aze.media, November 10, 2021, https://aze.media/free-shusha-the-road-to-victory-part-i/.

Amble: I know you have also highlighted how Shusha's high elevation offers a commanding position for artillery to control the lines of communication through the Lachin corridor and the capital city. At what point did Shusha become a factor in the war?

Spencer: Retaking Shusha was one of President Ilham Aliyev's strategic objectives—he wanted to reclaim the "cradle of Azeri culture" that his nation had lost nearly 30 years earlier—so it was a factor from the onset of the war, but it took several weeks for Azerbaijani forces to reach the city.[15] Shusha was also an operational objective due to its location and its ability to control the road through the Lachin corridor and commanding position over Stepanakert. Thus, the city became the target of artillery attack as early as the first or second week of the war.[16]

The war kicked off on September 27, 2020, when Azerbaijani forces crossed the line of contact. When Azerbaijan lost the first war, they ceded control of seven of its territories that surrounded Nagorno-Karabakh. The Armenians retained control of these territories following the first war to create a buffer zone around Nagorno-Karabakh. Since the Azerbaijani forces entered territory that was internationally recognized as belonging to them, yet controlled by Armenia, it was called a line of contact as opposed to a border.

Azerbaijani forces crossed the line of contact to the south of Nagorno-Karabakh. Within days they had penetrated 15 kilometers into the Armenia-controlled territory and had reached the cities of Jabyrili and Fazuli.[17] By October 7, Azerbaijani forces had reached Hadrut, a town in southern Nagorno-Karabakh that sits along one of the few roads that leads from the open plains in the south to Shusha. Although only 38 kilometers in a straight-line to Shusha, Hardut is 71 kilometers along the road through the wooded mountainous terrain.[18] The Azerbaijanis encountered significant resistance at Hardut, and it took the Azerbaijan forces until the end of October to finally advance to the outskirts of Shusha.

---

[15] "Ilham Aliyev was interviewed by Turkish A Haber TV channel," President.AZ, October 16, 2020, https://president.az/en/articles/view/43250.

[16] "Armenia: Journalists injured in alleged artillery strike on cathedral in Shusha (Nagorno-Karabakh) October 8 /update 9," *Crisis 24*, October 8, 2020, https://crisis24.garda.com/alerts/2020/10/armenia-journalists-injured-in-alleged-artillery-strike-on-cathedral-in-shusha-nagorno-karabakh-october-8-update-9?origin=fr_riskalert.

[17] Joshua Kucera, "As fighting rages, what is Azerbaijan's goal?" *Eurasianet*, September 29, 2020, https://eurasianet.org/as-fighting-rages-what-is-azerbaijans-goal.

[18] Dimitri Avaliani, "Karabakh: the battle for Hadrut and why it's important," *JAM News*, October 12, 2020, https://jam-news.net/karabakh-war-hadrut-azerbaijan-armenia/.

On their advance toward Shusha, the Azerbaijanis executed combined arms combat brilliantly, effectively using advanced technologies to engage air defense assets, tanks, and troops while also maneuvering their own tanks, troops, engineers, and other assets into position. The Azerbaijanis were the attackers so crossing the wide-open areas of the line of contact required very disciplined combined arms maneuver against an entrenched defending enemy. Artillery and armor specifically were critical to their ability to cross.

Amble: Eventually the fight for the city transitioned from artillery strikes to one that involved conventional forces fighting in the Lachin corridor outside Shusha. Special operations forces also contributed to the seizure of the city. Can you discuss that a little?

Spencer: In late October, a major battle was being waged for the terrain around Shusha, I call it the Lachin offensive. The Azerbaijani dismounted forces were slowly advancing north towards Shusha. If you step back and look at the war in late October, it shows an impressive campaign plan incorporating many of the elements of operational art that we teach in our professional military education institutions like West Point and the U.S. Army War College.[19] An entire army front was fighting north along the western edge of Nagorno-Karabakh towards the Lachin corridor, another massive force was facing a large Armenian force northeast of Nagorno-Karabakh near the city of Agdham, and there were still forces fighting in the central region near the city of Hadrut.[20] The Azerbaijanis were advancing along major lines of communications and seizing key terrain enroute to hold their gains and to continue their advance.

At some point between October 26 and October 30, the Azerbaijani military inserted at least two battalions of special forces soldiers north of Hadrut. Some reports show that one of these elements included approximately 400 special forces soldiers. All of them covertly infiltrated up the center of Nagorno-Karabakh and into the region around Shusha

---

[19] The ten elements of operational art include: end state and conditions, center of gravity, decisive points, lines of operations and lines of effort, basing, tempo, phasing and transitions, culmination, operational reach, and risk. Department of the Army, *Operations*, ADP 3-0 (Washington, DC: Headquarters, Department of the Army, 2019), https://armypubs.army.mil/epubs/DR_pubs/DR_a/ARN18010-ADP_3-0-000-WEB-2.pdf.

[20] Richard Giragosian, "Azerbaijan's next move will make or break Karabakh war," *Asia Times*, October 26, 2020, https://asiatimes.com/2020/10/azerbaijans-next-move-will-make-or-break-karabakh-war/.

and the Lachin corridor.[21] Over the next five days, this force split into four groups of approximately 100 personnel each. These groups traversed the rugged, mountainous terrain with heavy backpacks to occupy key positions to isolate and infiltrate Shusha City. As I mentioned earlier, the city is surrounded by cliffs on three sides with the road through the Lachin corridor, winding its way up to the city from the south.

One of the special forces elements walked, climbed, and scaled up the 300-meter-high cliffs on the southeastern side of the city's edge. It is a story as old as war itself. The Armenian-backed forces believed the cliffs and paths on the city's edge to be unscalable, even though they, the Armenians, had scaled the same cliffs in 1992 when they had captured the city, so they left them unguarded. The Shusha defenders seemed to believe that the Azerbaijani forces would mainly attack up the main road leading into the city from the Lachin corridor. They were unaware that this massive, dismounted commando force had encircled and was infiltrating the city. The Azerbaijani special forces used the cover of darkness, harsh terrain, ongoing Lachin offensive, and failed imagination of the defenders to make their decisive move and covertly isolate and pierce the Shusha defense through seemingly unpassable mountainous terrain.

Amble: It is striking that the defending force failed to remember its own history of scaling the cliffs and left them unobserved and undefended. Hindsight is 20/20, of course, and we have the luxury of analyzing this battle well after it was fought, but this tactical oversight produced strategic consequences. So, special forces appeared to have scaled these cliffs and occupied positions where they could isolate the city. What other operations were going on to support the attack of the city?

Spencer: On October 30, military clashes were reported just five kilometers east of Shusha signaling that the Azerbaijan forces were nearing the city.[22] Around November 4, Azerbaijan infantry forces reached Dashalti, a small village at the base of the cliffs, just south of Shusha.[23] The village controlled the road into Shusha, so both forces fought for control of the

[21] "We passed through Shusha through Armenians, they didn't know," qafqazinfo.az, December 3, 2020, https://qafqazinfo.az/news/detail/susaya-ermenilerin-icinden-kecib-getdik-xeberleri-olmadi-xtd-zabitivideo-307243.

[22] "On October 30, military clashes were reported just five kilometers east of Shush," *EurAsia Daily*, October 30, 2020, https://eadaily.com/ru/news/2020/10/30/minoborony-armenii-azerbaydzhanskie-podrazdeleniya-otbrosheny-ot-shushi.

[23] Tural Hamid, "Free Shusha: The road to victory – Part I." Aze.media, November 10, 2021, https://aze.media/free-shusha-the-road-to-victory-part-i/.

village. The Azerbaijanis also increased their artillery barrages. Azerbaijan engineers reportedly cut a path from Hadrut toward Shusha—the same route that special forces had taken—to bring artillery forward.[24]

This was an extremely challenging tactical fight for the Azerbaijanis. The Armenian forces still held the tactical advantage as they still controlled most of the wooded areas and mountains surrounding Shusha. As Azerbaijani forces advanced on Shusha from Dashalti, they were exposed to ambushes, artillery, and rocket attacks and suffered heavy losses.[25]

Within Shusha there were approximately 2,000 defenders who were fairly well armed with mortars, artillery, armor, and multiple launch rocket systems. They attempted to hold their ground but were forced to slowly withdraw through a series of ambushes to delay and attrit the advancing Azerbaijanis. A key moment in the 1992 battle of Shusha occurred when the attacking Armenians tricked the defending Azerbaijanis to move out of the city. The Armenians attacked the small villages surrounding Shusha, to include Dashalti, and the Azerbaijanis came out of the city to defend the villages. From a tactical perspective, it played into the Armenians' hand. The Azerbaijanis left their defensive positions in a difficult urban environment to counterattack in more open space.

Thus, in 2020, it was the Azerbaijanis turn to fight through the surrounding villages in their advance to Shusha. This time, however, the defending Armenians did not leave the city to counterattack. Instead, they laid a series of ambushes to attrit the advancing Azerbaijanis. It was a tough fight in which the Azerbaijanis no doubt suffered significant casualties but they slowly advanced, bringing their mortars and rockets to the edge of the town.[26]

Amble: It is the first week of November, the fifth week of the war, when the Azerbaijani forces reached Dashalti. When did the fight inside the Shusha begin?

Spencer: The first week of November was pretty intense. The reporting is fairly confusing, but it appears that Azerbaijani forces surrounded Shusha and seized control of the Lachin corridor, which prompted the Armenian

[24] Interview with Azerbaijan officer in Nagorno-Karabakh, November 7, 2021.

[25] "The battle for Shusha Fighting in Nagorno-Karabakh has reached a turning point. Here are the most recent developments in the conflict zone," *Meduza*, November 6, 2020, https://meduza.io/en/feature/2020/11/07/the-battle-for-shusha.

[26] There is almost no reporting and no official reports on casualties, so it is tough to know precise or even estimated casualty figures.

government to close the road through the pass. It also left 80 reporters, and no doubt many others, trapped inside Nagorno-Karabakh.[27]

Some of the Azerbaijani special operations forces reached the main road leading into Shusha from the north and blocked reinforcements attempting to aid in Shusha's defense. This, combined with the destruction of a key bridge over the Hakari River, meant little, if any, aid could come to Shusha as the town was now surrounded on at least three sides.[28] Once Dashalti fell, the Azerbaijanis had Shusha surrounded and completely cut off (see figure 15.3).

By November 5, Azerbaijani troops had reached the outskirts of the city. Reports surfaced that Argishti Kyaramyan, the former head of the Armenian National Security Service who had been appointed the military commander responsible for organizing the defense of Shusha, had left the city.[29] It was a precarious situation for the Armenian-backed forces.

On November 6, Azerbaijani forces finally entered the city and a close urban fight ensued. The special forces elements that had infiltrated up the cliffs started engaging the city's defenders using the urban terrain to mask their movement. By the time the defenders realized that the city had been infiltrated, it was too late. The Armenian-backed forces could not employ artillery, mortars, and other large guns against the infiltrators because the infiltrators were too close to their own troops. In urban warfare the tactic of getting so close to the enemy that they cannot use their heavier weapons is called "hugging the enemy." The battle of Shusha, like many urban battles, quickly turned into house-to-house fighting. Azerbaijan infantry forces cleared buildings while at times Armenian tanks were just 50 feet away in the courtyard of a neighboring building.

On the evening of November 7, fog rolled in which provided a temporary advantage to the Armenian-backed forces. By this point, the Azerbaijanis had firmly established air superiority which provided them

---

[27] "Turkish weapons, 80 stranded journalists – situation in Karabakh, November 5," *JAM News*, November 5, 2020, https://jam-news.net/turkish-weapons-in-karabakh-war-news-uavs-turkey-azerbaijan-armenia-aliyev/.

[28] Natalia Konarzewska, "Armenia and Azerbaijan sign Nagorno-Karabakh ceasefire deal brokered by Moscow," *New Eastern Europe*, November 20, 2020, https://neweasterneurope.eu/2020/11/20/armenia-and-azerbaijan-sign-nagorno-karabakh-ceasefire-deal-brokered-by-moscow/; and "Azerbaijan surrounded the city of Shushi from three directions overnight – the assault has already begun," *Avia.Pro*, November 5, 2020, https://avia-pro.net/news/azerbaydzhan-za-noch-okruzhil-gorod-shushi-s-s-tryoh-napravleniy-shturm-uzhe-nachalsya.

[29] "Media report: Argishti Kyaramyan left the city a few hours before the Azerbaijani sabotage group entered Shushi," *Armenian Report*, December 5, 2020, https://armenianreport.com/ru/pubs/267102/.

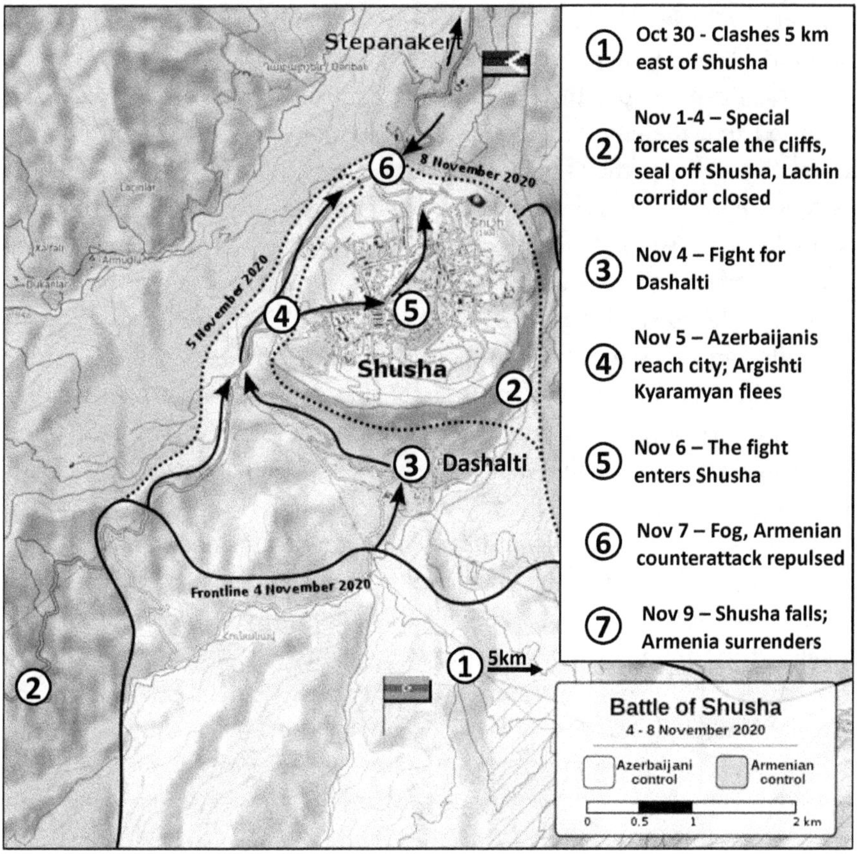

Figure 15.3. The battle of Shuhsa, November 2020.
Source: © Wikimediacommons, https://commons.wikimedia.org/wiki/File:Battle_of_Shusha_(2020).svg. Description of battle added to the figure by Liam Collins and John Spencer.

a tactical advantage with their unmanned aerial system fleet.[30] The fog neutralized this tactical advantage, because many of the Azerbaijani aerial strike, and intelligence, surveillance, and reconnaissance assets were not all-weather capable. Thus, the Azerbaijanis could not employ their TB-2 drones, for example, that had been so effective up until that point of the war. As a result, Armenian forces were able to maximize the use of their

[30] Air superiority is defined as being able to conduct air operations "without prohibitive interference by the opposing force." See, Phillip Meilinger, "Supremacy in the Skies," *Air Force Magazine* 99, no. 1 (2016), 46, https://www.airforcemag.com/PDF/MagazineArchive/Magazine%20Documents/2016/February%202016/0216supremacy.pdf.

armored vehicles—T-72 tanks and BMP-2 infantry fighting vehicles—to support major counterattacks attempts against the city.[31]

Despite the fog, the Azerbaijanis controlled more of the city and reported that the city was theirs. The Armenians, however, continued to claim the city was theirs, so it was very confusing. Around this time, the Armenian-backed forces launched three major counterattacks into the city from the heavily wooded areas outside the city. It was absolutely critical for them to regain the city, but it was equally important for the Azerbaijanis to repel the attacks, so the fighting was very heavy. Ultimately, the Azerbaijanis prevailed and after repulsing the counterattacks, the Azerbaijanis returned to clearing Shusha. They captured the building of Shusha's executive power, the prison, and a number of other key locations. They rooted out or destroyed the remaining Armenian defenders, deliberately clearing buildings throughout the city.[32]

On the morning of November 8, the Azerbaijanis once again claimed that they controlled the city, and once again, the Armenians refuted that claim.[33] The following day, Azerbaijan President Ilham Aliyev declared a total victory in Shusha and over the Armenians. This is why I argue that the battle of Shusha was a strategic battle, because once Shusha fell, the Azerbaijanis declared victory in the war. After Shusha fell, Armenia agreed to a peace deal that allowed Azerbaijan to keep all the land that it had seized during the war.[34] The Armenian-backed forces controlled little more than Nagorno-Karabakh's capital. Once they lost the key terrain of Shusha, which overlooks the capital city, they had no choice but to surrender or face what was certain to be a brutal urban fight that would destroy much of Stepanakert and cause significant civilian casualties.

The Armenian-backed government may have believed that the Azerbaijanis were going to start launching rocket attacks into the capital city and move forces in to attack it. They knew that they could not hold out if that happened and rather than risk devastation to the city, they

[31] Spencer and Ghoorhoo, "The Battle of Shusha City."

[32] Mircavidan Niftullayev, "Interview with Colonel Tehran Mansimov. First time," YouTube video, January 29, 2021, https://www.youtube.com/watch?v=P6l4R4-4QVo&t=2620s.

[33] Associated Press, "Azerbaijan forces seize control of city in conflict with Armenia," *New York Post*, November 10, 2020, https://nypost.com/2020/11/08/azerbaijan-forces-seize-control-of-city-in-conflict-with-armenia/.

[34] "Armenia, Azerbaijan and Russia sign Nagorno-Karabakh peace deal," BBC News, November 10, 2020, https://www.bbc.com/news/world-europe-54882564.

Figure 15.4. Armenian controlled territory in Nagorno-Karabakh after the 2020 war.
Source: Map modified by Liam Collins from original map created by Wikimedia Commons user Golden, © Wikimedia Commons, https://en.wikipedia.org/wiki/Nagorno-Karabakh_conflict#/media/File:Nagorno-Karabakh_conflict.png.

surrendered because Shusha had fallen. As a result, the Armenians lost the seven territories they had gained back in 1992, as well as portions of Nagorno-Karabakh (see figure 15.4).

Many analysts point to the Azerbaijani use of drones, anti-guided tank missiles, and other advanced military technologies as keys to their victory, but few to none discuss the strategic importance of Shusha. No doubt, those all contributed to the victory, but it was ultimately the loss of this urban battle that forced the Armenians to the negotiation table.

Amble: When we talk about centers of gravity—a center of gravity only becomes such if both sides essentially agree that it is, and that it is worth fighting over. And the capital, Stepanakert, clearly was a center of gravity for both sides. But then, by extension, Shusha's strategic location made it a proxy center of gravity—taking Shusha effectively made the defense of the capital untenable, which meant that for all intents and purposes, the war was over once Shusha was captured.

If we could go back in time to when we did not know the outcome of the battle, I think it is useful to look at certain factors that existed before the war. Azerbaijan had certain advantages. It had a defense budget nearly three times that of Armenia.[35] Azerbaijan's defense budget clearly dwarfs the resources Artsakh could marshal to an even greater degree. Azerbaijan also had arguably more highly trained special operations forces, as well as air superiority because of the fleet of unmanned aerial vehicles that it was able to fly. Azerbaijan had a number of advantages at the war's onset.

But the city's defenders had a number of advantages, as well, and there are many cases throughout history where a much larger force has been defeated by the smaller one.[36] The city sat atop a hill, so they had advantageous terrain. We can look at war throughout history and see how many lives have been lost trying to take the top of a hill. Moreover, this particular hill has almost sheer cliffs on three sides, a major obstacle for any attacker that would seemingly be insurmountable if the cliffs were covered with direct or indirect fire.

If I were to ask you which side was in a better position to win the fight for the city in September 2018, would it be the defending force or the attacking force?

Spencer: That is easy. The defense is a stronger form of fighting than the attack.[37] But knowing how the fight turned out, it is difficult to answer that without the bias of hindsight. Michael Kofman, a senior research scientist at the Center for Naval Analysis' Russia Studies Program, believes that the modernization of the Azerbaijani military played a critical role in their success. Turkey provided material support, in terms of weapon systems, and non-material support, such as training the Azerbaijani special forces, over the course of many years.[38] Given this modernization, I think the Azerbaijanis would have eventually won based on their qualitative advantages, but I think it would have been much costlier if the Armenian-backed forces had defended Shusha better.

Of course, one thing to consider is that sometimes nations feel pressure to end wars prematurely. So had the Armenian-backed forces defended

[35] In 2018, Armenia spent $523 million for defense while Azerbaijan spent $1,854. Data comes from the "SIPRI Military Expenditure Database," Stockholm International Peace Research Institute, https://sipri.org/sites/default/files/Data%20for%20all%20countries%20from%201988%E2%80%932020%20in%20constant%20%282019%29%20USD%20%28pdf%29.pdf.

[36] For example, see the 1973 Battle of Suez City. John Spencer and Jayson Geroux, "What We Can Learn From the 1973 Battle of Suez City: An Urban Warfare Project Case Study," Modern War Institute, January 13, 2022, https://mwi.usma.edu/what-we-can-learn-from-the-1973-battle-of-suez-city-an-urban-warfare-project-case-study/.

[37] Carl von Clausewitz, *On War* (Princeton, NJ: Princeton University Press, 1989), 84.

[38] Michael Kofman, "A Look at the Military Lessons of the Nagorno-Karabakh Conflict," *Russia Matters*, December 14, 2020, https://www.russiamatters.org/analysis/look-military-lessons-nagorno-karabakh-conflict.

better, and the battle dragged on longer, maybe the United Nations, the United States, or some other nation might have pressured the Azerbaijanis to halt the war before seizing Shusha. Or maybe Russia would have felt compelled to support Armenia. The Armenian-backed forces definitely could have designed a better defensive plan and held the city longer.

I turn to the 1991 Battle of Vukovar to consider how a small number of determined defenders in a city can impact a war. The city of Vukovar sits on a main supply route on the eastern border of Croatia. In May 1991, a large Yugoslav People's Army force taking part in a full invasion of Croatia hit a figurative wall in the small city of Vukovar when a force of 2,000-3,000 Croatian national guard, police officers, and civilian volunteers defended the town to prevent the Yugoslav People's Army from passing.[39] The defenders held out for over two months despite intense daily shelling and close urban combat. While the defenders eventually lost, they arguably exhausted the Yugoslav People's Army, and it was the culminating point of the war. A ceasefire was declared only a few weeks later.

It is hard to say what could have happened both tactically and strategically had the Armenian backed forces defended better, but I strongly feel after visiting the city that it could have been defended for a long time.

To be clear, the Azerbaijanis did not simply rely on superior technology and firepower, they designed and executed a superior plan. This was not a 2003 march to Baghdad: an overwhelming joint force that penetrated or bypassed anything that was in its way. The terrain in Nagorno-Karabakh would not allow that. The Azerbaijanis relied on deception, intelligence, and the sheer grit of dismounted and motivated soldiers to defeat a defender that occupied superior terrain. For the Azerbaijanis, the battle of Shusha was well-planned and well-executed.

Amble: Shusha is a city, but a relatively small one. It only had a population of about 5,000, so it is probably more accurate to call it a town.[40] Regardless, it is a built-up, urban area. Given its size, are there limitations on the lessons that can be learned from this battle relating to urban warfare?

---

[39] Mario Šebetovsky, "The Battle of Vukovar: The Battle That Saved Croatia," Thesis (United States Marine Corps Command and Staff College, 2002), https://apps.dtic.mil/sti/citations/ADA407751.

[40] There is no universally agreed size to distinguish between a town or village and a small city, for example. The is an effort to classify as containing at least 50,000 inhabitants, but this is not universally accepted. Lewis Dijkstra, "How do we define cities, towns, and rural areas?" World Bank Blogs, March 10, 2020, https://blogs.worldbank.org/sustainablecities/how-do-we-define-cities-towns-and-rural-areas.

Do you have to be careful about extracting too many lessons that cannot necessarily be applied to a city of a million people? Or are there enough structures and people to make the lessons important for urban warfare on a much larger scale?

Spencer: I often get pushback for calling Shusha a city, to include pushback from peers that reviewed early drafts of my initial report. First, it is important to remember that the Shusha region had a population of around 25,000 residents before the First Nagorno-Karabakh War when the city changed hands in 1992.[41] So while there were only 5,000 residents living in the city in 2020, and most evacuated when the fight began, there were still the buildings and infrastructure to support a small size city. And to be frank, the definition of what constitutes a city varies greatly in urban studies and around the world. For example, in the United Kingdom all that is required to denote an urban area as a city is a royal decree.[42]

There are many factors beyond population size that are used to distinguish a city from something that is not a city. Even a small urban area can have huge implications for a nation, region, and in war. What is important is that Shusha, whether you want to call it a town or a small city, was a strategic urban area that contained a population, dense man-made features like narrow streets, multi-story buildings, and complex infrastructure to support a population. So, yes, I do think there are urban lessons that can and should be learned from this battle, and they can be applied fairly universally.

The first lesson is that cities remain tactical, operational, and strategic objectives in war. If the Azerbaijanis did not believe that Shusha offered military value, they would have simply bypassed it. Throughout history, cities have been strategic objectives and that seems to be the case in this war as well. As long as cities have strategic value, armies will fight for them and in them. That does not mean that every city has strategic value, but some do. Many cities are located along major lines of communication—ancient crossroads and ports that still provide significant value to the civilization of the region and nation—and that is why they were built there and why they grew. Controlling a city, thus, often means controlling lines

[41] Elchin Amirbayov, *Shusha's Pivotal Role in a Nagorno-Karabakh Settlement* (Cambridge, MA: Belfer Center, 2001), https://www.belfercenter.org/publication/shushas-pivotal-role-nagorno-karabagh-settlement#session.content_root#documents/CSP6.pdf.

[42] John Spencer, "The Destructive Age of Urban Warfare; Or, How to Kill a City and How to Protect It," Modern War Institute, March 28, 2019, https://mwi.usma.edu/destructive-age-urban-warfare-kill-city-protect/.

of communication and this offers tactical and possibly operational value. In this case, Shusha controlled the route through the Lachin corridor that connected Armenia with Nagorno-Karabakh.

Cities often represent immense cultural value. Jerusalem is one obvious example, but even a city as small as Shusha may be culturally significant. As mentioned earlier, Shusha has been described as the "cradle of Azeri culture" yet houses Armenian cultural sites such as the Ghazanchetsots Cathedral, the church of Kanach Zham, and a monastic convent.[43] Thus, a city could offer little tactical value—in this case Shusha was tactically valuable—yet still offer significant cultural or strategic value. The Azerbaijanis lost the city in 1992, yet never gave up the idea of reconquering it. Shusha may be small, but the battle of Shusha demonstrates that control of a city, regardless of its size, may have tactical, operational, or strategic value. After the Azerbaijanis captured the city, the Armenian-backed government sued for peace.

A second lesson is that a full suite of modern, joint force capabilities is needed to seize and hold decisive urban terrain. While drones have a role in modern war, urban warfare still requires combined arms: infantry, armor, special forces, engineers, and artillery. The terrain dictates that urban warfare will be a close fight. My fear is that people will learn the wrong lessons from this war and overvalue the roles that unmanned aerial vehicles will play in future wars, just as we overvalued the role of networked systems, information management, and precision guided munitions following the Gulf War. It led to the concept of net-centric war, believing the United States could dominate its enemies with sensors and stand-off precision guided munitions.[44] Sure, that might work in the open desert, but those capabilities were of relatively little value when fighting in the streets, alleys, and buildings of Fallujah or Ramadi. I am afraid that may happen again with this war, an overemphasis on one particular technology.

A third lesson is that militaries must prepare for both urban offense and defense operations. The Azerbaijanis created a solid and bold plan to

[43] Francis Farrell, "The battle for Shusha: the cauldron of generational pain at the heart of the Nagorno-Karabakh war," *New Eastern Europe,* November 12, 2020, https://neweasterneurope.eu/2020/11/12/the-battle-for-shusha-the-cauldron-of-generational-pain-at-the-heart-of-the-nagorno-karabakh-war/.

[44] Christopher M. Schnaubelt, "Whither the RMA?" *Parameters* 37, no. 3 (2007), 95-107, https://press.armywarcollege.edu/parameters/vol37/iss3/24/?utm_source=press.armywarcollege.edu%2Fparameters%2Fvol37%2Fiss3%2F24&utm_medium=PDF&utm_campaign=PDFCoverPages.

seize the city, but they still had to execute it. They obviously trained and were prepared for the mission. Yet at the same time, they had to quickly transition from offense to defense to hold off the Armenian-backed forces counterattacks. Had they been unable to repel the counterattacks, all their gains would have been lost. By contrast, the battle demonstrated the Armenian-backed forces inability to build an effective defense that leveraged the existing natural obstacles like the cliffs and limited avenues into the city. Despite this extremely poor decision, the Armenian-backed forces were able to attrit the Azerbaijanis through a series of ambushes as they fell back to defend from inside the city. They did not make the same mistake that the Azerbaijanis made in 1992 by leaving the more defensible terrain of the city to fight in the villages.

Those are just three lessons that can be learned, or that are reinforced, from the battle of Shusha and I think they apply to almost any war.

Conclusion

This book has demonstrated that no setting is more challenging for militaries than the urban environment. No form of combat is more inherently destructive than urban combat. Yet militaries find themselves unable to avoid being pulled into brutal urban fights and, when they are, they often find that they are unprepared for the challenges of urban warfare.

We believe that this book justifies these claims. Yet, we also believe that this book will, at least to some extent, lessen the challenge that the urban environment poses to militaries by providing different ways to examine and understand it, from feral to smart cities. Likewise, we believe the cases that were examined illuminate some of the many lessons that, far too often, have to be relearned each time a unit finds itself engaged in urban combat.

Understanding the Complex Operational Environment of a City

The first part of the book is intended to provide readers with different ways to examine and understand cities. While each chapter has its own unique focus, some themes emerge.

The dual trends of population growth and urbanization mean that more and more people are living in cities, and as these metropolises grow, they are becoming increasingly more complex. By 2030 there will be 706 cities with more than one million residents and 43 urban areas with more than ten million residents.[1] This creates opportunities for many residents, but it also has the potential create problems that may be orders of magnitude greater than they have been in the past.

As municipalities grow, so do the challenges for militaries executing operations to achieve tactical, operational, and strategic objectives within them. Cities grow in size and density. They grow vertically both up

[1] Population Division, *The World's Cities in 2018* (New York: United Nations, 2018), https://digitallibrary.un.org/record/3799524?ln=en.

with skyscrapers and down with subterranean facilities and tunnels. As Dr. Richemond-Barak noted, the maze of underground spaces in cities pose tactical and even legal challenges that militaries must be prepared to deal with.

Cities are growing so large that some may have already eclipsed the importance of the nation in which they reside. States' monopoly of violence and their ability to govern and control their urban areas has eroded over time. As Dr. Sassen noted, the emergence of multi-national corporations and the corporate services complex has created global cities, which has resulted in power and wealth being transferred from the state to multinational corporations. There are 70 such organizations that have revenue that exceeds the gross domestic product of all but the top 30 nations.[2]

Globalization has produced "winners" like New York, Paris, London, and Tokyo that rank at the top of "globalness" lists. Yet not every municipality has been able to adjust to this rapid growth. Dr. Norton coined the term feral cities for population centers that have not adjusted to the challenges they have faced. Like Dr. Sassen, he finds that states have ceded power in the financial and economic sectors; yet he finds that in feral cities, they have also ceded power in governance, services, and security, and often lack a healthy relationship with civil society. His taxonomy examines these factors and provides a useful tool to classify cities on a feral spectrum.

Population centers on the healthier end of the spectrum tend to be smart cities—ones that leverage technology to improve government services and the quality of life for residents. Militaries must understand the four layers that comprise the smart city system: the physical systems layer, the communications layer, the data analytics layer, and the human and services layer. While global cities are often the "smartest" urban areas, even municipalities that trend toward the feral end of the spectrum have some smart city systems, albeit at a more limited level.

A megacity, a city with a population exceeding ten million people, is yet another classification of an urban area. While this might seem somewhat of an arbitrary population cutoff, Colonel (retired) Mahaney makes the case that megacities are much more complex than smaller cities, and thus, it is appropriate to create a term that acknowledges this difference. A city of ten million people is not simply ten times more complex than a city of one

[2] Milan Babic, Eelke Heemskerk, and Jan Fichtner, "Who is more powerful – states or corporations?" *The Conversation*, July 10, 2018, https://theconversation.com/who-is-more-powerful-states-or-corporations-99616.

million; it is orders of magnitude more complicated. Megacities do not just have more structures; the buildings are taller and denser, the ungoverned slums may be massive, the tunnel networks may not be mapped, and trying to build a network diagram to figure out where power resides is significantly more challenging. Yet despite the growing complexity of megacities over the past two decades, it is the small- and medium-sized municipalities that have become more problematic and this is where militaries have found themselves engaged.

Too often when people, especially soldiers, think of cities, they think of the urban triad: the complex man-made physical terrain, the population, and supporting infrastructure. Yet, with the emergence of global cities and the ability to transfer information, ideas, and funds almost instantaneously, it is more appropriate to think of a city as an urban quad that also includes information.

Another theme that emerges is the need to think of a city as a living organism, a complex adaptive system, or an open and incomplete system that ingests and processes critical flows such as commerce, energy, information, fuel, water, and waste. A city needs fresh air, water, food, and power and a way to expel its waste like sewage and trash. When populated areas are unable to effectively process these requirements, toxic buildup in the form of pollution, crime, violence—or flooding in the case of Fallujah—can result. It is important for militaries to have at least a basic understanding of these dynamic conditions because military operations tend to disrupt urban functions and, thus, risk increasing these toxic products, which make military operations significantly harder.

Military Application

Being armed with concepts that force cities to be thought of as organisms and taxonomies such as global, feral, smart, and megacities provide useful constructs to examine them. As such, they provide insights that militaries must consider when operating in urban environments.

Militaries Must Understand the Various Flows Within a City

Too often, militaries focus on the physical terrain aspects of urban areas, believing that understanding other flows belong to other entities, such as political or economic officers from the Department of State or Ministry of Foreign Affairs. Yet as has been demonstrated time and time again over the past two decades, there are not enough foreign service officers to support large scale deployments. Limited numbers often reside in embassies, but the company or battalion commander operating in a city—Ramadi or Fallujah for example—rarely has these experts at his or her level, so he or she has to at least have a basic understanding of the various flows.

Militaries must effectively apply tools like mission, enemy, terrain, troops available, time, and civilian considerations (METT-TC), political, military, economic, social, information, infrastructure, physical environment, and time (PMESII-PT), and areas, structures, capabilities, organization, people, and events (ASCOPE) to understand these urban environments from their three-dimensions to their flows and all the way down to sub-city units such as neighborhoods.

Colonel (retired) DeFrancisci's example from Fallujah perfectly illustrates why this is important. He was not a wastewater management or structural engineer, but he had to understand the sewage systems, the dam, and the power system to stop the flooding so that the assault against the enemy could continue. Likewise, Colonel (retired) Mahaney argued that it is important to have a micro-understanding of a city, because the basic function of security, rule of law, and area control can change from neighborhood to neighborhood or even block to block.

While not in this book, we will provide another example of why understanding the flows of a city can be important. When the United States transitioned from the liberation phase to the transition phase during the Iraq War in 2003, General Petraeus was the only division commander who had some understanding of these urban connections. He jump-started the political process by holding a mayoral election, opened the border to enable trade, aggressively pursued reconstruction projects, and held Ba'ath party renunciation ceremonies. As a result, Mosul had significantly fewer security incidents than other cities. Petraeus understood the need for the city to provide governance, services, and security, the need to reestablish a licit economy, and the requirement for the city to be engaged with its civil society. Thus, he prevented the toxic buildup that other cities experienced.[3]

[3] Liam Collins, "Military Innovation in War: The Criticality of the Senior Military Leader," PhD diss. (Princeton University, 2014), 141-3, http://arks.princeton.edu/ark:/88435/dsp01br86b3727.

Militaries Must Consider the Urban Quad: People, Infrastructure, Terrain, and Information

Militaries must understand not only their assigned area, but also how information can impact their operations. For example, when insurgents killed four American contractors in Fallujah in 2004, the immediate reaction from political leaders was to have the military launch an operation into Fallujah to find those responsible for the brutal killing and mutilation, despite the fact that the military leaders in Iraq warned against such an attack. General John Abizaid, the commander of U.S. Central Command acknowledged that, "Repeated play on the media has made the insurgents appear strong and the U.S. weak."[4] Thus, he was forced to launch the attack.

However, the assault was short lived as Arab media outlets sensationalized the high collateral damage and civilian casualties and sometimes misidentified fallen insurgents as civilians. Al Jazeera claimed 600 civilians had been killed and broadcast "images of dead children at the Fallujah hospital and elsewhere within the city."[5] As a result, the Iraqi interim government convinced American political leaders to end the operation, once again, in opposition to military leaders on the ground who felt that they needed to finish the mission that they had started.[6] This example illustrates how quickly information can transmit across the globe and impact a company or battalion operating in a foreign city.

On the positive side, Dr. Kilcullen noted that militaries may be able to impact operations by leveraging a diaspora population half a world away to influence operations without risking troops on the ground. The point is that when considering operating in a modern city, militaries must consider how they can leverage information to help or how others may leverage information to hurt their operations.

Lieutenant General Noble also noted the importance of information operations during the 2016 Iraqi-led battle of Fallujah. Instead of attempting to surround the city and clear it, the Iraqi military used a multi-faceted information operations campaign to convince Islamic State of Iraq and al Sham (ISIS) fighters to leave the city or surrender.

---

[4] Joel Rayburn and Frank Sobchak, *The U.S. Army in The Iraq War, vol. 1* (Carlisle Barracks, PA: U.S. Army War College Press, 2019), 283, https://publications.armywarcollege.edu/pubs/3667.pdf.

[5] Ibid, 291.

[6] Ibid., 291-2.

Militaries Must Consider Existing Smart Technologies Within the City

All cities are employing at least some of the layers of the smart city system even if they have not yet reached full sophistication. Every city has at least rudimentary physical and communications systems. More advanced cities also have the data analytics layer and the human and services layer. These systems or stratum may be controlled by the government, militias, criminal elements, or an enemy military. Militaries must consider how others could use these systems against them or how they can neutralize or seize control of the systems for their benefit.

One example would be a camera system. Obviously, controlling a city's cameras offers one side a marked advantage. Another example might be a city's traffic systems. Consider a humanitarian mission in a semi-permissive environment with many rogue actors. If each traffic light represents a potential ambush location for a humanitarian convoy, then controlling the traffic lights to facilitate the rapid movement of a convoy through the city is something that a military would want to do.

Militaries Must Prepare to Operate in High-Risk Cities

Dr. Norton provides a taxonomy to identify high-risk cities, those tending to the feral end of the spectrum. These are the cities in which militaries are most likely to find themselves engaged, either for humanitarian operations or for more traditional combat missions. Militaries must prioritize understanding these cities, so that they are better prepared to function in them.

The clearest examples are the 2004 battles of Fallujah. While the death of four civilian contractors in April 2004 triggered the first battle, it was the city turning completely feral that forced the coalition to initiate the second battle. The Iraqi national government had lost control, influence, and the ability to support local governance and to provide basic services to the city's residents. The coalition was forced to clear the city because attacks emanating from the city impacted the entire region.

Learning From Operational Case Studies

In his examination of the Battle of Marawi, Knight noted that "the use of mouseholes, like most of the lessons of Marawi, was not new at all, rather the method can be retrieved from the history books of earlier conflicts."[7] Like Knight, we find that there are few, if any, new lessons to be discovered after examining urban battles over the past several decades. Yet the fact that militaries seem to have to learn the same lessons over and over again, means it is still a worthwhile exercise to provide them. What follows are some of these lessons.

Militaries Must Stop Thinking They Will Ever be Able to Avoid Cities

In every case that was examined, with the exception of Mogadishu, the city was the objective. Shusha was a strategic objective. Many others were operational objectives: liberating Mosul from ISIS, clearing Fallujah and Ramadi of al Qaeda in Iraq, and dislodging German soldiers from Ortona to continue the advance up the Italian peninsula.

Urban Warfare Requires Combined Arms Operations

The cases demonstrated the importance of combined arms operations in urban warfare. This is by no means a new observation as Major Geroux noted that the Canadians in Ortona experienced success when they employed infantry, tanks, anti-tank guns, mortars, and engineers together with artillery in support from positions outside the city.

In the battle of Sadr City, Lieutenant Colonel (retired) MacMillan described how 1-68 Armor created company teams that included a mix of tanks and infantry for urban fighting. He also illustrated the limitations of the Stryker battalions. When the enemy situation forced units in Baghdad to quickly transition from stability operations to offensive operations, the Stryker battalion needed to be reinforced and ultimately give up parts of its sector to 1-68 Armor because it lacked heavily armored tanks that were more survivable against the enemy's explosively formed penetrators. He demonstrated that urban combat requires armored vehicles.

7 See chapter 14, "The Battle of Marawi," in this book.

Mogadishu also demonstrated what can happen when forces lack adequate armor. Had Task Force Ranger possessed sufficient armor assets within its exfiltration convoy or quick reaction force, the battle likely would have been shorter and had fewer casualties. The assault force was only able to exfiltrate after the Malaysian armored personnel carriers and Pakistani tanks arrived.

Marawi is yet another example that demonstrates the importance of having mobile protected firepower. Lacking tanks and heavily armored vehicles, the police and army teams that immediately responded after the arrest operation were easily ambushed by the Islamic State in the Philippines militants because their thin-skinned vehicles were vulnerable to militants armed with older rocket propelled grenades. If the Armed Forces of the Philippines (AFP) had tanks, they could have rolled through the ambushes.

A military that is trained, manned, and equipped to fight in combined arms formations that include artillery, armor, infantry, engineers, and aviation has the greatest odds at overcoming the equalizing effects of urban terrain to achieve objectives.

Urban Battles Often Take Longer Than Expected

The raid into Mogadishu was supposed to last 20-30 minutes yet it lasted 15 hours. Ortona was one of the shortest battles examined in this book, lasting only eight days; the duration, nevertheless, was four times what the Canadians expected. It is unrealistic to expect urban battles to last only eight days. The battle was relatively short because the city was small and because the Canadians exercised combined arms maneuver against a German airborne unit that lacked tanks. A city ten times its size might take ten times longer to clear, or perhaps exponentially longer. Also, an enemy that plans to fight to the bitter end is likely to take longer to clear than one that is willing to leave the city once the outcome of the battle is no longer in doubt.

Some thought it took the AFP too long to clear Marawi, but they had to conduct repeated deliberate stronghold assaults, which is by no means a quick or easy military mission. Likewise, Colonel (retired) DeFrancisci noted that the Marines had to clear over 35,000 buildings in Fallujah.[8] Thus,

[8] DeFrancisci remembers it being 35,000 buildings, but some reported as many as 50,000 buildings. See, "U.S. Troops Occupy Fallujah," CBS News, November 14, 2004, https://www.cbsnews.com/news/us-troops-occupy-fallujah/.

it is important to remember that the removal of enemy fighters from urban terrain takes more time than it would in other environments. The 2016 battle of Mosul was planned over a year in advance, and it was originally planned as a 3-month mission. Yet it took over 100,000 Iraqi and coalition security forces nine months to liberate the city.

There is No Single Best Approach to Clear, Hold, and Build a City

To clear Ramadi, Colonel MacFarland's brigade established company outposts throughout the city using the counterinsurgency "ink spot" method—slowly expanding security outward from the combat outposts until the entire city was secure. It required fewer troops and took longer than the second battle of Fallujah but was also less destructive. In Fallujah, the coalition forces conducted a deep penetration to Jolan Park that forced the enemy to leave its defensive positions and fight the coalition forces on less favorable terms. In Marawi, the AFP conducted a systematic clearance from three directions, constricting the enemy into a smaller and smaller space. By building the wall around Sadr City, 1-68 Armor forced the Jaish al-Mahdi militants to expose themselves in a fight more favorable to the American forces.

All these were successful operations, demonstrating that no single approach works in all situations. The best methodology depends on the number and variety of available forces, the type of enemy that is being engaged, the willingness of the local population to support, and many other factors.

Information Operations Must be Part of the Plan From the Onset

Too often, at least in the U.S. military, information operations are often viewed as an afterthought. Units often develop a tactical plan and then figure out how information operations can support a plan that has already been developed. Yet Fallujah and Marawi demonstrated the significant impact that information operations can have on mission success when they are considered throughout planning and execution.

Before the second battle of Fallujah, coalition information operations convinced most of the population to flee the city before the battle started. They also contributed to the deception plan, which prevented the enemy from massing its defenses along the actual avenue of attack. Dr. Knight described how the AFP made information operations a priority, equal in importance to the lethal operations. The AFP backed up its words with actions by creating the Hijab troopers, reassigning combat engineers to construct shelters for homeless residents, and creating a joint task group to focus on information operations. These efforts helped turn the population from one that largely distrusted the military and government into one that started supporting it.

Civil Affairs Operations Must Support All Phases of Combat Operations in Cities

The chapter on rebuilding Fallujah demonstrated how civil affairs operations supported all phases of the operation. During the shaping operations, civil affairs constructed lists of key individuals, mapped out critical infrastructure, supported displaced residents, and supported the division's deception operations. During the assault, civil affairs conducted humanitarian assistance, infrastructure assessments, and helped stop the city's flooding. After the assault, civil affairs helped facilitate the recovery of more than 2,000 dead enemy fighters and helped get the city functioning again.

Urban Warfare Can Negate or Reduce the Effectiveness of Some Superior Technology

A dense urban area greatly reduces the engagement range of all weapons systems; thus, a hand thrown Molotov cocktail may have the same range as a 120mm tank round. Buildings and tunnels provide cover and concealment for enemy forces to maneuver, thus negating the effects of strike and intelligence, surveillance, and reconnaissance assets. Fighters can drop charges from the upper stories of buildings onto vehicles while only being vulnerable to hand-fired weapons since they are above the maximum angle of elevation for tank or armored personnel carrier's main battle guns. Thus, urban terrain can be a great equalizer. It offers an advantage to the force that understands how to use it effectively.

During the Battle of Mosul, ISIS used the dense urban terrain to reduce the effectiveness of coalition intelligence, surveillance, and reconnaissance, and strike capabilities. The Iraqi military would have made quick work of the estimated 3,000 to 12,000 minimally trained terrorist fighters in nearly any other environment.[9] In Mogadishu, the militias figured out how to volley fire rocket propelled grenades to take down hovering helicopters. This greatly reduced the ability of the command-and-control aircraft to hover and provide directions to the exfiltration convoy. As a result, the convoy was unable to evacuate the assault force.

Urban Warfare Requires Innovation—Especially When Previous Lessons Have Not Been Learned

Mouseholes and loopholes were present in nearly every battle, but unfortunately, many soldiers learned by experience during the battle as opposed to implementing these techniques from the start. Dating back to at least Ortona, existing entrances such as doorways and windows often become death traps, armed with improvised explosive devices (IEDs), or used by the enemy as kill zones for direct gunfire. In Ortona, soldiers learned to breach walls and clear from the top-down.

Sometimes innovation is characterized by the employment of weapon systems in nontraditional ways. In Ortona, Major Welsh brought his six pounder anti-tank guns forward to help reduce the rubble obstacles. In Sadr City, tank personnel deduced that they could clear IEDs from the road using canister shot. Thus, urban combat often involves employing existing weapons in non-traditional means.

In Marawi, the AFP built giant ramps to maneuver their M113 armored personnel carriers onto the second floor of buildings so they could employ them as mobile gun platforms. While this technique may not have widespread application since many structures may not be able to handle the vehicle's weight, it demonstrates how troops can be innovative.

It is also important to consider how the enemy might innovate. In Mosul, an engineer was smart enough to consider that the enemy might float IEDs down the Tigris River to blow up the bridge. An enemy might rig entire buildings to detonate as the Germans did in Ortona.

---

[9] The Mosul Study Group estimated that the number of fighters ranged from 3,000 to 5,000, but others estimated them to be 12,000 or more. See, Mosul Study Group, *What the Battle of Mosul Teaches the Force* (Fort Eustis, VA: U.S. Army Training and Doctrine Command, 2017), 5, https://www.armyupress.army.mil/Portals/7/Primer-on-Urban-Operation/Documents/Mosul-Public-Release1.pdf; and "Over 12,000 ISIS militants fighting for Mosul," *ARA News*, June 27, 2016, https://web.archive.org/web/20170516043248/http:/aranews.net/2016/06/12000-isis-militants-fighting-mosul/.

Urban Warfare is Likely to Result in Significant Collateral Damage and a Large Number of Civilian Casualties

Although we must now consider the urban quad to understand urban terrain, militaries must also consider an urban risk triad for urban operations that includes risk to force, risk to civilians, and risk to infrastructure. Militaries can often mitigate the risk to one or two, but rarely, if ever, all three. To minimize collateral damage, the force must often assume a higher risk to itself. To reduce risk to the force, the city often suffers higher levels of collateral damage, and the case studies seem to bear this out.

The Battle of Marawi illustrated this trade-off fairly well. The AFP attempted to fire a 105mm artillery round or detonate a high explosive in every room prior to entering. While not always possible, when employed, this technique greatly reduced the risk to the force, but it also resulted in significant destruction. But the AFP was forbidden from using high explosives in mosques, and, as a result, suffered heavier casualties.

Thus, it is not surprising that the level of destruction in urban combat is so high. In Marawi, an estimated 95 percent of the buildings in the main battle area were damaged and more than 50 percent destroyed.[10] In Fallujah, an estimated 60 percent of the city's buildings were damaged or destroyed.[11]

Militaries and their governments must be ready to accept high levels of destruction, a high number of friendly casualties, or they should avoid the city altogether. It is simply a reality of urban warfare. The first battle of Fallujah was terminated prematurely because the Iraqi and the United States governments were unwilling to accept intense collateral damage. Over time, both realized it was necessary and a reality of war; the learning curve, however, resulted in additional devastation because during that time, the enemy only strengthened its position. The end result was that the coalition forces likely produced a higher level of damage during the second battle than it would have had they not prematurely ended the first battle.

The greater the population density, the more likely there will be civilian casualties. But trying to minimize noncombatant fatalities often puts the force at greater risk because it often requires limiting firepower.

---

[10] Ervin Malicdem, "Aftermath of the Battle of Marawi," *Shadow Expeditions*, November 30, 2017, https://www.s1expeditions.com/2017/11/223-marawi-battle-structures.html.

[11] Kendall Gott, ed., *Eyewitness to War Volume I: The US Army in Operation AL FAJR: An Oral History* (Fort Leavenworth, KS: Combat Studies Institute Press, 2006), 14, https://apps.dtic.mil/sti/pdfs/ADA460458.pdf.

Additionally, the more civilians that exist, the easier it is for an enemy to use them as human shields or to take hostages to limit the opponents use of firepower. Thus, it is often preferable to get the inhabitants to evacuate the city prior to assaulting it.

In Ortona, the Germans evicted the city's residents, yet 10 percent, or roughly 1,000, remained at the start of the battle. In Fallujah, an estimated 70–90 percent of the city's 300,000 residents fled, leaving a minimum of 30,000.[12] In Marawi nearly 98 percent of the city's population in the main battle area fled within the first week following the raid.[13]

All three were still tough fights, but the assaulting force, by and large, did not have to consider civilians when making tactical decisions regarding firepower. As mentioned earlier, the AFP was able to conduct pre-assault fires on nearly every room after most of the hostages had been recovered. In Fallujah, the United States killed an estimated 2,175 insurgents and detained an additional 2,052, while suffering only 82 killed.[14] It is reasonable to conclude the United States would have suffered dozens more if it had to consider civilians and withhold firepower.

Civilians in Urban Warfare Are Not Obstacles to be Overcome, Rather They Are Part of the Urban Environment That Can Help or Hinder Operations and Must be Considered During Planning

It is important to understand the interconnectivity of a city and the local population understands these flows more than anyone. Thus, it is imperative to influence and leverage the population whenever possible.

In Fallujah, the coalition leveraged local leaders to convince the population to leave before the battle; this made the fight easier than it otherwise would have been. Also in Fallujah, Colonel (retired) DeFrancisci had to convince municipal workers to enter a city with a raging war to identify sewage lift stations to stop the flooding. In Ramadi, coalition forces gained the support of local sheiks who were necessary to build local police

[12] Though other sources reported as few as 500. Anne Barnard, "Fallujah refugees describe ordeal of life in crossfire," *The Boston Globe*, November 15, 2004, http://archive.boston.com/news/world/articles/2004/11/15/fallujah_refugees_describe_ordeal_of_life_in_crossfire/.
[13] "Marawi Crisis," UNHCR, https://www.unhcr.org/ph/marawi-crisis.
[14] Rayburn and Sobchak, *The U.S. Army in the Iraq War*, 354; and Timothy S. McWilliams and Nicholas J. Schlosser, *U.S. Marines in Battle: Fallujah, November-December 2004* (Quantico, VA: U.S. Marine Corps History Division, 2014), 67, https://www.usmcu.edu/Portals/218/FALLUJAH.pdf.

forces capable of holding areas that had been cleared by coalition forces. In Marawi, the Philippine government gave unprecedented political authority for non-military decisions to local leaders. This helped them further the narrative that the government was working with the local community.

Civilians are always part of the battlefield environment and especially so in cities. The military can use acronyms like METT-TC (Mission, Enemy, Troops, Terrain, Time, Civil Considerations) to implement population factors in planning, but they are often considered merely as an afterthought as opposed to something that can facilitate offensive and defensive operations. The chapter "Rebuilding Fallujah" demonstrated why it is important to implement civil considerations during all phases of an operation.

It is important to remember that civilians remain a consideration even if most have fled. Even if only 95 percent of a city's population has evacuated, for a city of 300,000 that still leaves 15,000 noncombatants in the battle area. There will be no urban operations in the future completely void of civilians and militaries must not train or plan like there will be.

Recommendations

Perhaps the most important lesson is that cities often hold tactical, operational, or strategic value and cannot be avoided in war. We often hear the argument that cities are too big and too difficult to fight in, so the military should avoid them. We do not deny that cities pose an extremely challenging operational environment, and they should be avoided, but this book is a testament to the fact that this is not always possible.

The United States and other professional militaries have a long history with urban warfare. Many of the most well-known battles in the past 20 years were urban and they ranged from Iraq to Nagorno-Karabakh, and the Philippines to Ukraine. During its initial invasion of Iraq, the U.S. military bypassed Fallujah and Ramadi but eventually found itself fighting battles in both.

The U.S. military gained experience in urban warfare during the Iraq War but has chosen to push much of that knowledge aside and reset to prioritizing for operations in other environments. It is not a matter of if militaries will have to conduct attacks and build defenses in urban areas during major combat operations, but when. One only needs to look at Russia's invasion of Ukraine in 2022 to see this reality play out. Thus, militaries must do more to understand urban terrain and prepare to fight and win there.

Acknowledging the fact that cities often cannot be avoided in war and given the dual trends of population growth and urbanization, it is important for militaries to be prepared to conduct a full range of operations, both combat and noncombat, in cities. As such, this final section provides some recommendations to militaries to be better prepared for urban combat.

Militaries Must Allocate Considerably More Resources to Understanding and Preparing for Urban Warfare

When he was Chief of Staff of the Army, General Mark Milley, now chairman of the Joint Chiefs of Staff, stated: "In the future, I can say with very high degrees of confidence, the American Army is probably going to be fighting in urban areas…We need to man, organize, train and equip the force for operations in urban areas, highly dense urban areas, and that's a different construct. We're not organized like that right now."[15]

Yet the U.S. military's actions do not match the former chief's words. One can examine where the U.S. military spends its funding, the types of environments in which the military trains, or the allocation of time in professional military education courses to see that the urban environment is far from a priority. The U.S. military is simply not manning, organizing, training, or equipping the force for operations in urban areas. There are many reasons for that, but it must change. Other nations, often following the United States' lead when it comes to defense investment, are also underinvested in preparing for urban warfare.

Assign the U.S. Army as the Executive Agent to Ensure the U.S. Military is Adequately Preparing for Urban Operations

This is a U.S. centric recommendation, but the point is to figure out the process within a nation's political-military system to force change. For the United States, one way to ensure that the military increases the resources dedicated to preparing for urban warfare is to assign the U.S. Army as the

[15] Michelle Tan, "Army Chief: Soldiers Must Be Ready to Fight in 'Megacities'," *Defense News*, October 5, 2016, https://www.defensenews.com/digital-show-dailies/ausa/2016/10/05/army-chief-soldiers-must-be-ready-to-fight-in-megacities/.

executive agent for urban operations. An executive agent is "the Head of a [Department of Defense] Component to whom the Secretary of Defense or the Deputy Secretary of Defense has assigned specific responsibilities, functions, and authorities to provide defined levels of support for operational missions, or administrative or other designated activities that involve two or more of the [Department of Defense] Components."[16]

The U.S. Department of Defense establishes an executive agent when "no existing means to accomplish the [Department of Defense] objective exists," "resources need to be focused on a specific area(s) of responsibility to minimize duplication or redundancy," or "Required by law, Executive order, or Governmentwide regulation." In lay terms, the Department of Defense establishes an executive agent to tackle important issues that impact multiple services.

When the U.S. Department of Defense dissolved United States Joint Forces Command in 2011, the U.S. military lost its only office, the Joint Urban Operations Office, solely focused on urban operations.[17] No office or like organization has been created since. The U.S. Army's proponent for urban operations is the Combined Arms Center, but urban operations is among a long list of critical areas against which the center's time and resources must be prioritized. To date, urban operations have clearly not been a priority.

Assigning the army as the executive agent would make urban operations a priority for the army and the Department of Defense and require the army to provide periodic updates to an Office of the Secretary of Defense Principal Staff Assistant which is designated to oversee its activities.[18] It would allow the army to develop standards, develop training programs, and other activities for all aspects of urban doctrine, organization, training, materiel, leadership and education, personnel, facilities, and policy development, advocacy, and integration.[19]

---

[16] Department of Defense, *DoD Executive Agent*, DoD Directive 5101.1 (Washington, DC: U.S. Department of Defense, 2003), 2, https://www.esd.whs.mil/Portals/54/Documents/DD/issuances/dodd/51011p.pdf?ver=2019-01-14-091137-860.

[17] Kenneth Goedecke and William Putnam, *Urban Blind Spots: Gaps in Joint Force Combat Readiness* (Cambridge, MA: Belfer Center, 2019), https://www.belfercenter.org/sites/default/files/2019-11/UrbanBlindSpots.pdf.

[18] Government Accounting Office, *Defense Management: DoD Needs to Improve its Oversight of Executive Agents* (Washington, DC: U.S. Government Accounting Office, 2017), 6, https://www.gao.gov/assets/gao-17-601.pdf.

[19] Ibid., 12; and John Spencer, "Four Transformational Steps the U.S. Army Should Take to Get Serious About Urban Operations," Modern War Institute, July 21, 2020, https://mwi.usma.edu/four-transformational-steps-the-us-army-should-take-to-get-serious-about-urban-operations/.

Militaries Must Create Urban Warfare Centers

This book has highlighted the need to understand the urban environment and the requirements of urban operations across the range of military operations. It has, however, only scratched the surface of the work that needs to be done to prepare armies for urban warfare in the future. One of the best ways to improve the training, education, leader development, and readiness across an entire army is to create a fully resourced and staffed urban warfare center.

Congress attempted to push the U.S. Army to do more to prepare for urban warfare when it passed the 2019 and 2020 National Defense Authorization Acts.[20] Members of the House Armed Services Committee were concerned about the army's lack of prioritization and resourcing to address the challenges of conducting military operations in complex, densely populated urban terrain. The 2020 act stated the army should create both an "Army urban warfare center dedicated to the study and research of urbanization, mega-cities, urban warfare, and military operations in dense urban environments" and an "urban warfare training center that focuses on advanced skills to fight, survive, and win in urban operating environments at the brigade level and higher." The latter organization, the bill says, "address the challenges associated with vertical, subterranean, and dense urban terrain, and the inclusion and integration of joint and interagency enablers."[21]

We agree that the U.S. Army must create an urban warfare center that is fully resourced and staffed. A topic so broad requires longitudinal research from experts across multiple fields to work together to address the military challenges related to operating in the complex operational environment of urban terrain. The army must develop research, historical analysis, doctrine, curriculum, wargames, and a cadre of experts and expertise within the force. A true urban warfare center is not just about constructing a bunch of buildings, it requires building a cadre of trained personnel to research, analyze, instruct, and execute urban operations in multiple contexts.

---

[20] For 2019 and 2020 National Defense Authorization Acts, see, https://www.congress.gov/115/crpt/hrpt676/CRPT-115hrpt676.pdf, and https://www.govinfo.gov/content/pkg/CRPT-116hrpt120/html/CRPT-116hrpt120.htm.

[21] See, 2020 National Defense Authorization Act.

An urban warfare center could be housed within the U.S. Army's active or National Guard component. Using a National Guard unit to develop expertise in a single environment has precedence. The Vermont National Guard is home to the army's mountain warfare experts and the U.S. Army Mountain Warfare School that includes multiple courses in how to survive, lead, operate, and fight in mountainous terrain.[22] In 2021, John Spencer worked with the California National Guard's 40th Infantry Division to create the army's first urban warfare planners course.[23] This course was the largest single urban military education event in decades, but it was a unit as opposed to an army initiative. Rather than the army funding the course, the unit funded the course out of its internal budget. The 40th Infantry Division's efforts should be commended, but this effort must be institutionalized with a permanent home.

Lastly, preparing for the future of urban operations is a global and international effort. An urban operations center would allow international partners, militaries, and academia to combine efforts to improve the understanding of the requirements and impacts of military operations in urban environments. It is time to focus energies and expertise to do more. It is vital to ensure that future soldiers do not have to relearn the same lessons over and over again the hard way.

[22] Whitney Huges, "Making Mountaineers: U.S. Army Mountain Warfare School," NationalGuard.mil, February 17, 2022, https://www.nationalguard.mil/News/Article/2938143/making-mountaineers-us-army-mountain-warfare-school/.

[23] John Spencer, "What We Learned Creating the Army's First Urban Planners Course," Modern War Institute, November 29, 2021, https://mwi.usma.edu/what-we-learned-creating-the-armys-first-urban-planners-course/.

Index

Page numbers in italics refer to figures and tables.

We hope that you have enjoyed this book.
Please consider posting a review for future readers.

Thank you.

Printed in the USA
CPSIA information can be obtained
at www.ICGtesting.com
LVHW020828040824
787236LV00003B/33

9 781912 440351